Handbook of Special and Remedial Education

Research and Practice

Second Edition

Titles of related interest

WANG *et al.*
Handbook of Special Education, Volume 1: *Learner Characteristics and Adaptive Education*

WANG *et al.*
Handbook of Special Education, Volume 2: *Mildly Handicapped Conditions*

WANG *et al.*
Handbook of Special Education, Volume 3: *Low Incidence Conditions*

WANG *et al.*
Handbook of Special Education, Volume 4: *Emerging Programs*

HELLER, MONKS & PASSOW
International Handbook of Research and Development of Giftedness and Talent

Handbook of Special and Remedial Education
Research and Practice
Second Edition

Edited by

MARGARET C. WANG
Temple University, Center for Research in Human Development
and Education, Philadelphia, USA

MAYNARD C. REYNOLDS
University of Minnesota, Minneapolis, USA

and

HERBERT J. WALBERG
University of Illinois at Chicago, USA

Pergamon
An imprint of Elsevier Science

UK   Elsevier Science Ltd, The Boulevard, Langford Lane,
Kidlington, Oxford OX5 1GB, UK

USA   Elsevier Science Inc., 660 White Plains Road, Tarrytown,
New York NY 10591-5153, USA

JAPAN   Elsevier Science (Japan), Tsunashima Building Annex,
3-20-12 Yushima, Bunkyo-ku, Tokyo 133, Japan

First edition: Vol. 1, 1987; Vol. 2, 1988; Vol. 3, 1988;
Vol. 4, 1991
Second edition 1995

Library of Congress Cataloging-in-Publication Data
Handbook of special and remedial education : research and
practice / Margaret C. Wang, Maynard C. Reynolds,
Herbert J. Walberg.—2nd ed.
   p. cm.
Updated ed. of: Handbook of special education.
Includes indexes.
   1. Special education—Handbooks, manuals, etc. 2. Special
education—United States—Handbooks, manuals, etc.
I. Wang, Margaret C. II. Reynolds, Maynard Clinton.
III. Walberg, Herbert J., 1937–   . IV. Handbook of
special education (Oxford, England)
LC3965.H263   1995
371.9'0973—dc20   95-37011

ISBN 0-08-042566-6

Printed and bound in Great Britain by BPC Wheatons Ltd, Exeter

Contents

Section 2: Distinct Disabilities

Section 3: Support Systems

Preface

A decade ago work began on the four-volume *Handbook of Special Education: Research and Practice*, published over several years by Pergamon Press beginning in 1987. A fifth volume summarizing the *Handbook*'s four volumes was published in 1990. The *Handbook* achieved a foremost position as a professional statement in the field of special education. It provided a comprehensive summary of the well-confirmed knowledge in the field, along with commentary on the status of practice in the various domains and some discussion of needed "next steps," in research, policy, and practices. The field has been active over the past decade, thereby establishing a clear need for an updated version of the *Handbook*. These few remarks introduce the second edition. The general commitment to summarize the knowledge base or the "state of the art" and to comment on the "state of practice" are held in common over the two editions. But some changes have been made.

A change in title, adding the words *and remedial*, signals a quite pervasive shift in orientation. That is, to see special education in the context of other categorical programs and of the important continuing renegotiation of the relations among categorical education programs with the general education programs. The full title for the second edition is *Handbook of Special and Remedial Education: Research and Practice*. There is growing recognition that the rewards of narrowly framed categorical programs are few and limited and that there is need for a broadly integrated school system even as students with special needs are considered.

The increasing numbers of "special" or categorical programs has itself become problematic, by causing an extreme disjointedness of programs within schools. The need for repair of the disjointedness in favor of a broad and systematic approach to education for *all* students was represented explicitly in recent revisions in the United States of the Chapter 1 programs (for economically disadvantaged students). Chapter 1 programs can now be operated on a schoolwide basis without identifying and labeling particular students and can be integrated with special education programs. The change in title to *Handbook of Special and Remedial Education* is intended to reflect this kind of melding of programs.

For similar reasons, the second edition is presented in a single volume.

There is a growing concern about separation of exceptional students in various ways, often with accompanying labels and stigmatization. Even the professional literature has been separated by categories, resulting in narrow streams of awareness and reading by educators. By reducing from 45 chapters, as in the first edition, to 15 in this edition, and by presentation of them in a single volume, we have intended to help diminish some of the negative divisiveness of the literature.

In some parts of the world finding and enrolling exceptional students in the schools is no longer a major problem. Continuing efforts will be required, of course, to maintain universal school enrollments, but this no longer need be a major element in rallying energies to school reform. There are, however, four correlated issues: (a) the extent to which severely disabled children should be placed within general education classes; so-called *full inclusion* is a controversial matter; (b) the increasing use of suspension and expulsion procedures to rid the schools of students who are troublesome; (c) the omission of disabled students in assessment systems that otherwise involve all students; (d) the high costs of special programs. These concerns are reflected in several chapters of this volume.

Special education over at least the past two decades has led the way in forming partnerships with parents to plan and evaluate school programs. Now, even broader coalitions linking schools with social and health agencies are being formed. This pattern of "school–parent–community" linkages grows from acknowledgment that no one agency can meet the complex needs of many students and their families. Special educators have more reasons than most others to be involved, as broad forms of service to children and families are designed; the story of this rapid development appears at several places in this volume. Other topics receiving increased attention include early education and school-leaving, with special emphasis—in the latter case—on transitions from school to employment and other aspects of adult life.

Special and remedial education over the past decade has accomplished much, but it has not been a story of uniform, steady progress. Indeed, the life situation and the learning of too many students has eroded. This seems especially true for the increasing numbers of children living in poverty, those whose families live outside the complex, global, information-based economy. Such students are concentrated in inner cities and in the poorest rural areas. These are not favored places for teachers and other staff of the schools; "burn-out" rates are high in them—for students and teachers alike. But these are areas of growing challenge that must be addressed by researchers and practicing educators.

Programs of special and remedial education now draw a large share of the funds provided to the schools; so much so, in fact, that policymakers are asking for two things: (a) better evidence that the outcomes of the programs are positive as a condition for continued funding; and (b) cutbacks in categorical funds in favor of improved support for general education. These are hard demands upon special fields already overburdened with problems.

In general, the field of special education does not produce its own researchers. Two exceptions, it may be argued, exist in the cases of speech impairments and severe cognitive impairments. These two fields have been quite remarkable in producing researchers who show persistent dedication to research in them. In the field of impaired vision quite an opposite condition may be observed; it "borrows" researchers when possible from nearby disciplines, but with only mixed, occasional success. Too often, it appears, people who might have made useful contributions to research in the various fields have been overwhelmed by practical problems and find themselves drawn into program administration. The lack of researchers and strongly supported research centers are serious problems, especially in some of the low-incidence areas. On the other hand, there has been steady improvement in the central recording of research findings, as in the ERIC system, which has aided the assembly of information for the several authors involved in producing this volume.

The *Handbook* is divided into three major sections. The first deals with that very large part of special education sometimes referred to as teaching the "mildly disabled" plus those who are at risk for school failure and need remedial help. This section also includes chapters on accelerated learning by students who show gifts and talents and on resilient students who "beat the odds" in learning even though their life situations show much adversity.

It is in programs described in this first section that major shifts have occurred and still are occurring in student classification practices, and it is here that much embarrassment has surfaced concerning racial disproportionalities in child-labeling practices. Some professionals will wish to continue the use of narrower categorical groups and labels, but we believe the time has arrived for a new formulation in this large aspect of education.

The second major section includes the several categorical areas of special and remedial education in which distinctive practices are required for selected pupils. The third and final section encompasses general support structures and functions for the several fields. Here, as elsewhere, there is research to be reviewed, practices to be evaluated, and a knowledge base to be presented. Each of the three sections begins with a brief introductory statement.

As a final note, we wish to acknowledge the contributions of others who have had important roles in producing this work. To the 25 authors we are extremely grateful. We would also like to thank Ellen Weinstein of the Temple University Center for Research in Human Development and Education for her coordination and editorial support for the successful completion of the final revisons of the chapters and Bonnie Warhol, who provided efficient and much-appreciated secretarial services. Finally, we repeat a note of gratitude, offered in the preface of the first edition and now in the second, to Barbara Barrett at Elsevier for her encouragement and support for this work and to Michele Wheaton, also at Elsevier, for her caring hands through the entire editing and production process.

Margaret C. Wang
Maynard C. Reynolds
Herbert J. Walberg

Section 1

Learners at the Margins

The focus of this first section is on knowledge that might have been dealt with in the past in narrow categorical terms such as mild mental retardation, learning disabilities, and acceleration. It is estimated that at least 75% of students now enrolled in special education programs and many others served by remedial and at-risk programs are encompassed by this topical area. Today, the most rapidly developing category of special education in the United States appears to be "cross-categorical." Reports show, for example, that the number of special education teachers employed in cross-categorical programs increased by more than 130% in the United States in the two-year period from 1987 to 1989.

From another perspective, researchers have demonstrated that there is no disordinal interaction between traditional special education categories at the "mild" levels and the types of instruction needed. The special panel of the National Academy of Science that studied placement practices in special education in the United States concluded that there was no instructional justification for categorical distinctions between mild retardation, learning disabilities, and Chapter 1. Research findings also indicate that there is no instructional justification for categorical distinctions between students in compensatory or remedial education programs for children from economically disadvantaged circumstances and those students with mild or moderate disabilities. Children assigned to such programs may need very intensive help, but not of different kinds associated with the common categorical delineations.

This large domain is dealt with in six chapters in this section as described below. For each subtopic/chapter, the authors summarize the knowledge base, but also comment on current practices in special and remedial education.

A first chapter summarizes approaches to assessment, classification, placement, and individualized planning for students who show problems of learning in basic academic areas, such as in reading. The discussion encompasses aspects of prevention and treatment as well. The authors have been active in examining practices in applied situations and are well prepared to offer suggestions on policy issues in this field.

A second chapter covers comparable topics in programs of early education for disabled and at-risk children. This is one of the rapidly developing fields of the past decade. In a major way, these developments have helped to raise issues and propose possible solutions to general issues, such as: classification and labeling, prevention of learning problems, and working collaboratively with parents.

A third chapter covers organizational, curricular, and instructional approaches at the elementary school level for students with mild disabilities or learning difficulties. A fair amount of research has been conducted in this domain and the exchange of ideas at an international level has been quite evident. Because so much of special and remedial education, and related research, is centered at the elementary school level, this chapter is likely to be of wide interest. But it is also an area full of issues, covered in the chapter with much care.

A fourth chapter covers elements of curriculum, instruction, and support activities at the secondary school level. Attention is given also to transition programs designed to link school programs to post-school employment and other aspects of adult life. It is at the secondary school level that issues concerning school demissions (suspensions, exclusions, expulsions, and "drop-outs") become most prominent and troublesome. Points of consensus and conflict are noted, along with a summary of well-confirmed knowledge.

The fifth chapter turns to the contrasting topic of accelerated learning. Students whose rates of learning are especially high need adapted programs, just as do those who lag in their learning and development. Finally, a sixth chapter provides a discussion of resilience. It brings focus to students who might well be considered to be "at risk" for educational failure, but who somehow "beat the odds" and succeed well in school. The research in this area may be helpful in designing programs that serve students whose life situations show much adversity.

Without doubt, some educators will prefer to deal with students who show "mild" learning problems in the traditional categorical terms and to approach issues of rapid learning in terms of "giftedness" and special talents. Our judgment, however, is that treating this total set of challenges in terms of dimensional rather than typological differences will be most useful; all the more so as the negatives of crude typologies and labels are discredited and that trend, we think, is already here.

1

Individualized Planning

RICHARD L. ALLINGTON and ANNE MCGILL-FRANZEN
State University of New York at Albany, School of Education, Department of Reading, 1400
Washington Avenue, Albany, NY 12222, USA

We trace the history of educational interventions for harder-to-teach children with an emphasis on the historical pattern of segregation of these children from their peers. The expanding federal role in funding compensatory and special education programs for harder-to-teach children is examined. We conclude that several unintended effects of federal education policy require a substantial rethinking of the current efforts which primarily perpetuate a segregation of harder-to-teach children—primarily children of low-income families. Current policies foster continuing expansion of a "second system" of education that has failed to provide substantial educational benefit to participating children—at least when educational benefit is defined in terms of acceleration of academic achievement. We call for a renewed emphasis on enhancing regular classroom instruction and building the capacity of regular education programs to better meet the needs of all children.

Introduction

Different children find learning different things more or less difficult. This holds true even when we consider learning other than academic learning. In this chapter we will be concerned with academic learning, and in particular literacy learning. We narrow the focus because academic learning is the responsibility of schools and literacy learning is critical to other academic learning. In this chapter, we discuss, principally, the evolution of remedial programs and the development and expansion of the "second system" of American education. It is the remedial programs, both compensatory education and special education efforts, that sit outside the regular education system and comprise the "second system."

The notion that children differ in the rates at which they acquire control over academic learning is rarely challenged. What is vigorously debated is the etiology of these differences and the potential of alternative instructional environments to ameliorate differences in rates of learning. For most of the Twentieth century, the educational profession (and the larger public) believed that differences in learning rates were attributable to differences in intellectual abilities. Children who found school learning more difficult were simply less intelligent than the others. This view held that differences in learning rates were not only inevitable but largely irremediable. Some children would simply lag behind their peers regardless the of interventions offered.

More recently, some have argued (c.f. Allington, 1991; Bloom, 1976; Clay, 1985; Lareau, 1990) that differences in learning rates can be attributed to a number of factors beyond intellectual capacity including the quality, quantity, and intensity of the instruction experienced in school and the familial resources available to support children's learning (and protect children from ineffective school instruction). According to this argument, differences in learning rates are viewed as indications that some children will need more experience with intensive and higher quality instruction in order to achieve the same outcomes. But the schools we have were organized during an earlier era when such views were uncommon. Thus, most schools are not organized in ways that easily allow for some children to receive more and better instruction than others. Most schools are organized to offer children a standard curriculum with standard amounts of standard quality instruction. Most schools are organized so that some children learn more and others learn less. Unfortunately, in most schools, children who begin school with fewer experiences with books, stories, and print, remain behind their more advantaged peers throughout their school careers.

Educating All Children: A Recent Idea

Remedial programs, whether compensatory or special education, are relatively recent additions to the educational systems of developed nations. It is only after universal compulsory school attendance was achieved (in the U.S. this occurred around the turn of the last century) that issues of underachievement and learning difficulties became relevant and reflected in the professional discourse. Of course, some children have always required more effort than others to educate. But actually educating these "harder to teach" students was a relatively late developing feature of educational systems (Windham, 1992). In other words, as educational systems develop and mature, the focus shifts from educating the more advantaged and easier to teach children, to educating all children, including those who—because of difference, disadvantage, or disability—are difficult to teach.

In the U.S., and some other developed nations, the effort to educate those harder-to-teach children has largely involved the development and expansion of a second system of categorical education programs. That is, extra-ordinary educational efforts targeted to enhance the learning of harder-to-teach children have been organized as special programs with specialized personnel added to the educational workforce to deliver special instruction to targeted children. This approach has dramatically altered the character of American schools and the educational workforce.

Since 1950, the number of professional staff in American elementary schools has very nearly doubled while class size has hardly changed. Over half of the adults working in American schools today are not classroom teachers (U.S. Department of Education, 1992). The importance of the "first system," the regular classroom teacher and the core curriculum, has been gradually but steadily eroded as federal and state educational policies constructed a "second system" of education comprised of specialist teachers and other staff supported by special categorical funding streams (Wang, Reynolds, & Walberg, 1988). Some are, at first, skeptical about assertions of the size and influence of the "second system". However, consider the professional staff listed in Figure 1.1. In almost any elementary school enrolling 500 children (and say 20 classroom teachers) we can find all of the non-classroom professionals listed in the first column and several of those listed in the second column. In

Principal	Asst. Principal
Librarian	Gifted & talented teacher
Speech teacher	Guidance counselor
Reading teacher	Social worker
Special education teacher	Bilingual/ESL teacher
Art teacher	Remedial math teacher
Music teacher	Computer teacher
Physical education teacher	Foreign language teacher
Physical therapy	Home/school coordinator
School psychologist	Mentor teacher
School nurse	Sex/health/drugs teacher
	In-school suspension teacher
	Transitional-grade teacher
	Migrant education teacher
	Paraprofessional

Figure 1.1 Non-classroom professional staff commonly employed in American elementary schools (from Allington, 1994).

some elementary schools we can find all of those listed in the both columns and other staff whose roles are not listed here (e.g. school attendance offcer, at-risk coordinator, community affairs coordinator, an so on). In most elementary schools with 500 children there seem to be 20–25 classroom teachers and 15–25 nonclassroom professionals and paraprofessionals employed and, in some schools, these numbers rise appreciably until these personnel substantially outnumber classroom teachers.

There are two factors, generally related, that influence the number of "second system" personnel found in schools. The concentration of children from poor families that attend the school is the first and most influential factor. This is because schools with many poor children are eligible for a variety of federal and state special funding streams, most especially federal Title 1 program funds, to support hiring additional professional and paraprofessional staff. In addition, poor children are substantially more likely to be identified as disabled than are more economically advantaged children (Wagner, 1995). Thus, schools with many poor children generate far more participants in compensatory and special education programs than schools enrolling more economically advantaged children.

The second, and related, factor is the literacy achievement of the children who attend the school. Currently, many state programs provide additional funding to schools where the literacy development of many children is deemed unsatisfactory. These funds have often been one part of individual states' efforts to improve instruction in low-achieving schools. In most communities, there exists a fairly strong correlation between levels of poverty and children's school achievement (Cooley, 1993; McGill-Franzen & Allington, 1991). Thus, schools enrolling more poor children generally have lower average achievement levels than schools enrolling few poor children. Thus, such schools most often employ more non-classroom personnel using the categorical monies.

Because schools have been less successful in fostering academic achievement of poor children, much of the expansion of the "second system" has been based in efforts to improve educational achievement generally, and literacy achievement particularly, of children from low-income families. It may be useful, then to briefly review the history of the development of this "second system" and note how the current organizational structure of remedial education into a dual system of compensatory and special education came about.

A Brief History of the "Second System"

American education before 1900 was primarily available to the easy to teach, easy to reach, and the children of more affluent families. As the nation moved from a predominantly rural society with a largely agricultural economy

to an urban society with a manufacturing economy, higher levels of educational proficiency were deemed desirable and schools were required to educate
a larger number and a greater variety of children. But schools did not find
this easy.

> Though turn-of-the-century compulsory attendance laws challenged
> exclusionary practices, many school systems were unwilling to relinquish
> them ... Schools required an alternative that would enable them to comply
> with the law while supporting the sentiment for exclusion. Special classes
> provided that alternative. (Tropea, 1993, p. 59)

Both compulsory attendance laws and child labor laws worked to increase
the number of harder-to-teach children attending schools. Many of these
children had parents who would have preferred that the children remained
employed; these children were little interested in the regimen of the school
day and previously had been excluded because their behavior or learning
capacity was judged unacceptable. The educational profession has used a
variety of labels in referring to these children. The labels typically reflected
responses to legislation and judicial decrees concerning the school's responsibility to educate harder-to-teach students. What has held constant has been
the creation of special programs for labelled children that segregated them
from other children and excluded them from the regular classroom for all or
part of the day.

In their analyses of special programs, both Tropea (1993) and Skrtic (1991)
conclude that the perpetuation of a largely separate and segregated system
for harder-to-teach children allowed schools to maintain their traditional
organizational structure and emphasis on educating the easier-to-teach while
de-emphasizing the need to adapt either curriculum or instruction to better
educate the harder-to-teach students. However, with the heavy emphasis that
has long been placed on achieving efficiency in education, it is not difficult
to understand why schools are often organized in ways that limit the likelihood that harder-to-teach children will be well-served. The least expensive,
most efficient instruction is some standard program that does reasonably well
in educating most of the students.

The failure of standard programs to actually educate the harder-to-teach
student was the raison d'être for special schools, special programs, and special
teachers. By shunting harder-to-teach children into these special programs, the
traditional organization of schools—the regular education curriculum, classrooms, and lessons—could remain largely unaffected. In the latter half of this
century, especially, federal education policy has emphasized the creation of
categorical education programs to better serve harder-to-teach children. Those
policies, however, seem to reinforce traditional school organization and,
perhaps, slow the pace of restructuring schools. And traditional school
organization must be altered if harder-to-teach children are to attain high

levels of academic proficiency. An unintended effect of these federal "second system" policies is that the regular education program (the "first system") has not only been reluctant to change but has, over the past thirty years, shifted the responsibility for educating the harder-to-teach children to the "second system" (McGill-Franzen, 1994).

However, it is not wholly fair to conclude that segregating harder-to-teach children is a recent phonemon brought largely about by the creation of federally-funded remedial programs beginning in the 1960s. Exclusionary and segregationist education has a longstanding American educational history. Some argue (Baker, 1994) that the second system prevailing in schools today is not so much the result of apriori planning and research but rather the result of educational policies that historically have "drifted" into place. One gets a historical sense of segregated education for harder-to-teach children from the labels presented in Figure 1.2. Listed in the first column are a number of the labels found in the literature since 1900 describing marginalized students. The second column lists programs and practices described in the literature as appropriate educational responses to student differences, disadvantagedness, and disability (from Allington, 1994). The lists are not comprehensive and are only loosely organized by historical use, but they do illustrate of the popularity that both labels and exclusionary practices have long enjoyed in American education.

In examining the lists, there appears to be a trend in the labeling of children. The labels move from descriptions based on social aspects of school life to labels based upon presumed psychological deficits of children. But, as Cummins (1986) notes, it is no surprise that educational psychologists develop psychological explanations for failures to learn. And it is educational psychology that has dominated theorizing about learning and failure to learn across this century.

The educational programs and practices seem to move from an exclusionary toward a more inclusionary focus. That trend, though, is generally a more recent phenomenon, largely the result of various legislative and judicial activities in the past twenty-five years (Singer & Butler, 1987). While there seems to be a trend toward more inclusionary education, currently, most schools still seem to segregate harder-to-teach students for at least part of the day (Sawyer, McLaughlin, & Wingless, 1994). The democratic ideal of educating all children together on a common curriculum has not yet been implemented in American schools. However, Osborne and Mattia (1994) note that a trend toward judicially mandated inclusion has begun. Historically, the U.S. courts have deferred to educators' judgements about the appropriateness of segregation but more recent court decisions require s   districts to provide substantial evidence that placement of handicapped students in a general education classroom with support and adaptations had been attempted and evidence that the handicapped child failed to benefit from these efforts.

Labels	Programs
laggards	truant schools
unmanageables	disciplinary classes
neglected	home visitors
defectives	vocational schools/classes
incorrigibles	suspension from school
backwards	mentally handicapped schools
truants	Negro schools
immigrants	child guidance clinics
subnormals	crippled, blind & deaf schools
low-grade children	ungraded schools/classes
irregular attendants	pupil personnel services
feeble-minded	home tutors
high-grade morons	adjustment classes
dull normals	achievement tracking/grouping
immatures	retention in grade
maladjusted	remedial classes/courses
coloreds	fresh air classes
slow learner	social promotion
reluctant learner	individualized instruction
minimal brain damage	summer schools
dyslexic	special education schools
educable mentally retarded	paraprofessional assistance
educationally disadvantaged	special education classes
remedial readers	extended day programs
emotionally disturbed	drug therapies
developmentally delayed	consultant teachers
learning disabled	in-school suspension rooms
language impaired	self-esteem training
behaviorally disordered	multi-age groupings
attention deficit disordered	magnet schools
exceptional	mainstreaming
differently abled	inclusionary education

Figure 1.2 Labels and programs for marginalized students in American elementary schools, 1900–2000 (from Allington, 1994).

The Medical Model

The education profession's early and continuing fascination with medicine prompted general acceptance of a pathological approach to learning difficulties. Even the term, remediate, for instance, is drawn from the medical notion of remedy and thus linked to the notion of disease or illness (Johnston & Allington, 1991). Historically, remediation has largely involved completing a diagnostic assessment of children who were experiencing difficulties in academic learning (note that *diagnosis* also is drawn from medical terminology). Several underlying assumptions have guided this diagnostic emphasis. One was that the source of interference with learning could be identified through psychological assessment. A second was that the source of interference was typically located in the child or in the child's home environment. A third was that harder-to-teach children required a different curricular emphasis and specialized instruction. A fourth assumption was that significant numbers of children would be unlikely to ever develop anything other than minimal academic proficiencies because of intellectual limitations.

Early literature (1910–1930) on difficulties in literacy development among children illustrates such assumptions. This literature was often focused on the newly identified *slow learner* (defined in terms of below average scores on the increasingly popular norm-referenced tests of intellect). The source of slow learners' learning difficulties was then located in their limited intellectual capacity. Not surprisingly, it was children of the poor and the poorly educated who were most often found to have intellectual limitations that interfered with learning (see Gould's [1981] classic treatise on the role biological determinism played development of psychological assessments of intellectual ability for a quite compelling summary).

At the same time, a view of the reading process as a hierarchical progression of subskills to be mastered was also emerging (Langer & Allington, 1992). Thus, diagnosis of reading difficulties became, largely, a search to identify which subskills were yet unmastered. Thus, children who found learning to read difficult were often diagnosed as being intellectually limited and having a long list of reading subskills that need to be mastered.

Because slow learners were viewed as intellectually limited, the design of the instruction recommended can be best characterized as "slow down the pace of instruction and make instruction more concrete and littered with drill and repetition" (Allington & McGill-Franzen, 1994). Oddly, it now seems, memorizing sight words in isolation and individual letter-sounds were, for example, viewed as concrete activities, while reading books was viewed as an abstract intellectual activity. Remedial instruction then developed into a diagnostic search for the isolated skills elements yet to be learned, followed by lessons that offered primarily a focus on rote memorization of rules and isolated skill elements of literacy learning. Actual reading and writing activity were not often the focus of the intervention. Thus, remedial instruction, even when not intended for children identified as exhibiting impaired intellectual capacity, came to follow the "slow it down and keep it concrete" plans. Unfortunately, slowing the pace of instruction and focusing on rote learning of isolated elements resulted in perpetuating a slow rate of literacy acquisition.

The early history of remedial instruction seems important because it was so influential in shaping the design of subsequent remedial practices. In other words, while we have moved away from explaining reading difficulties as based in limited intellect, we still often design interventions that slow the pace and increase the drill and repetition of isolated skills. Skills-based instruction is still often viewed as more concrete and more appropriate for children experiencing difficulty than instruction that engages children in actual reading and writing activities. While few speak of the slow learner anymore, teachers do still commonly use the words "slow" and "low" to describe children experiencing difficulty. Thus, it seems, conventional educational wisdom still bears the legacy of the roots of remedial instruction (Allington, 1994).

Also important in the history remedial instruction is the common root of both compensatory and special education. In the 1940s both Sam Kirk, the father of the American learning disabilities movement, and Edward Dolch, the reading expert who developed the famous Dolch 220 high-frequency sight word list, both authored books with a focus on the remediating the slow learner (Dolch, 1948; Kirk, 1940). Similar recommendations are offered by both authors and both suggest "slow it down and keep it concrete" instructional plans. Within their lifetimes the professional careers of these two stalwarts would diverge as two distinct professional fields were created—compensatory reading and learning disability. This professional separation was nurtured and sustained by federal program regulations and funding streams that created two different sets of interventions to assist low-achieving readers.

Expanding the Federal Role in Education

In the first half of the twentieth century individual states largely controlled public education in the U.S. with little regulation or support from the federal government (although the passage of the federal child labor laws did, for instance, have a dramatic effect on school attendance). In the second half of the century the federal educational role expanded, first in civil rights activities following the 1954 Supreme Court ruling in the Brown v. Board of Education case that undermined the "separate but equal" school systems for African-American students that had long existed in some regions.

At about the same time, general public dissatisfaction with American schools and fears of Russian educational and technological superiority spurred passage of the National Defense Education Act of 1958 (NDEA). This Act marked the beginning of a federal push to reform American schooling. The NDEA provided funds to train math and sciences teachers, to train teachers to work in low-income communities, to expand testing and guidance services to better sort children into academic and non-academic career tracks, and for projects to produce improved curriculum materials. The NDEA added curriculum specialists and guidance counselors to many schools as well as funding for expanded psychological services (Spring, 1989).

Further expansion of the federal role came with the passage of the Elementary and Secondary Education Act (ESEA) of 1965 with its emphasis on funding supplemental educational services to overcome the negative effects of cultural deprivation and economic disadvantage. The focus of the ESEA was the development of basic skills in economically and educationally disadvantaged students. The law required that educational service provided under the Act be supplemental, that is, in addition to, the services normally provided students. It was illegal to supplant, or replace, routine educational services with services funded under ESEA. In addition, students might be eligible for services but there was no mandate to provide the services to all eligible children and adolescents (Allington, 1986).

ESEA also funded training for basic skills teachers, especially reading teachers, creating another new professional role. This funding resulted in the creation of graduate programs to prepare diagnostic-prescriptive reading teachers to work with disadvantaged children. While ESEA did not require schools to employ special reading teachers in order to provide the supplemental instruction, most school districts found this strategy the simplest way to meet federal program audit requirements. Thus, the specialist teacher, pull-out model for compensatory educational services was widely implemented. The ESEA also provided funding for creating and staffing school libraries. By the late 1960s, remedial reading teachers, paraprofessionals, librarians, psychometrists, media specialists, and a variety of program administrators were thus added to the American education workforce.

Coincident with the ESEA implementation, the field of learning disabilities was emerging. Kirk and Bateman (1962–63) coined the term "learning disability" to describe the unexpected learning difficulties of those children who exhibited normal intellectual abilities but significant underachievement in one or more academic areas. In 1966 the U.S. Office of Education modified the formal definition of "minimal brain dysfunction" thereby producing the category of learning disabilities based upon an achievement deficit in children with normal intellectual abilities (Fletcher et al., 1994).

In 1975 the Education of Handicapped Children Act (EHA) was passed. Federal law now entitled handicapped children to a "free appropriate public education" in the "least restrictive environment." Children with disabilities could no longer be denied access to educational services nor could they be routinely shunted into special schools. In addition, another group of children who were having difficulty with literacy learning were codified into law and special services mandated. The "learning disabled" child was not, by legislative definition, disadvantaged, language different, emotionally unstable, behaviorally disruptive, intellectually limited, or an irregularly attending student. Learning disabled children were having difficulty learning to read and write in American schools but did not easily fit any of the categorical program guidelines previously implemented (McGill-Franzen, 1987).

The EHA also provided funds for training additional special education teachers to meet the requirements of the special education law and included a regulatory structure adding school psychologists, speech therapists, physical therapists, and special education administrators as well as special education teachers to the professional staff of American schools (Singer & Butler, 1987).

Compensatory and Special Education

By 1975, then, remedial education was legislatively and fiscally divided into two broad but separate strands: compensatory education, with a focus on economic and environmental disadvantage as primary sources for literacy learning difficulties and special education focusing more on children whose learning capacity was deemed impaired. Special education itself included "moderately and profoundly handicapped" children—children with fundamental intellectual and physiological impairments—as well as "mildly handicapped children" with seemingly normal intellectual capacity who were experiencing learning difficulties. This latter group included children identified under a number of categorical labels who had not often participated in special education programs or been identified as handicapped prior to the passage of EHA (e.g., children identified as learning disabled, language impaired, behaviorally disordered). Children who had historically been identified as "dyslexic" were subsumed, largely, under the new and broader umbrella of the "learning disability" label.

The traditional concept of the slow learner was split into *disadvantaged* and *disabled* orientations (McGill-Franzen, 1987). Children experiencing difficulties with literacy learning were fit into one of these two broad explanations. Two sets of remedial programs evolved. Compensatory education programs were designed to provide instruction that would alleviate the negative effects of family economic disadvantage, with federally funded *Head Start* and *Title 1* programs but two examples. Children who grew up in poverty often lacked the book, story, and text experiences of their more advantaged peers—experiences valued in the early school curriculum. These children started school "behind" their more advantaged peers. So, preschool compensatory education emerged, programs designed to compensate for the economic and environmental disadvantages experienced by poor children (McGill-Franzen, 1993).

New special education programs and categories emerged for children. The categories of "educable mentally retarded," "language impaired," and "learning disabled" became the labels given children who seem to function normally outside of school settings but who experienced substantial difficulties with literacy learning. Children identified with these labels now comprise the majority of all children with disabilities. These disability labels are part of the broadly defined mildly handicapped classification, which largely replaced the older label, "slow learner." While the the notion of intellectual limitations as a source of learning difficulties was perpetuated in the concept of "educable mentally retarded", children considered "language impaired" or "learning disabled" were those whose intellect fell within average or above-average ranges but who still experienced difficulties in school. Although some educators believe that some sort of neurological malfunctioning is the source of these difficulties, supporting evidence is largely unavailable (Tashman, 1995).

Thus, two sets of programs and professionals have evolved primarily for the purpose of remediating difficulties in language and literacy learning. Although learning difficulties occur in other academic areas, it is literacy learning difficulties that most often precede referral to compensatory or special education programs. Compensatory education programs usually employ curriculum specialists (e.g., reading teachers, writing teachers, math teachers), while special education programs for the mildly handicapped usually employ categorical specialists (e.g., learning disabilities teachers, speech and language pathologists, occupational therapists). Although compensatory education programs have usually adopted (but rarely met) goals of accelerating literacy development in an attempt to normalize achievement, special education programs have more often adopted more educationally limited goals because the presumed disabilities are expected to limit the learning potential of individual children.

These differences in program orientations have grown in importance as large numbers of children have been identified as mildly handicapped in recent years, a trend that is as yet unabated. Few differences have been found in the characteristics of children participating in compensatory education programs and special education programs for the mildly handicapped (Birman, 1981; Fletcher et al., 1994; Shepard, 1989; Ysseldyke, 1987). Few differences have been observed in the instruction offered in the two programs (Allington & McGill-Franzen, 1989; Gelzheiser & Meyers, 1991; Turnbull, 1995). These findings run contrary to conventional wisdom and current educational policy, both of which rest on the assumption that there are two quite different populations of children having difficulty in school.

The net effect of redefining the sources of difficulties in literacy learning has been to shrink the number of children served in compensatory education programs and increase the numbers served in special education programs for the mildly handicapped. Since the 1960s participation in Title 1 compensatory education has decreased substantially from about 8 million children served annually to about 5 million (U.S. Department of Education, 1993). At the same time the number of children identified as learning disabled or language impaired has risen dramatically from about 1 million to over 3 million children. The identification of children as learning disabled has expanded dramatically over the past 20 years increasing 183% and growing from about a quarter to half of all children identified as handicapped (Office of Special Education and Rehabilitative Services, 1993).

Similarly, the number of compensatory education teachers funded through the federal Title 1 shrunk, between 1980 and 1990, from 87,000 to 69,000 (Moskowitz, Stullich, & Deng, 1993). Between 1976 and 1990 the number of special education teaching positions nearly doubled from 180,000 to 300,000 (Office of Special Education and Rehabilitative Services, 1985; 1993). There were also 65,000 Title 1 paraprofessionals and 300,000 other special education staff members employed, including 155,000 paraprofessionals, in 1990. These numbers suggest an additional trend common to both compensatory and special education efforts—increasing rates of employment for paraprofessional staff. This trend is of concern given the relatively low level of education and professional preparation of these paraprofessionals, especially given the evidence that paraprofessionals often offer instruction with little or no professional supervision (Millsap, Moss, & Gamse, 1993).

As McGill-Franzen (1987) noted, recasting the source of difficulty in literacy learning from disadvantage to disability has had a substantial impact not only on the funding of educational programs but also on teacher training and the professional and paraprofessional staff employed in American schools. Between 1960 and 1990 there was an enormous expansion in the numbers of persons employed in American schools. Large numbers of these additional staff arrived in schools as a result of federal educational initiatives that gave birth to a second system of education.

Compensatory and Special Education Today

The situation today, then, is that children experiencing difficulty in literacy may be served in programs developed from either of these two broad perspectives about learning difficulties. The federal education initiatives that created these two categorical programs now represent the largest federal financial contributions to education. But while the initiatives that created these programs (actually funding streams for programs) were enormously well intended, their very presence may have fostered quite unintended effects—especially the creation of a huge segregated second system of education.

This brief history of the "second system" has offered only the major federal education initiatives that have added special program staff to American elementary schools but a daunting array of other smaller, federal, state, and local initiatives have also contributed to the expansion (e.g., Bilingual Education Act, Drug Free Schools Act, dropout prevention projects, magnet schools initiatives, gifted and talented programs, computer technology initiatives, community school projects, Eisenhower math and science initiatives, migrant education programs, sex and health education initiatives, career ladder projects, and so on).

The expansion of the "second system" has added new personnel to our schools at an ever-increasing rate since the passage of the NDEA some 35 years ago. The addition of these personnel has not had much impact on the average class size because most of these new staff members are not classroom teachers. The "second system" expansion has, however, absorbed much of the new money available to educate children and created substantial fragmentation of the school day and of the professional responsibility for educating children who find learning to read difficult (Janney, Snell, Beers, & Raynes, 1995; Lankford & Wyckoff, 1995; McGill-Franzen & Allington, 1991).

The current state of the "second system" in American schools seems well-described in a recent federal policy statement concerning "reinventing government."

> It is almost as if federal programs were designed not to work. In truth, few are "designed" at all; the legislative process simply churns them out, one after another, year after year. (Ivins, 1993, p. 6)

It has been roughly 100 years since universal compulsory school attendance was mandated and American schools were required to serve everyone. Across this 100-year period remedial education has been legislated, judicially mandated, or simply funded in an attempt to entice or coerce schools into providing a good common education for all children. But most previous reforms have had but small impacts on the actual instruction offered in the classrooms (Cohen & Spillane, 1992). Perhaps the major accomplishment of American education in the twentieth century has been the creation of the "second system" of educational services.

American schools can no longer deny some children entry nor can they suspend students who fail to maintain normative behavior. Schools cannot let the uninterested drift away and cannot ignore those students who fail to thrive academically. Nonetheless, Tropea (1993) concludes that while labels have changed in response to legislative mandates and court rulings, schools have always managed to maintain the segregation of harder-to-teach children from the regular classroom program primarily through the maintenance of special programs. Federal education policy has largely, though unintendedly, served to perpetuate such segregation and undermine a sense that the regular education system has the primary professional responsibility for educating all children.

Perhaps in recognition of these unintended effects, recent federal education policy, in both compensatory and special education, has moved towards an emphasis on more inclusionary and more collaborative approaches to addressing the instructional needs of harder-to-teach children and adolescents.

The Future of Compensatory and Special Education

While there is a general consistency in the evidence and opinion concerning the ineffectiveness of current approaches to educating hard-to-teach children (Skrtic, 1995), a full-blown debate literally rages within the profession as to the most appropriate solution. The notion that the primary purpose of special programs is to accelerate participants' academic achievement is a relatively recent phenomemon (Allington, 1991). While few educators would likely feel comfortable describing compensatory and special education programs as "warehousing" efforts, truth be told, few special programs either set or achieved academic acceleration goals until 1985 or later. Perhaps Marie Clay (1985) deserves as much credit as anyone for providing the impetus for accelerating learning and for providing the evidence that a dramatic acceleration of literacy learning could be achieved when particular instructional conditions were met. She brought her Reading Recovery program with its emphasis on intensive teacher training and tutorial instruction to the U.S. in the early 1980s. The various reports of the success of this early intervention effort, even with learning disabled children (Lyons, 1994), drew the attention of the professional community immediately. However, others (c.f. Allington & Walmsley, 1995; Slavin, Karweit, & Madden, 1994; Hiebert & Taylor, 1994) have also provided good evidence that restructuring traditional schooling can result in accelerated academic achievement and in a significant decrease in the number of children who are retained in grade or identified as handicapped.

Acceleration of academic achievement has become a potentially defining purpose for compensatory and special education programs (though we will note that the impact on practice generally seems yet quite small). For instance, LeTendre (1991, p. 328), head of the U.S. Office of Compensatory Education, has written that, "The explicit purpose of Chapter 1 [now Title 1] is to enable low-achieving students to catch up and keep up." This statement would seem to provide substantial clarity concerning the primary mission of efforts funded under this federal program.

More recently, the U.S. Supreme Court has upheld the U.S. Court of Appeals interpretation that the role of special education services, at least for students identified as learning disabled, also fit the "catch up and keep up" definition (Walsh, 1993). The lower court's decision in the Carter v. Florence County School District criticized the school district's practice of preparing annual goals for a learning disabled student, Shannon Carter, that reflected only four months growth in literacy development. In the decision that was upheld, the lower court had argued, "Clearly, Congress did not intend that a school district could discharge its duty under the Act by providing a program that produces some minimal academic achievement, no matter how trivial." (United States Court of Appeals for the Fourth Circuit, Carter v. Florence County School District, November 26, 1991, p. 7a)

Thus, the lower court redefined the concept of educational benefit, moving away from relying on examining simple educational program inputs (e.g. teacher certification, class size, existence of IEP) and placed the question of academic growth clearly on the agenda (McGill-Franzen, 1994).

Historically, of course, both compensatory and special education programs have been largely evaluated not from an educational benefit perspective but from a regulatory compliance perspective. In the case of the federal Title 1, compensatory education began to change in 1988 with the passage of the Hawkins-Stafford amendment to the Educational Consolidation and Improvement Act. This amendment mandated an annual evaluation of Title 1 program effects on participating children. Basically the new regulation created a "do no harm" provision by mandating that participants in compensatory education should, at the very least, not fall further behind their peers academically as a result of receiving services. Still, a significant number of local programs failed to meet even this minimal standard for producing an educational benefit.

Currently, there is much interest in the issue of assessing the educational benefit of participation in compensatory and special education. The panel of educational researchers who provided much of the framework for the recent Improving America's Schools Act of 1994 (which reauthorizes the Elementary and Secondary Education Act, including Title 1 and other compensatory education programs), perhaps best summarized the current mood when it noted, "Accountability systems must focus more on outcomes than on regulation of process and inputs." (U.S. Department of Education, 1993, p. 19)

We expect that the issue of evaluating educational benefits will be an important aspect of the discussion as Congress considers the reauthorization of the legislation supporting current special education programs. This is hardly a case of crystal-balling on our part since several influential groups have called for inclusion of students with disabilities in the administration of the National Assessment of Educational Progress (e.g., Shriner, Ysseldyke, Thurlow, & Honetschlager, 1994). In addition, several meetings of scholars and practitioners have been held to discuss inclusionary testing and to discuss appropriate modifications to testing procedures when needed.

Much recent policy anaysis has focused on reconceptualizing compensatory and special education programs (e.g. Allington & Walmsley, 1995; Commission on Chapter 1, 1993; Congressional Budget Office, 1993: McGill-Franzen, 1994; Schrag, 1993; U.S. Department of Education, 1993; Wang,

Reynolds, & Walberg, 1993). There is mounting evidence that current programs not only provide few instructional benefits to targetted children but
such programs are also more often a barrier to educational reform than a
catalyst for accomplishing reform.

Two reports to Congress pointed out that the "second system" design simply
could not overcome fundamental deficiencies in regular education.

> The "add-on" design, ... cannot work when the regular school program
> itself is seriously deficient ... If [compensatory education] is to help
> children in poverty to attain both basic and high-level knowledge and
> skills, it must become a vehicle for improving whole schools serving
> concentrations of poor children. (Commission on Chapter 1, 1993,
> pp. 6–7)
>
> Federal funds should be used to reform and improve the whole school
> program ... supplementary services for 30–40 minutes a day cannot be
> expected to compensate for regular educational services with low
> expectations for students, ineffective curricula and instructional practices,
> and inadequately trained staff and professional leadership. (U.S.
> Department of Education, 1993, p. 10)

As soon as these statements were released they were attacked for being
focused only on compensatory education programs. "We believe the panel
made a major mistake by excluding from its proposals [other] categorical
programs ... such as programs for migrant workers' children, handicapped
children, neglected and delinquent children, limited English proficient
children, and Native American children." (Wang, Reynolds, & Walberg, 1993,
p. 64)

A central theme in each of these statements is the need to focus on building
the capacity of schools—regular education—to better serve children with
special educational needs. The statements reflect a shift away from the
tradition of increasing categorical funds to employ ever more special program
personnel. The current focus of educational reform efforts seems to be one
of strengthening regular education and shrinking the "second system" to pay
for some of the capacity-building that is needed. If this is the future, then
substantial changes will need to be made in current routines and practices,
perhaps none more so than the testing routines that have been constructed
for identifying children who will participate in the "second system"
programs.

Identification and Assessment in Compensatory and Special Education

Virtually all remedial education programs, whether compensatory or special
education, have used some form of testing to identify students who will

participate. Testing, then, has been big business in remedial education with much time and effort spent in establishing eligibility and estimating educational progress after admittance to the programs. In fact, children who participate in compensatory and special education are undoubtedly the most frequently tested children in schools. As a result of widespread use, the assessment practices and tools have been widely studied.

In both compensatory and special education, testing routines have been developed to identify children eligible for program participation and to evaluate participants' instructional needs. There has been an on-going debate about the adequacy of these testing routines and the actual role the routines play in defining either participant eligibility or instructional needs.

Testing in Compensatory Education Programs

Eligibility for participation in compensatory education has usually been restricted to students exhibiting substantial achievement deficits. Thus, standardized achievement testing has been the most common method of establishing eligibility for participation in compensatory education programs.

The extra instructional support has typically been available only when academic achievement fell below some predetermined standard. However, this standard often shifted depending on the level of funding available. Few compensatory programs have been fully funded—funding was, and is now, simply insufficient to provide instructional support services to all children who qualify. For instance, Title 1, the largest federal compensatory education program, has been underfunded such that only about half of all children and youth who would be eligible to participate receive any services whatsoever. Similarly, Head Start programs have been available to only about one in four eligible children (U.S. General Accounting Office, 1995). Thus, rather than determining who might need extra-ordinary instructional support, the assessment routines generally have served simply to ration limited compensatory education services.

Under Title 1, the regulations have required the children in greatest need be the first served and schools with the largest enrollments of poor children to be first funded. The common assessment routine to establish educational "neediness" has been to administer a standardized norm-referenced reading achievement test to eligible disadvantaged children and then to rank order performances to establish need for compensatory education services. Common practice or not, such procedures based on norm-referenced group achievement test scores typically violate accepted use of these measures.

School districts have often simply set a standard cut-off score on norm-referenced reading achievement tests to determine who will participate in compensatory education. However, the level of these cutoff scores varies from district to district. In schools serving the largest numbers of eligible children, only those with very low test scores are served, while in schools with fewer eligible children, those with much higher measured achievement have been selected participate in the programs (Moskowitz, et al., 1993)

Once selected for compensatory education services, children have often undergone a second round of testing. Such testing is considered diagnostic and is to provide information about the specific information to guide instruction offered the participating children. Diagnostic testing has often included oral and silent reading and testing of word and phonics knowledge. Most of the testing instruments used in these diagnostic sessions have minimal, if any, demonstrated reliability (Salvia & Ysseldyke, 1985). The diagnostic judgements made by those administering these tests were usually quite variable and unreliable. However, since little evidence is available that teachers actually used such diagnostic information to plan instruction for the children tested, this lack of evidence of the reliability is, perhaps, not a major cause for concern.

Testing is also done in compensatory education programs as part of an effort to estimate program effectiveness. After a quarter-century of such testing it has been established that most interventions have small impacts on the achievement of participating children (LeTendre, 1991).

The testing that is driven by categorical programs is expensive and widely viewed as ineffective. Not surprisingly, these criticisms have resulted in serious reexamination of traditional testing practices, especially those involving group administered norm-referenced tests and the use of many of the traditional diagnostic measures (Garcia & Pearson, 1991; Johnston, 1992). It is generally recommended that traditional testing routines be replaced with measures that are more curriculum-based and more naturalistic assessments of academic learning (Darling-Hammond, 1994). However, to date little real progress has been made in this area.

Testing in Special Education Programs

Identifying a child as disabled has suggested some physical or psychological impairment that interferes with learning. Thus, much of the testing has been designed to provide evidence for inferences about the extent to which learning capacity is impairedand the source of that presumed impairment. Much of the initial testing has been done by school psychologists or licensed psychometrists. While achievement testing has been included in these routines, more emphasis has been placed on attempting to assess academic aptitude and to isolate psychological factors that seem to interfere with learning. Of course, for many low-incidence conditions such as deafness, visual impairments, severe mental retardation and so on, other types of tests are also administered. As with testing in compensatory education programs, there is little evidence of the reliability of most of the commonly used assessment instruments and routines (Coles, 1987; Salvia & Ysseldyke, 1985). Likewise, there is little evidence that special education teachers use these test results in instructional planning or delivery.

For instance, Smith (1990) provided a comprehensive review of the research on the development of IEPs. He found substantial documentation of general inadequacies including little evidence that the substantial amounts of testing contributed much to the content of the IEP. He also noted the growing use of computer technology in the development of IEPs and argued that this practice undermined the very notion of individual planning. Likewise, Wesson and Deno (1989) found a high degree of uniformity in the IEPs developed by the same teacher for different children suggesting little attention to the results of individual assessments. Smith (1990) concluded that the weight of the research suggests that it is time to acknowledge that the IEP process is not viable and is an impractical notion that should be replaced with alternatives more attuned to fostering specifically designed instruction.

Further support for the relative impotence of the IEP and test results as instructional planning supports is found in studies of regular education teachers planning for mainstreamed students with disabilities (Schumm & Vaughn, 1992). Both were ranked among the least used sources of information by elementary, middle school, and high school classroom teachers. The classroom teachers reported that other teachers, teacher manuals, and professional journals were more frequently used than either IEP information or test results.

Nonetheless, testing plays an enormous role in reifying the decision to identify children as handicapped. Most children referred by classroom teachers for consideration as handicapped are identified as handicapped after testing. There is evidence that testing usually continues until test results support the decision to identify as handicapped (Mehan, Hartweck, & Meihls, 1986). The very nature of the most commonly used tests suggests why these results are so seldom used in planning educational services. The focus on isolating psychological factors seen as relevant to the disability provides little useful information about participant's instructional needs or instructional histories.

Hutton, Dubes, and Muir (1992) and Ganschow, Sparks, & Helmick (1992) report that school psychologists most commonly use the *Illinois Test of Psycholinguistic Abilities*, the *Bender Visual-Motor Gestalt Test*, the *Wide Range Achievement Test*, the *Brigance Diagnostic Inventory*, the *Woodcock-Johnston Psycho-educational Battery*, and the *Visual Motor Integration Test* in assessments of children referred for evaluation for the existence of a handicapping condition. These are assessment tools that Salvia and Ysseldyke (1985) identify as failing to meet even minimum standards of demonstrated reliability and validity. Thus, the testing for identification fails to even approach the minium standards of adequacy.

Large-scale achievement testing has seldom been conducted in special education programs. Currently, federal and state agencies rarely include disabled children in public educational accountability testing programs and rarely require any routine reports of academic progress once a child is identified as disabled (Allington & McGill-Franzen, 1992; McGrew, Thurlow, & Spiegel, 1993). This pattern of exclusion of disabled children from achievement assessments has been attributed to the psychometric difficulties in testing low-achieving children. But such difficulties are hardly insurmountable.

Another explanation is that schools have low expectations for the academic potential of these children and academic achievement testing is seen as largely irrelevant. Finally, it has been reported that some schools remove low-achieving children from the accountability roles by identifying them as handicapped (McGill-Franzen & Allington, 1993). This allows schools to report only the reading achievement of students not identified as handicapped—scores that invariably produce an illusion of a higher level of achievement in the school.

Students who receive remedial instruction in either compensatory or special education settings are the most frequently tested persons in schools. Yet much of this testing is done with instruments and procedures that lack both demonstrated validity and reliability. Only rarely, it seems, do these students participate in evaluative sessions that produce rich sources of information about their academic development or their instructional needs. Too often the achievement of these students is buried or removed from public summaries of student achievement. In short, one wonders why such testing continues and why no one seems much interested in dramatically improving the situation or eliminating the ruse of reliable classification of children and reliable identification of instructional needs.

Shepard (1993) argues that educational testing should be subject to tests of "consequential validity." In other words, she argues that demonstrations of traditional sorts of validity (e.g., content or construct validity—the tests measures what it purports to measure) though necessary, are insufficient. Most educational testing has intended consequences. The test results might, for instance, be intended to inform teachers or administrators in ways that lead to improved instruction. Or the testing might be intended to determine whether a child will participate in a compensatory or special education program, in order to enhance educational opportunity. In her argument Shepard uses the particular example of testing children as part of the process of determining whether placement in special education will occur. While establishing the content validity of a reading achievement test is important, it pales in importance compared to the larger intended consquences of the test—placement in special education. Of particular importance is whether the intended result, improved educational opportunity, is achieved through the test administration. In other words, Shepard argues that we need to evaluate the tests more on the larger consequences while remaining concerned with traditional measures of test reliabilility and validity. As Shepard (1993) notes, "The most critical requirement is that ... groups must be better off in their respective treatments than they would have been without the test-based placement." (p. 441) Given what we know about the importance of testing in identifying children who participate in compensatory and special education programs, and what we know about the unlikelihood that participation substantially enhances educational opportunity, it would seem that most current testing schemes utterly fail to meet the consequential validity standard.

There are substantial problems with current testing routines for identifying students who will participate in compensatory and special education programs. Problems of misuse and abuse of tests and test information dominate discussions in the professional literature. However, few studies of testing have even considered the longer-term educational consequences of testing routines used to identify participants in compensatory and special education programs. How might testing change if schools were required to demonstrate the consequential validity—the longer-term educational benefits—of their assessment practices?

Nonetheless, widespread testing of students continues. Which students, then, do the testing routines identify to participate in compensatory and special programs?

Who Participates in Compensatory and Special Education Programs?

It is largely the children of poverty that populate compensatory and special education programs. Of course, the federal Title 1 effort has always been targeted to serve children from economically disadvantaged backgrounds. But special education programs also enroll primarily poor children. Singer and Butler (1987) noted that almost half of the children enrolled in special education were from families with incomes below 150% of the federal poverty line.

Wagner (1995), reporting on data from the National Longitudinal Transition Study of Special Education, notes that both poor and minority students were represented among special education students at rates higher than they occurred in the general population. Children from low-income families (incomes of less than $12,000 in 1986 dollars) were substantially overrepresented, while economically more advantaged children were underrepresented. For instance, over a third of the children were from poor families in the special education population analyzed by Wagner, whereas poor children represented only 18 percent of the general school population. She reported that over two-thirds of the special education students were from families with incomes below $25,000 while fewer than a third of the children from families with incomes above that level were identified as disabled (even though these children represented almost two-thirds of the school-age population). These figures indicate that special education programs primarily serve children of low-income families, much like the federal Title 1 program.

While reports of the overrepresentation of minority students in special education has been a longstanding issue (Heller, Holtzman, & Messick, 1982) and the basis for several successful lawsuits (Reschly, 1988), Wagner's (1995) analysis suggests that such overrepresentation occurs primarily because minority students are disproportionately from poor families. Minority over-representation is largely limited to one ethnic minority—African-American students. African-American students accounted for 24 percent of the special education population, double the proportion of African-American students in the school population. But Wagner argues that African-American children are similarly overrepresented in the population of poor children (53% came from household with incomes of less than $12,000). Thus, according to Wagner's analysis, childhood poverty, rather than ethnicity, seems to be the key variable in explaining the overrepresentation of African-American students in special education programs.

Gregory (1993), on the other hand, provides an analysis of special education students in one large, mid-western state and arrives at a different conclusion. He argues, for instance, that of every 100 students classified as emotionally impaired, 50 students were African-American males while only 9 were Caucasian girls. He presents similar analyses of other categories that illustrate both the overrepresentation of males generally, but African-American males particularly. He characterizes the phenomonon as the M &M (male and minority) syndrome. Similar findings are reported by Walsh (1988) in his analysis of flunking in the primary grades.

The available evidence suggests an additive structure whereby the likelihood of children being referred for identification as disabled will be increased with each additional factor. The most powerful factor seems to be family socioeconomic-status (and accompanying parental educational attainment). Poor children are far more likely to have difficulties with academic learning upon entering school. Gender may be the next critical factor, since boys are typically overrepresented in both compensatory and special education even though the evidence that boys are more likely to experience literacy learning difficulties is contradicted by at least one well-designed, large-scale study (Shaywitz, Shaywitz, Fletcher, & Escobar, 1990). Finally, ethnicity also seems a contributing factor since African-American students are overrepresented and Latino and Asian students are underrepresented in special education programs.

Are Compensatory and Special Education Designed to Accelerate Learning?

The evidence suggests that compensatory and special education programs have had mixed success in mitigating literacy learning difficulties. Large-scale evaluations of the federal Title 1 program, for instance, suggest that the average program promotes improvement in achievement but the gains are small enough to disappoint most proponents(Anderson & Pellicier, 1990; Birman, et al., 1987; Puma, Jones, Rock, & Fernandez, 1993). There has been no large-scale evaluation of the effects of participation in special education on academic achievement (Gartner & Lipsky, 1987). Nonetheless, the available evidence (e.g. Englert, 1983; Kavale, 1988; Lytle, 1988; Singer & Butler, 1987) suggests that special education only rarely provides services sufficient to remediate the academic difficulties of participating students.

However, there is good evidence that programs can be designed and implemented so that far more encouraging results are achieved (Allington & Walmsley, 1995; Slavin, et al., 1994). Nonetheless, such programs have not yet been established in most school systems. Most compensatory and special education programs operate during the regular school day and rarely extend children's opportunity to learn, even in terms of instructional time (Allington & McGill-Franzen, 1989; Haynes & Jenkins, 1986; O'Sullivan, Ysseldyke, Christianson, & Thurlow, 1990).

The most effective programs seem to be those that (1) offer early intervention, usually at ages 4–7, (2) increase the opportunity to learn by making more instructional time available to children, (3) increase the intensity of instruction as in tutorials, and (4) provide children with expert instruction delivered by well-trained professionals. Programs that typify such designs offer good evidence that most children can be successful in learning to read and write alongside their peers, including especially disadvantaged children and children now identified as learning disabled and language impaired (Lyons & Beaver, 1995; Knapp, 1995; Slavin, 1994). Unfortunately, compensatory and special education programs often match none of these criteria, particularly when the instruction is delivered by paraprofessionals to small groups of children during the school day (often during classroom language arts instruction).

Special education interventions for children labelled as mildly handicapped have not been evaluated for achievement gains in the same spirit as compensatory programs. Perhaps because these children are viewed as "damaged" and not capable learners, few have set accelerating literacy development (or catching up to peers) as a goal for such students. With no such goal few evaluations examine program effects in this area. Nonetheless, special education programs for the mildly handicapped only rarely improve on regular classroom instruction. For instance, special education teachers offer reading and language arts lessons of shorter duration than either regular or compensatory teachers, the focus is more often on low-level task completion, and participating children are no more likely to be using instructional materials of appropriate difficulty in the special class setting than in the regular education classroom (Allington & McGill-Franzen, 1989; McGill-Franzen & Allington, 1990).

In short, then, there is currently too little evidence that either compensatory or special education programs actually provide substantial academic benefits to most participating children. Research on such programs has noted that the lack of effects on literacy development is more easily explained by program designs than by characteristics of the children who participate (Allington & Johnston, 1989). When the design of compensatory and special education interventions is examined carefully, fundamental problems are often obvious.

For instance, since both types of programs typically operate during the school day, they simply cannot actually expand the instructional time available. To produce an academic benefit, then, the instruction offered would have to be significantly more appropriate and more intensive than that available in the classroom. As noted earlier, that is too often not the case. In addition, participation in either compensatory or special education is typically a source of fragmentation of curriculum for the children who participate. Often children who attend compensatory or special programs, as well as participate in "regular" classroom lessons, are exposed to different literacy curricula developed from different philosophical positions and delivered by teachers who rarely meet to plan instruction collaboratively. These children, then, often have to make greater adjustments to varying curricular demands and instructional emphases across the school day than do achieving children. As the children move from location to location in the school to participate in special instructional programs they lose potentially valuable instructional time in the transitions.

Both compensatory and special education personnel are often largely unaware of the nature of the instruction offered in regular education classes and regular education teachers have little knowledge of instruction offered in special classes (Johnston, Allington, & Afflerbach, 1985). This lack of shared knowledge about the content and nature of the instruction creates an unlikely situation for delivering instruction in either setting that is coherent, consistent, and coordinated. There have been several changes in program regulations meant to foster greater collaboration between regular education and special programs teachers over the past decade. Nonetheless, Turnbull (1995) notes, "The resource room represented sort of a Bermuda Triangle for the instructional program: The classroom teachers tended to know little or nothing about the instruction that took place there." (p. 109)

A few schools are experimenting with alternative program designs and moving toward a more unified effort, attempting to undo much of the categorization of children that has accumulated over time (Bean, Hamilton, Zigmond, & Morris, 1994; Gelzheiser, Meyers, & Pruzek, 1992; Goatley, Brock, & Raphael, 1995; Hiebert, Colt, Catto, & Gury, 1992; Jenkins, Pious, & Petersen, 1988; Jenkins & Leicester, 1992). New federal policy empha- sizes extended-time programs, with before- and after-school, or Saturday school, and summer school remedial instruction. The potentials of early intervention, tutorial support, and family-based programs are also receiving renewed emphasis. Nonetheless, compensatory and special education pro- grams typically remain quite traditional, often bound in organizational constraints that limit possible responses to learning difficulties.

While educational policy has supported expansion of the "second system" of compensatory and special education programs there is growing concern about the issues raised above. Some children do need access to more and better instruction in order to maintain development alongside their peers. Accelera- tion of academic development can be achieved and at an expense no greater than that required to support the second system now in place. But the keys to achieving such a result is the development of early, intensive, and responsive educational interventions.

Summary

Perhaps it is time, as Allington and Cunningham (1996) suggest, to refocus our efforts on enhancing the quality of classroom core curriculum instruction. They recommend that all school staff who are not classroom teachers be evaluated annually against a single criteria: Evidence that classroom instruc- tion has been enhanced as a direct, or indirect, result of the "second system" staff member's presence and professional efforts. This emphasis harkens back to Reynold's (1976, cited in Skrtic, 1995, p. 624) early call for using the extra- ordinary funding provided under various federal and state initiatives to create "more diverse educational environments [in the regular classroom], which would diminish the need to develop and use separate specialized educational environments ... through the redistribution of resources and energies, through training, and finally, through the redistribution of students."

Such a shift would require a massive change of attitudes, beliefs, programs, and practices in both regular education and compensatory and special educa- tion. But such a shift seems, perhaps, the most appropriate reconceptualization of an appropriate public education in the least restrictive environment. Rather than continuing to design compensatory and special education programs as if the source of the problem was some difference, disadvantage, or disability inherent in the student, it is now time to consider the restructuring of regular education to better meet the needs of all students. Instead of perpetuating past practice and the segregationist tradition of American education, the time has come to change the regular education system in a dramatic fashion.

References

Allington, R.L. (1986). Policy constraints and effective compensatory reading i review. In J. Hoffman (Eds.), *Effective teaching of reading: Research* (pp. 261–289). Newark, DE: International Reading Association.

Allington, R.L. (1991). The legacy of 'slow it down and make it more concret. & S. McCormick (Eds.), *Learner factors/teacher factors: Issues in literacy research and instruction* (pp. 19–30). Chicago: National Reading Conference.

Allington, R.L. (1994). What's special about special programs for children who find learning to read difficult? *Journal of Reading Behavior, 26,* 1–21.

Allington, R.L. & Cunningham, P. (1996). *Schools that work:All children readers and writers.* New York: HarperCollins.

Allington, R.L., & Johnston, P.A. (1989). Coordination, collaboration, and consistency: The redesign of compensatory and special education interventions. In R. Slavin, N. Karweit, & N. Madden (Eds.), *Effective programs for students at risk* (pp. 320–354). Boston: Allyn-Bacon.

Allington, R.L., & McGill-Franzen, A. (1992). Unintended effects of educational reform in New York State. *Educational Policy, 6,* 396–413.

Allington, R.L., & McGill-Franzen, A. (1989). Different programs, indifferent instruction. In A. Gartner & D. Lipsky (Eds.), *Beyond separate education: Quality education for all* (pp. 75-98). Baltimore: Brookes.

Allington, R.L., & McGill-Franzen, A. (1994). Compensatory, remedial, and special programs in language and literacy. In A. Purves (Ed.), *Encyclopedia of English Studies and Language Arts* (pp. 225–228). New York: Scholastic.

Allington, R.L., & Walmsley, S.A. (1995). *No quick fix: Rethinking literacy programs in America's elementary schools.* New York: Teachers College Press.

Anderson, L.W., & Pellicier, L.O. (1990). Synthesis of research on compensatory and remedial education. *Educational Leadership, 48,* 10–16.

Bean, R.M., Hamilton, R., Zigmond, N., & Morris, G.A. (1994). The changing roles of special education teachers in a full-time mainstreaming program: Rights without labels. *Reading and Writing Quarterly: Overcoming Learning Difficulties, 10,* 171–185.

Birman, B.F. (1981). Problems of overlap between Title I and P. L. 94–142: Implications for the federal role in education. *Educational Evaluation and Policy Development, 3,* 5–19.

Birman, B.F., Orland, M.E., Jung, R.K., Amon, R.J., Garcia, G.N., Moore, M.T., Funkhouser, J.E., Morrison, D.R., Turnbull, B.J., & Reisner, E.R. (1987). *The current operation of the Chapter I program: Final report from the National Assessment of Chapter I.* Washington, DC: U.S. Government Printing Office.

Bloom, B.S. (1976). *Human characteristics and school learning.* New York: McGraw-Hill.

Clay, M.M. (1985). *The early detection of reading difficulties: A diagnostic survey with recovery procedures.* Portsmouth, NH: Heinemann.

Cohen, D.K. & Spillane, J.P. (1992). Policy and practice: The relations between governance and instruction. In G. Grant (Ed.), *Review of Research in Education, vol. 18.* Washington, DC: American Educational Research Association (pp. 3–49).

Coles, G. (1987). *The learning mystique: A critical look at learning disabilities.* New York: Pantheon.

Commission on Chapter 1. (1993). *Making schools work for children of poverty: A new framework prepared by the Commission on Chapter 1.* Washington, DC: American Association for Higher Education.

Congressional Budget Office. (1993, May). *The federal role in improving elementary and secondary education.* Washington, DC: Congress of the United States.

Cooley, W. (1993). The difficulty of the educational task: Implications for comparing student achievement in states, school districts and schools. *ERS Spectrum, 11,* 27–31.

Crawford, J. (1989). Teaching effectiveness in Chapter I classrooms. *Elementary School Journal*, *90*, 33–46.

Cummins, J. (1986). Empowering minority students: A framework for intervention.*Harvard Educational Review, 56*, 18–36.

Darling-Hammond, L. (1994). Performance-based assessment and educational equity. *Harvard Educational Review, 64*, 5–30.

Dolch, E. (1948). *Helping handicapped children in school*. Champaign, IL: Garrard.

Dyer, P., & Binkley, P. (1995). Estimating cost-effectiveness and educational outcomes: Retention, remediation, special education, and early intervention. In R.L. Allington & S. Walmsley (Eds.), *No quick fix: Rethinking literacy programs in America's elementary schools*. New York: Teachers College Press. (pp. 61–77)

Fletcher, J.M., Shaywitz, S.E.,Shankweiler, D.P., Katz, L., Liberman, I.Y., Stuebing, K., Francis, D., Fowler, A.E., & Shaywitz, B.A. (1994). Cognitive profiles of reading disability: Comparisons of discrepancy and low achievement definitions. *Journal of Educational Psychology, 86*, 6–23.

Ganschow, L., Sparks, R., & Helmick, M. (1992). Speech/language referrals by school psychologists. *School Psychology Review, 21*, 313–326.

Garcia, G.E., & Pearson, P.D. (1991). The role of assessment in a diverse society. In E.H. Hiebert (Ed.), *Literacy for a diverse society: Perspectives, practices, and policies*. New York: Teachers College

Gartner, A., & Lipsky, D.K. (1987). Beyond special education: Toward a quality system for all students. *Harvard Educational Review, 57*, 367–395.

Gelzheiser, L.M., & Meyers, J. (1991). Reading instruction by classroom, remedial, and resource room teachers. *Journal of Special Education, 24*, 512–526.

Gelzheiser, L.M., Meyers, J., & Pruzek, R.M. (1992). Effects of pull-in and pull-out approaches to reading instruction for special education and remedial reading students. *Journal of Educational and Psychological Consultation, 3*, 133–149.

Goatley, V.J., Brock, C.H., & Raphael, T.E. (1995). Diverse learners participating in a regular education book club. *Reading Research Quarterly, 30*, 352–380.

Gould, S. J. (1981). *The mismeasure of man*. New York: Norton.

Gregory, J.F. (1993). *The M & M syndrome*. Paper presented at the conference of the International Association of Special Education, Vienna, Austria.

Haynes, M.C., & Jenkins, J.R. (1986). Reading instruction in special education resource rooms. *American Educational Research Journal, 23*, 161–190.

Heller, K.A., Holtzman, W., & Messick, S. (1982). *Placing children in special education: A strategy for equity*. Washington, DC: National Academy Press.

Hiebert, E.H., Colt, J.M., Catto, S.L., & Gury, E.C. (1992). Reading and writing of first-grade students in a restructured Chapter 1 program. *American Educational Research Journal, 29*, 545–572.

Hiebert, E.H., &. Taylor, B.M. (1994). *Getting reading right from the start: Effective early literacy interventions*. Boston: Allyn-Bacon.

Hutton, J.B., Dubes, R., & Muir, S. (1992). Assessment practices of school psychologists: Ten years later. *School Psychology Review, 21*, 271–284.

Ivins, M. (1993, September 14). Here's to the bureaucrats. *Times-Union*, p. 6

Janney, R.E., Snell, M.E., Beers, M.K., & Raynes, M. (1995). Integrating students with moderate and severe disabilities: Classroom teachers; beliefs and attitudes about implementing an educational change. *Education Administration Quarterly, 31*, 86–114.

Jenkins. J.R., & Leicester, N. (1992). Specialized instruction within general education: A case study of one elementary school. *Exceptional Children, 58*, 555–563.

Jenkins, J. R., Pious, C., & Peterson, D. (1988). Categorical programs for remedial and handicapped students: Issues of validity. *Exceptional Children, 55*, 147–158.

Johnston, P. (1992). *Constructive evaluation of literate activity*. New York: Longmans.

Johnston, P.A., & Allington, R.L. (1991). Remediation. In R. Barr, M. Kamil, P. Mosenthal & P. D. Pearson (Eds.), *Handbook of Reading Research, Vol. II* (pp. 984–1012). New York: Longmans.

Johnston, P., Allington, R.L., & Afflerbach, P. (1985). The congruence of classroom and remedial reading instruction. *Elementary School Journal, 85,* 465–478.

Kavale, K.A. (1988). The long-term consequences of learning disabilities. In M.C. Wang, M. C. Reynolds, & H. J. Walberg (Ed.), *Handbook of special education research and practice: Mildly handicapped conditions.* (pp. 303–344). New York: Pergamon.

Kirk, S.A. (1940). *Teaching reading to the slow learning child.* Boston: Houghton-Mifflin.

Kirk, S.A., & Bateman, B. (1962–62). Diagnosis and remediation of learning disabilities. *Exceptional Children, 29,* 73–78.

Knapp, M.S. (1995). *Teaching for meaning in high-poverty schools.* New York: Teachers College Press.

Langer, J.A., & Allington, R.L. (1992). Curriculum research in writing and reading. In P. W. Jackson (Ed.), *Handbook of research on curriculum.* (pp. 687–725). New York: Macmillan.

Lankford, H., & Wyckoff, J. (1995). Where has the money gone? An analysis of school district spending in New York. *Educational Evaluation and Policy Analysis, 17,* 195–218.

Lareau, A. (1989). *Home advantage: Social class and parental intervention in elementary education.* Philadelphia: Falmer.

LeTendre, M.J. (1991). The continuing evolution of a federal role in compensatory education. *Educational Evaluation and Policy Analysis, 13,* 328–344.

Lyons, C.A. (1994). Reading Recovery and learning disability:Issues, challenges, and implications. *Literacy, Teaching and Learning, 1,* 109–120.

Lyons, C.A., & Beaver, J. (1995). Reducing retention and learning disability placement through Reading Recovery: An educationally sound, cost-effective choice. In R.L. Allington, & S.A. Walmsley (Ed.), *No Quick Fix: Rethinking literacy programs in America's elementary schools.* New York: Teachers College Press.

Lytle, J. (1988). Correspondence: Is special education serving minority students? *Harvard Educational Review, 58,* 116–120.

McGill-Franzen, A.M. (1994). Is there accountability for learning and belief in children's potential? In E.H. Hiebert &. B.M. Taylor (Ed.), *Getting reading right from the start: Effective early literacy interventions.* Boston: Allyn-Bacon. (pp. 13–35)

McGill-Franzen, A.M. (1993). *Shaping the preschool agenda: Early literacy, public policy and professional beliefs.* Albany: State University of New York Press.

McGill-Franzen, A.M. (1987). Failure to learn to read: Formulating a policy problem. *Reading Research Quarterly, 22,* 475–490.

McGill-Franzen, A.M., & Allington, R.L. (1993). Flunk'em or get them classified: The contamination of primary grade accountability data. *Educational Researcher, 22,* 19–22.

McGill-Franzen, A.M., & Allington, R.L. (1991). The gridlock of low-achievement: Perspectives on policy and practice. *Remedial and Special Education, 12,* 20–30.

McGill-Franzen, A., & Allington, R.L. (1990). Comprehension and coherence: Neglected elements of literacy instruction in remedial and resource room services. *Reading, Writing, and Learning Disabilities Quarterly, 6,* 149–182.

McGrew, K.S., Thurlow, M.L., & Spiegel, A.N. (1993). An investigation of the exclusion of students with disabilities in national data collection programs. *Educational Evaluation and Policy Analysis, 15,* 339–352.

Mehan, H., Hartweck, A., & Meihls, J.L. (1986). *Handicapping the handicapped: Decision making in students' educational careers.* Stanford, CA: Stanford University Press.

Millsap, M. A., Moss, M., & Gamse, B. (1993). *The Chapter 1 implementation study: final report.* Washington, DC: U.S. Department of Education.

Moskowitz, J., Stullich, S., & Deng, B. (1993). *Targeting, formula, and resource allocation issues: Focusing federal funds where the needs are the greatest.* Washington, DC: U.S. Department of Education.

Office of Special Education and Rehabilitative Services. (1993). *To assure a free and appropriate public education of all children with disabilities: Fifteenth annual report to Congress on the implementation of the Individuals with Disabilities Education Act.* Washington, DC: U.S. Department of Education, Division of Innovation and Development.

Office of Special Education and Rehabilitative Services. (1985). *To assure a free and appropriate public education of all handicapped children: Seventh annual report to Congress on the implementation of the Education of Handicapped Act.* Washington, DC: U.S. Department of Education.

Osborne, A.G., & DiMattia, P. (1994). The IDEA's least restrictive environment: Legal implications. *Exceptional Children, 61,* 6–14.

O'Sullivan, P.J., Ysseldyke, J.E., Christenson, S.L., & Thurlow, M.L. (1990). Mildly handicapped elementary students' opportunity to learn during reading instruction in mainstream and special education settings. *Reading Research Quarterly, 25,* 131–146.

Puma, M.J., Jones, C.C., Rock, D., & Fernandez, R. (1993). *Prospects: The Congressionally mandated study of educational growth and opportunity—The interim report* (GPO 1993 0–354–886 QL3). U.S. Department of Education.

Reschly, D.J. (1987). Learning characteristics of mildly handicapped students: Implications for classification, placement, and programming. In M.C. Wang, M.C. Reynolds, & H.J. Walberg (Ed.), *Handbook of special education: Research and practice: Learner characteristics and adaptive education, vol. 1.* (pp. 35–58). New York: Pergamon.

Reynolds, M.C. (1976). *New perspectives on the instructional cascade.* Paper presented at the conference, The least restrictive alternatives: A partnership of general and special education. November 20–22, Minneapolis, MN.

Salvia, J., & Ysseldyke, J.E. (1985). *Assessment in special and remedial education.* Boston: Houghton-Mifflin.

Sawyer, R.J., McLaughlin, M.J., & Winglee, M. (1994). Is integration of students with disabilities happening? An analysis of national trends over time. *Remedial and Special Education, 15,* 204–215.

Schrag, J.A. (1993). Restructuring schools for a better alignment of general and special education. In J.I. Goodlad, & T.C. Lovitt (Ed.), *Integrating general and special education.* (pp. 203–228). New York: Merrill.

Schumm, J.S., & Vaughn, S. (1992). Planning for mainstreamed special education students: Perceptions of general classroom teachers. *Exceptionality, 3,* 81–98.

Shaywitz, S.E.; Shaywitz, B.A.; Fletcher, J.M., & Escobar, M.D. (1990). Prevalence of reading disability in boys and girls: results of the Connecticut Longitudinal Study. *Journal of American Medical Association, 264,* 998–1002.

Shepard, L.A. (1989). Identification of mild handicaps. In R.L. Linn (Ed.), *Educational Measurement* (pp. 545–572). New York: Macmillan.

Shepard, L.A. (1993). Evaluating test validity. In L. Darling-Hammond (Ed.), *Review of Research in Education, vol. 19.* (pp. 405–450). Washington, DC: American Educational Research Association.

Shriner, J.G., Ysseldyke, J.E., Thurlow, M.L., & Honetschlager, D. (1994). All means all–including students with disabilities. *Educational Leadership, 51,* 38–42.

Singer, J.D., & Butler, J.A. (1987). The Education for All Handicapped Children Act: Schools as agents of social reform. *Harvard Educational Review, 57,* 125–152.

Skrtic, T.M. (1991). The special education paradox: Equity as the way to excellence. *Harvard Educational Review, 61,* 148–206.

Skrtic, T.M. (1995). The special education knowledge and tradition: Crisis and opportunity. In E. L. Meyen & T. M. Skrtic (eds.), *Special education and student disability: Traditional, emerging, and alternative perspectives.* Denver: Love Publishing. (pp. 609–672)

Slavin, R.E. (1994). Preventing early school failure: Implications for policy and practice. In R. Slavin, N. Karweit & B. Wasik (Ed.), *Preventing early school failure: Research, policy & practice* (pp. 206–231). Boston: Allyn-Bacon.

Slavin, R.E., Karweit, N.L. & Wasik, B.A. (1994). *Preventing early school failure: Research, policy, and practice*. Boston: Allyn-Bacon.

Smith, S.W. (1990). Individualized Education Programs (IEP's) in special education: From intent to acquiescence. *Exceptional Children*, 6–13.

Spring, J. (1989). *The sorting machine revisited*. New York: Longman.

Tashman, B. (August, 1995). Misreading dyslexia: Researchers debate the causes and prevalence of the disorder. *Scientific American*, 14–16.

Tropea, J.L. (1993). Structuring risks: The making of urban school order. In R. Wollons (Ed.), *Children at risk in America: History, concepts and public policy*. Albany, NY: SUNY Press.

Turnbull, B. (1995). Supplementing classroom instruction: Implications for meaning-oriented instruction. In M. Knapp (Ed.), *Teaching for meaning in high-poverty classrooms*. (pp. 104–123). New York: Teachers College Press.

U.S. Department of Education (1993). *Reinventing Chapter 1: The current Chapter 1 program and new directions*. Washington, DC: Office of Policy and Planning.

U.S. Department of Education. (1992). *The condition of education*. Washington, DC: National Center for Educational Statistics (p.141).

U.S. General Accounting Office (1995). *Early childhood centers: Services to prepare children for school often limited*. Health, Education, and Human Services Division, Washington, DC.

Wagner, M. (1995). *The contribution of poverty and ethnic background to the participation of secondary school students in special education*. Menlo Park, CA: SRI International.

Walsh, D.J. (1988). *The two-year route to first grade: The use of testing to validate decisions based on class and age*. Paper presented at the annual meeting of the American Educational Research Association, New Orleans.

Walsh, M. (November 17, 1993). In special education case, justices back parents. *Education Week*, *13*, 1 & 16.

Wang, M.C., Reynolds, M.C. & Walberg, H.J. (March 24,1993). Reform all categorical programs. *Education Week*, *12*, 64.

Wang, M.C., Reynolds, M.C., & Walberg, H.J. (1988). *Handbook of special education research and practice: Mildly handicapped conditions*. New York: Pergamon.

Wang, M.C., Reynolds, M.C. & Walberg, H.J. (1988). Integrating the children of the second system. *Phi Delta Kappan*, *70*, 248–251.

Wesson, C.L. & Deno, S.L. (1989). An analysis of long-term instructional plans in reading for elementary resource room students. *Remedial and Special Education*, *10*, 21–28.

Windham, D. (1992). The role of basic education in promoting development: Aggregate effects and marginalized populations. In D. Chapman & H. Walberg (Eds.), *International perspectives on educational productivity*. London: JAI Press (pp.91–108).

Ysseldyke, J.E. (1987). Classification of handicapped students. In M.C. Wang, M.C. Reynolds, & H.J. Walberg (Ed.), *Handbook of special education research and practice: Learner characteristics and adaptive education, vol. 1*. New York: Pergamon.

2

Early Education for Disabled and At-Risk Children

REBECCA R. FEWELL

Department of Pediatrics, University of Miami, Miami, FL 33136, USA

This chapter conveys the historical foundations of the field of early intervention, tracing the roots of today's services to the pioneers who developed successful programs for economically disadvantaged children and to the first models of service funded under the Handicapped Children's Early Education Program. The growth in services can be seen in the figure of 598,221 children birth through five years of age who were served in 1992.

In order to serve children today, a process of identification; assessment; and program development, implementation, and evaluation must be in place which is most frequently seen in Individualized Family Service Plans. This chapter reviews the common assessment tools, the models of programs that have emerged, and the program implementation issues.

The final focus of this chapter conveys what research has said about the effectiveness of services and models when children and parents have different needs. The chapter concludes with a summary of the contributions of early intervention to the field of special education and the issues that face early childhood service providers and researchers as they move into the next century.

Introduction

The movement to expand the concept of special education to include infants and other very young children in specific and direct services began in earnest about three decades ago. The timing of many events contributed to this new focus. One of the strategies of the War on Poverty was to provide enrichment experiences for children who lacked early exposure to school-valued learning experiences. Reports of successful outcomes with these children, considered at that time to be "at risk" or "disadvantaged," led to the expansion of services to children known to be delayed when they entered school, that is, children with established disabilities. Political changes called for an emphasis on the civil rights of all persons, with equal opportunities for all to have free and appropriate education and access to public places and experiences. Medical and technological advances were increasing the life and health of individuals previously denied services, and with each new technological advancement, full participation in appropriate services was becoming a reality. The curriculum provided to instruct children during their earliest experiences in special education, the methods and procedures used to foster development and learning, the support services provided to children and their families, and the organizational arrangements that facilitate the ebb and flow of services have changed swiftly and dramatically in this short time period.

This chapter will review the brief history of this field, note some of the early and continuing issues in defining the population and determining not only what services to provide, but also other parameters of the service question. After noting the general demographic trends, we will address identification and assessment issues, service models, service personnel, theoretical and curricula models and close with research findings and some possible future directions.

Early Intervention Services: An Evolving Field

The phenomenal growth in services to young children with special needs is enabling increased numbers of children to enter public schools ready to learn and to be included in classes with their peers. Because the need existed and the political and fiscal support came forth to galvanize the movement, progress came quickly and, at times, advanced on theories and feelings rather than on rigorous empirical evidence. The field continues to evolve and in the minds of some (Guralnick, 1993) is just entering into a second generation of services and research.

Historical Foundations

In order to put into perspective today's early intervention services, it is helpful to reflect on a few significant events that contributed to this movement. In the 1960s school administrators began to voice concerns that children were arriving at the doors of their elementary schools ill-prepared for the education awaiting them. In Tennessee, a response to that concern was developed by Susan Gray and her colleagues (Gray, Ramsey, & Klaus, 1982). They developed a program of early intervention that was a center-based program during the summer months followed by programs of home visits during the remainder of the year. With early intervention teachers in short supply in the area, Gray used a strategy that remains an important lesson today. Program implementors identified the community women who were trustworthy, had good listening skills, and were highly valued by young parents for their advice. This might lead them to a "Miss Sally," and the staff would hire her to be the home visitor to implement the program with parents. The critical skills sought in these visitors are as important in the 1990s as they were 30 years earlier.

In Michigan, Schweinhart and Weikart (1993) began the Perry Preschool Program, a program of high-quality, center-based, preschool education for children born to families in poverty. These and the other members of the now famous Consortium of Longitudinal Studies worked predominantly with children in economically depressed areas and demonstrated impressive gains in the early IQ scores of the children enrolled in those programs. These findings were the basis for the massive Head Start effort. A logical conclusion was that, through the provision of early intervention programs, one could ameliorate, if not prevent, mental retardation.

Legislation and Significant Public Laws

The time was right for advocates of early intervention services to fuse the social, political, economic, and ethical arenas and garner support for programs for young children with disabilities. The work of the early researchers and the Head Start efforts were critical to the passage in 1968 of the Handicapped Children's Early Education Program (HCEEP) Act. This was a remarkably successful act that established 25 programs to develop, implement, evaluate, and replicate models for early intervention services for children under the age of eight years with disabilities.

Three programs illustrate the contributions: The Portage Project was established as a home-based service model for rural areas; the Model Preschool Program for Children with Down Syndrome, at the University of Washington, became a strong example of center-based services; and, the Precise Early Education of Children with Handicaps, developed at the University of Illinois, became the classic program combining center- and home-based models. The timing of these efforts was helped by the political and legislative changes that were taking place in the United States; the equal rights of persons with disabilities was gaining strength, and by 1975, the access to equal education became the now famous Public Law 94-142.

P.L. 94-142 was Part B: Assistance for Education of All Handicapped Children of the Education of the Handicapped Act. This law provided authority for federal assistance with states' excess costs for providing equal education opportunities to children who were handicapped. This law was followed by P.L. 99-457, passed on October 8, 1986. This law was the most important single piece of legislation ever enacted for serving children with disabilities who are under the age of six years. Specifically, it contained two major components, Part B and Part H. Part B was actually a revision of Part B of the Education of the Handicapped Act (P.L. 94-142). This section provides assistance for states and local education agencies to extend services to children down to age three.

Part H of P.L. 99-457 law was new and represented the most recent thoughts about effective intervention. Specifically, the law addresses early intervention as opposed to special education. The concept of early intervention is much broader than that of special education; it includes family training, counseling, therapy services, psychological services, limited medical and health services, screening and assessment services, and special instruction. P.L. 99-457 mandated 14 components that states were required to have in their fully operational comprehensive early intervention system. Unlike P.L. 94-142, P.L. 99-457 can be administered by any agency within a state that the governor designates. Only 18 states designated their departments of education as the lead agency. As expected, states have varied considerably in how they administer services under this law.

In 1991, P.L. 94-142 was reauthorized as P.L. 102-119, the Individuals with Disabilities Education Act (IDEA). This legislation further strengthened family involvement by including service coordination and the Individualized Family Service Plan (IFSP) as principal components. The family directs the IFSP, and importantly, the assessment of their needs, resources, priorities, concerns, and supports (Early Intervention Programs for Infants and Toddlers with Handicaps, 1992). With the changes in this reauthorization, there is no question: family systems are the focus of services and the family-focused model is now required. While changes have been made, more will be forthcoming. The orientations that guided services three decades ago are much different today, and before the turn of the century, we can expect further directions in service legislation.

Definitional, Incidence, and Prevalence Issues

When services began for very young children in the late 1960s it was common to see programs focusing on specific and serious disabilities such as Down syndrome, cerebral palsy, hearing and vision impairments. As services began expanding, more children with mixed and moderate degrees of impairment began to be included. Today, we find some states, such as Hawaii, including in their definition of children eligible for services those who are at risk environmentally. States set their own definitions, and, thus, considerable variability exists.

Definitional Issues

The law requires that each state set eligibility criteria. Typical of state definitions and criteria are those established by the State of Florida.

Established Conditions.

(1) Definition. A child with an established condition is defined as a child from birth through two years of age with a diagnosed physical or mental condition known to have a high probability of resulting in developmental delay or disability. Such conditions shall include genetic disorders, metabolic disorders, neurological abnormalities and insults, or severe attachment disorder.

(2) Criteria for eligibility. A child is eligible for the special program for children who have established conditions when the following criteria are met:

    (a) The child is below the age of 36 months; and

    (b) A licensed physician(s), qualified to assess the child's physical or mental condition, makes a diagnosis or suspected diagnosis of a condition that has a high probability of resulting in developmental delay or disability. (Chapter 6A-6.03030, 1993, p. 329)

Developmentally Delayed.

(1) Definition. A child who is developmentally delayed is defined as a child from birth through two years of age who has a delay in one or more of the following areas:

    (a) Adaptive or self-help development;

    (b) Cognitive development;

    (c) Communication development;

    (d) Social/emotional development;

    (e) Physical/motor development.

(2) Criteria for Eligibility. A child is eligible for the special program for children who are developmentally delayed when the following criteria are met:

    (a) The child is below the age of 36 months; and

    (b) There is documentation of one of the following:

    1. A score of 1.5 standard deviations below the mean in at least one area of development. For children below the age of 24 months, the delay shall be defined in accordance with the child's corrected age; or

    2. A 25% delay on measures yielding scores in months in at least one area of development. For children below the age of 24 months, the delay shall be defined in accordance with the child's corrected age; or

    3. Based on informed clinical opinion and the observation of atypical functioning, the multidisciplinary team makes a recommendation that a developmental delay exists and exceptional student educational services are needed. (Chapter 6A-6.03031, 1993, p. 329)

Like Florida, many states have opted for eligibility criteria for services that permit the use of age-referenced, nonstandardized tests. Team members in a local area can develop their own test, provide the age references they feel are appropriate for skills, and then declare children eligible or not eligible for services. Quite common are practices that permit decisions to be made on the basis of scores from curricula, age-referenced tests, and programs that lack precise scoring procedures. It is on the basis of these tests and checklists that teams apply a state's criteria such as the 25% delay in one area or, in some cases, 20% delays in two areas. Given the variability across states, it is quite likely that a child who is eligible in one state is not eligible for services in another state.

Unlike services to school-aged children that are delivered through state and local administrative units and according to stringent criteria, preschool and early intervention services across and within states differ tremendously in how they determine eligibility for services and in how they deliver services. Some states do not serve children who are developmentally delayed, or they provide them with a different level of services than those provided to children with established conditions. Even within states local service agencies differ vastly in the amount of services they provide, this being a function of many things, including history, availability of professional staff, fiscal resources, and empowered parents.

Incidence and Prevalence Issues

Issues of incidence and prevalence are confusing to many seeking information on numbers of children currently served. These numbers, represented by prevalence data, are those most meaningful for discussions in this chapter. States report prevalence numbers in very different ways, making it difficult to ever know the actual number of children served during a given time period. Nonetheless, the most recent numbers available are from the Annual Report to Congress on the Implementation of The Individuals with Disabilities Education Act (U.S. Department of Education, 1994). According to this report, the 50 states, the District of Columbia, and Puerto Rico served 138,493 infants and toddlers, birth through age two, in Chapter 1 and other programs in 1991. In addition, under Part B, which covered children three through five years, 459,728 children were served, bringing the total through five to 598,221 in 1992. Given the differences in the year of these counts, it is reasonable to assume that approximately 600,000 children under the age of six were receiving some kind of early intervention services through federal, state, and local service agencies.

Identification, Child Assessment, and Diagnostic Practices

As noted, states decide how they will determine who will be served and some of the specifics that will be used in the process of identification and delivering services. This section describes the identification process and required practices, then identifies some of the assessment and diagnostic practices that prevail in states and local education agencies charged with conducting the child valuations, and in some cases, delivering the services.

Identification and Required Practices

The process and parameters of child assessment were outlined quite succinctly in P.L. 99-57. Each state must have in place a child-find system. Once children are found, the law requires that each be evaluated by qualified, multidisciplinary team members, in conjunction with family members. The child's level of functioning in the following developmental areas must be reported: (1) cognitive development; (2) physical development, including vision and hearing; (3) language and speech development; (4) psychological development; and (5) self-help skills ("Early intervention programs," 1992). The findings of the assessment are used to determine eligibility and, in some cases, lead directly to the development of the IFSP within 45 days of referral.

States differ widely in how they conduct the referral and assessment process and in the tests they use. While some states let local agencies determine tests and procedures, others have state approved test lists. Some of the typical tests and brief descriptions of new trends in assessment in the major domains will be described.

Assessment Measures and Practices

In the domain of **cognitive development,** the most widely used measure for infants and toddlers is the Bayley Scales of Infant Development (BSID) (Bayley, 1969). A new edition of the BSID, published in 1993, is rapidly replacing the earlier edition. This edition includes tables for younger children, and because of the new standardization, more children will meet eligibility criteria. When children are above 2.5 years in development, psychologists turn to either the Kaufman Assessment Battery for Children (Kaufman & Kaufman, 1983) or the McCarthy Scales of Children's Abilities (McCarthy, 1972).

If a child has a serious sensory or physical disability, the aforementioned, traditional tests are clearly inappropriate and should not be used. In such cases the child's cognitive abilities must be assessed on a measure that does not penalize a child for the particular sensory or physical disability. Among the tests and procedures preferred are the Ordinal Scales of Psychological Development (Uzgiris & Hunt, 1975) and the modification by Dunst (1980). Today, examiners use computers, communication boards, and other assistive devices to ensure that physically impaired children's abilities are adequately assessed.

New techniques in cognitive assessment testing frequently use children's play as a means for gathering assessment data. Glick (1994) reported moderate to high correlations between the Play Assessment Scale (Fewell, 1991) and the BSID.

Gross and fine motor development, referred to in earlier laws as physical development, was clarified in P.L. 102-119. The measurement of these skills is usually done by physical and occupational therapists. The test that has been used most widely is the Peabody Developmental Motor Scales (Folio & Fewell, 1983) and the motor portion of the BSID.

In recent years, a number of new tests have become available. These include but are not limited to the Toddler and Infant Motor Evaluation (Miller & Roid, 1994), a standardized test for infants from 4 months to 3.5 years, and the Pediatric Evaluation of Disability Inventory (Haley, Faas, Coster, Webster, & Gans, 1989), a test of multiple skills that indicates the level of assistance and the environmental modifications required. This test also provides a structure from which one can evaluate motor skills. New trends in motor assessment, both gross and fine include observing children performing motor movements in natural environments, and then scoring the behaviors on observational protocols.

Language and speech development tests are almost always included in batteries given to young children suspected of having some kind of delay. Among the scales commonly used to assess receptive and expressive language and communication skills in young children are the Sequenced Inventory of Communication Development (Hedrick, Prather, & Tobin, 1984), the Receptive Expressive Emergent Language Scale (Bzoch & League, 1978), and the Preschool Language Scale—3 (Zimmerman, Steiner, & Pond, 1992).

More recently, some therapists have found the criterion-referenced, Rossetti Infant-Toddler Language Scale (Rossetti, 1990) useful because of the comprehensive mix of interaction-attachment, pragmatics, gesture, play, and other language-related skills. New trends in communication assessment are often associated with the videotaping of interactions during play (Fewell & Glick, 1993). Language samples are taken during play and then scored on protocols. In addition, parent child interaction scales that measure the contributions of both members of the dyad are becoming routine. Scales like the Maternal Behavior Rating Scale (Mahoney, 1992) are becoming widely shared by professionals; however, few such measures are available commercially.

Multidomain, Curriculum Referenced, and Multipurpose Measures

Very popular for a number of years, multidomain and curriculum-referenced measures are easy to use and lead directly to intervention goals and objectives. Among the frequently used models are the Battelle Developmental Inventory (Newborg, Stock, Wnek, Guidubaldi, & Svincki, 1984), and the Early Intervention Developmental Profile (Schafer & Moersch, 1981), despite the difficulties in determining specific age scores. New trends are multifunction tools that have added functions such as procedures to plan services and evaluate progress and program intensity. One example of this is the System to Plan Early Childhood Services (Bagnato & Neisworth, 1991).

Family Assessment. Prior to 1986 a only a few research and service groups were engaged in the study of families of young children with special needs. Ann Turnbull and her colleagues at the University of Kansas (Turnbull, Summers, & Brotherson, 1984), Don Bailey and others at the Frank Porter Graham Center in Chapel Hill, North Carolina (Bailey & Simeonsson, 1984), and Keith Crnic and fellow psychologists at the University of Washington (Crnic, Friedrich, & Greenberg, 1983). The Washington group captures through the title of one of its studies the focus of early family assessment: "Adaptation of families with mentally retarded children: A model of stress, coping and human ecology." Researchers sought to learn how families coped with what was perceived to be a negative stressor. As we reflect back on those early studies, particularly the studies of researchers who were actually engaged in interventions to reduce stress and increase coping skills (e.g, Vadasy, Fewell, Meyer, & Greenberg, 1984), we glimpse roots that led to the heightened interest in the role of families in the lives of very young children and the need to measure the effectiveness of such interventions. This timely movement, at sites across the nation, was instrumental in the enhanced role for families being written into P.L. 99-457 in 1986.

The typically used family assessment measures are the Family Support Scale (Dunst, Jenkins, & Trivette, 1984) and the Family Needs Survey (Bailey & Simeonsson, 1988). The more common practice in early childhood programs is to simply ask parents to tell them about their needs, resources, and supports, and then to record these on the IFSPs.

As we look at family assessment almost 10 years after the first programs began serious family assessment, we see the impact of maturity, new knowledge, and broader perspectives. The shift is away from family pathology to the use of comprehensive conceptual models that identify and value successful family adaptation. One group of researchers whose work exemplifies the more recent thinking is that of Gallimore and his colleagues (Gallimore, Weisner, Kaufman, & Bernheimer, 1989; Nihira, Weisner, & Bernheimer, 1994) at UCLA. Based on the assumption that families create their own ecology and, by studying the process, one can identify the resources and constraints through examination of their daily routines, Nihira et al. investigated the construct and concurrent validity of 12 ecocultural factors within the context of families of young children. Although scales do not exist that conform to these new directions, we can expect these to follow and be shared across programs as soon as they are released from the researchers.

Service Models

Agencies have considerable freedom in where and how often they deliver services. Given the impressive services delivered to children in child development and school settings and in home and hospital settings, local agencies have many models from which to choose. Not only has precedent been a factor in what an agency selects, but space availability and cost are also major considerations. Several models that were among the first funded under the Handicapped Children's Early Education Program Act of 1968 continue to be the most widely used systems for delivering services.

Home-Based Services

Home-based services are those in which service providers go to a child's home to work with the child and family on the goals and objectives of the child's education or therapy. While some programs send providers into the homes as many as two or three hours a week, others visit families once a month or less.

Center-Based Services

Center-based services are those provided at hospitals, schools, day-care centers, or other community agencies. Family members either transport children to these sites or the centers provide transportation. Again, services vary widely in frequency, from daily services of 5.5 hours per day of early intervention to about one hour per month.

Consultative Services

Consultative services are used in conjunction with some services or in lieu of direct services. In these cases, a specialist gives advice or assistance to someone who provides the actual service. For example, physicians and nutritionists advise staff concerning issues they raise.

Support Services

It is very difficult to pinpoint or define early intervention services even within one state because a child and family are counted as receiving these services as long as they receive one of the required services. Thus, for one family it may mean assessment and an IFSP carried out by the family, whereas with another child and family it may mean five days a week of intensive early intervention, multiple therapies, and transportation to a center for these services.

Arrangements reflect local and state agency proclivities, funding sources, histories, the availability of personnel in a locale, and the strength of advocacy groups. In cases in which departments of education deliver these services, we see a heavy emphasis on the child-focused special instruction component as delivered through early intervention special education teachers as the major service providers and others serving in important ancillary roles. These systems look predominantly to funds through education channels to cover the costs of services. In contrast, in states in which the lead agency is a medical or human service structure, we see a greater emphasis on aspects of services that reflect the priorities of these agencies. For example, children may be given several therapies a month at a hospital or clinic and be visited once a month by a social worker or early intervention teacher. There are no required services or number of hours of services. These programs use a combination of Part H, state, Medicaid, and insurance funds. As children get older and enter into more traditional educational programs, schools are required to provide "educationally relevant" therapies, that is, therapies that are necessary to enable chhildren to participate in educational services. Therefore, therapy that is needed for medically related problems is seldom found on children's educational plans. Parents must seek such services from their health care professionals, using health care funds.

Service Personnel. As services opened up for young children and their families, we have seen a large increase in the demand for service providers. This has been acute in all service areas, but especially in some of the related services such as physical and occupational therapy. Regardless of the kind of service, there is a shortage of persons trained, licensed, or certified to deliver services to infants and toddlers.

The personnel shortage is being met in several ways. First, institutes of higher education are increasing their training efforts and thus the supply of professionals is greater. Second, in the case of special instruction, programs hire persons who are not certified or hire assistants who must work under the supervision of persons licensed or certified in the discipline. Third, systems with more fiscal resources meet the increasing salary demands linked to personnel shortages. The result is increased program costs and a search for funds from other sources. Given the shortages and the variability in the services delivered across programs, it is reasonable to assume that shortages of personnel and the costs of such services, rather than child needs, are influencing the services provided.

Sources of Fiscal and Community Support. Today early intervention programs are strongly urged to work with other community agencies to provide services and support for the various needs of children and families. Among the federal, state, and local programs that are known to work together to provide early intervention services for children under the age of three years in the Miami, Florida, area are the Dade County Public Schools; Children's Medical Services (one program provides free primary medical care, not only to children with special needs, but also to their siblings); Medicaid; health maintenance organizations; the Women, Infants, and Children Program; Department of Agriculture Nutritional Supplements; Community Committee for Developmental Handicaps; Chapter 1; Part H; the Department of Transportation; and a number of private organizations. These community supports will differ by location and need, but what is evident today is that the responsibilities for meeting child and family needs are beyond the professional expertise and fiscal means of a single provider, and interagency collaboration is more than a good idea; it is a necessary service model.

Theoretical, Curricula, and Instructional Models

Best practices require that early intervention services be based on sound theory of how children develop and learn and how families nurture their young. Because service programs tend to evolve over a period of years, program changes are gradual and service modifications emerge from new requirements or ideas rather than continuous study of theory and research.

Practices that exist in early intervention programs are heavily influenced by the thinking of individuals who pioneered the programs or joined the program at some point in time. While most programs acknowledge the importance of following developmental theories in determining instruction goals and objectives for very young children, many others draw heavily from other theoretical constructs. For example, with very young children, many programs use ideas espoused by Piaget (see Dunst, 1980). With children who have very severe disabilities, service providers often use applied behavior analysis techniques to target and change very precise behaviors. Some frequently used examples of curricula that combined these models are the Portage Project curriculum and the Hawaii Early Learning Program. This kind of detailed, one-to-one instruction is difficult to conduct when children are included in programs for normally developing children; thus, such programs as the High Scope Curriculum, developed at the Perry Preschool Program for normally developing, but "at-risk" children (see earlier description), have become popular and typical of what is happening in programs for three- to five-year-old children.

An example of new trends in curriculum and instruction can be seen in the Assessment, Evaluation, and Programming System (Bricker, 1992). This curriculum uses developmentally sequenced instructional targets, instructs within daily routines, measures effectiveness in teaching, and provides raw scores rather than traditional age-related scores. While the testing and the curriculum is viewed positively by staff, administrators have been hesitant to accept it because it is difficult to interpret progress or to compare it to gains made on other curricula.

What Research Has Revealed: Descriptions of Models and Services, Effectiveness, and Program Variables

What do we know about effective models and services and how strong are the findings? To fully appreciate the expansion of early intervention services over the last three decades, it is important to examine the research that has been instrumental in changing practices within the field and what we teach our future field leaders. This section will focus briefly on some of the important milestones in research and then provide a closer look at more recent studies that are changing practices and policies now and in the future.

While it is possible to locate early intervention studies prior to the 1980s, the conditions and procedures under which the interventions were provided have little relevance for today. To understand today's practices, it is appropriate to begin with the results from the first funding of the HCEEP programs.

Descriptions of Models and Services

Many of the professionals funded to develop the HCEEP model programs documented their procedures and findings in an effort to gain approval from the Joint Dissemination and Review Panel. This approval allowed them to gain access to additional funding sources for the replication of their programs through the National Dissemination Network. In 1981, the *Journal of the Division of Early Childhood*, in collaboration with the Office of Special Education, devoted an issue to "Efficacy Studies in Early Childhood Special Education." The programs described were considered among the best models of effective early intervention. Characteristic of programs at that time, Simmons-Martin (1981) used an instrument developed by the project (Early Education Project of the Central Institute for the Deaf). Programs developed at three other sites (Bricker & Sheehan, 1981; Moore, Fredericks, & Baldwin, 1981; Rosen-Morris & Sitkei, 1981) used a State of Oregon developed curriculum guide, the Student Progress Record, to measure change. Bricker and Sheehan were among the first to use standardized tests of intelligence and an analysis of educational significance. Interestingly, they eliminated from consideration children whose impairments were so substantial they could not perform on the tests. By today's standards, the procedures and methods used to validate these programs may not merit peer-review publication; nevertheless, taken collectively, these early studies laid the groundwork for the important conclusion that early intervention with children who have disabilities was effective in changing their developmental trajectories. The policy changes and new laws that followed can be traced to the case made by these important early studies.

Multiple Studies: The Solution to the Effectiveness Dilemma

In the decade of the 1980s, a number of reviews of multiple studies and meta-analyses were published (e.g., Bricker, Bailey, & Bruder, 1984; Casto & Mastropieri, 1986; Dunst, McWilliam, & Trivette, 1985; Dunst & Rheingrover, 1981; Guralnick, 1988; Odom & Fewell, 1983; Simeonsson, Cooper, & Scheiner, 1982; White & Casto, 1985). These researchers recognized the considerable problems faced by the early program staff and the pressure on them to publish findings that could impact social policy and provide a basis for the large-scale establishment of early intervention programs (Casto, 1988). These reviewers were consistent in their criticisms of the studies; nevertheless, they included many of these studies in their analyses. In general, these large-scale studies came to similar conclusions: intervention was moderately effective in producing short-term benefits for young children with special needs (Shonkoff, Hauser-Cram, Krauss, & Upshur, 1992).

Second Generation Research: Findings upon Which Policies Can Be Enacted and Programs Developed

In 1993, Michael Guralnick wrote in a commentary at the end of the White and Boyce monograph, "...the field is moving rapidly through a transition, from what might be referred to as first generation research, toward second generation research" (p. 366). No longer was the focus on the global efficacy question but the "...far more demanding task of identifying the child characteristics, family characteristics and program features that interact to optimize one or more outcomes within the framework of contemporary early intervention issues" (p. 367). The need for valid, reliable findings that could direct policies, models, and services became the impetus for many agencies, public and private, to fund significant programs of early intervention research. A number of these studies have produced multiple findings that are likely to shape the future directions of policies and services.

Low Birthweight and Mildly Impaired Young Children. Certainly one of the best designed research projects of this new generation was the Infant Health and Development Program (IHDP) (1990), a randomized clinical trial to determine the efficacy of an intensive early intervention program for preterm, low birthweight infants and their families. Two-thirds of the 985 infants were equal to or less than 2,000 grams and the remaining third were between 2,001 and 2,500 grams. In most cases children with special needs were served in the program; however, in some cases, children were provided therapies outside of the program. It is estimated that approximately 15% of the children in the study would have been eligible for some early intervention service.

The program was in effect for three years at eight sites in the United States (Boston, New Haven, New York, Philadelphia, Little Rock, Miami, Dallas, and Seattle). The intervention program consisted of weekly home visits for the first year, then biweekly visits in the next two years. In addition, all intervention children attended child development centers for a minimum of five hours per day for two years. At corrected age 36 months, the intervention group's subjects had significantly higher mean IQ scores than did the follow-up subjects (13.2 points for the heavier weight subjects and 6.6 points for the lighter weight subjects). When these subjects were retested at age five (Brooks-Gunn, et al., 1994), the intervention group had full-scale IQ scores similar to children in the follow-up group. However, significant differences continued to be present for the heavier children when compared to the follow up children (3.7 points), but no differences were observed when the lighter weight groups were compared. As would be expected, more children with serious disabilities were in the lower weight groups. In commenting on their findings, the researchers suggest these enrichment experiences in the first few years of life are not enough to protect children against biological disadvantages over extended periods of time. Further, they concluded it would be necessary to extend intervention beyond age three in order to attain maximum effectiveness during the school years.

As with many of these larger, well-designed studies, there are numerous subsequent studies that have been and will continue to be forthcoming. Many of these studies are able to address some of the more puzzling questions that previous studies simply did not have the power to address. Three of the many IHDP studies address questions related to important programmatic and policy issues: curriculum and model intensity. Sparling, Lewis, Ramey, Wasik, Bryant, and LaVange (1991) found the rate at which curriculum was delivered contributed significantly to the prediction of children's IQ scores at 36 months. For the lighter weight babies, those most likely to be candidates for typical early intervention services, the researchers found that five activity episodes from their *Partners* curriculum experienced during daily visits to the child development centers would be considered a basic minimum (predicting an IQ of 88), with the possible use of up to 10 opportunities per day (predicting a 98 IQ) would be the maximum. However, for the heavier children, most of whom are developing normally, the story is quite different. These children were predicted to receive an IQ in the 90s regardless of the number of activity episodes received per day. The results of this study inform us that what we do and how much we do it in our classrooms and home intervention programs is not only important, but also directly related to outcomes. While many have said this previously, few have had the data to state it convincingly.

With the shift to family focus intervention the field has sought to determine directions and effects of involving families in early intervention services. Two of the IHDP studies report findings on this issue. Ramey, Bryant, Wasik, Sparling, Fendt, and LaVange (1992) collapsed three forms of intervention (home visits, parent group meetings, and child development center days attended) and labeled this a Family Participation Index (FPI). They used multiple linear regression models to determine the significance of the variation in the FPI on IQ scores at 36 months and on raw scores from the total problem section of the Behavior Problem Checklist after controlling for initial status variables and study design variables. More frequent participants received significantly higher IQ scores than did less frequent participants. By further examining other factors associated with these differences, the researchers reported the probability of a child's functioning in the borderline intellectual range or lower decreased significantly with increases in family participation. No linear relationship was found as a function of participation. This finding offers some support for family participation; however, it is reasonable to question whether the child's participation at the centers (family members were rarely present as children were provided with transportation) should be considered in the index.

More recently, Bradley, Whiteside, Mundfrom, Casey, Caldwell, and Barrett (1994) examined the impact of the IHDP program on the home environments of the children. Using the Home Observation for Measurement of the Environment as the outcome measure, they found no differences in the scores between intervention and follow up at 12 months, but found differences on five of the eight subscales favoring the intervention group at 36 months. The researchers concluded that short-term intervention with parents will not necessarily impact the child's outcome. However, after the children came to the centers on a daily basis and the home visitors continued to visit the homes, support the families, and coordinate services, the program's effects appear to be real. Moreover, the subscales on which the two groups differed (e.g., Learning Materials and Learning Stimulation) appeared to be mediated by the curriculum delivered during the home visits. The researchers also raised the question of child age and program intensity: they suggested the intervention was more limited before age one (children continuing to recover from prematurity and less likely to be actively involved in the intervention), and, thus, the intervention may not have been as effective as when the children were older and an entire set of program components was in place. Taken together, these two studies suggest that family participation is a critical element that needs further study. We need to study the impact of the program on the child's performance, on the family, and on the home environment. Studying the impact of early intervention services on multiple elements is complex and challenging. We are just beginning to understand what is happening and how it can be measured.

Children with Mild and Moderate Impairments. Also addressing early intervention concerns with mildly impaired children are studies of the effects of contrasting theories and curricula, conducted at a single site, the University of Washington's Experimental Education Unit. In 1989, Cole, Mills, and Dale reported findings for years one and two of a study that randomly assigned mild and moderately impaired children (mean age at entry was 4.9 years; mean IQ was 76.7) to either a direct-instruction, with a strong emphasis on achievement of academic and behavioral competencies, or a cognitive-mediated learning model program. At that time main-effect differences were observed during these early years, but they faded by two years, postintervention (Mills, Dale, Cole, & Jenkins, 1995). Interestingly, the researchers found aptitude-by-treatment interactions at the end of year one (an eight-month intervention) and have continued to follow the children over time: they reported persistence in these interactions between child characteristics and preschool curricula. In their most recent article, they report findings when the children were nine years of age. Although the researchers found no main-effect differences as a result of the preschool program, they found the same aptitude-by-treatment interactions they had reported after one year of intervention. Out of a possible 28 interactions, 7 were found to be significant: higher performing students at pretest also performed higher at posttest if they had been given the direct instruction. Likewise, lower performing students at pretest performed higher at follow up if they had received the mediated-learning intervention. These findings are similar in direction and magnitude to those the researchers (Cole et al., 1993) reported at the end of the intervention. The researchers found the interactions to be a preservation of the initial student/curricula interactions as they did not get stronger after the programs ceased.

The Cole et al. (1993) study has three other findings that clearly inform the field and need to be shared. First, the researchers found differences in school placements: the direct-instruction students were significantly more likely to be in self-contained special education programs, whereas the mediated learning students were more likely to be in either special education resource programs or a more difficult to define, specialized self-contained service. Secondly, the researchers found support for the increasing stability of performance scores as children age: Early measures of language and cognitive development, obtained during children's first year in the intervention program, continued to predict achievement at age nine some four years later. The full McCarthy Scales of Children's Abilities was the best predictor, but the language measures (McCarthy Verbal and the Peabody Picture Vocabulary Test–Revised) were also good predictors. Finally, these researchers reported that the earliest posttest scores, taken 6–10 months after entry into the program, consistently predicted long-term outcomes much better than pretest scores. The authors suggested this finding may result from the children's ability to learn from early intervention. Additionally, they proposed, the early posttest score may have additional power to predict the children's later achievement.

Seriously Impaired Children. When children are born with obvious impairments or established conditions, it is very likely that they will be referred early in life to intervention and therapy services. Research with these children continues to be difficult to conduct and weak in findings. The heterogeneity of the children, the low numbers at any one site, the rapid changes and unstable profiles during the early months of life, and the paucity of valid ways to measure abilities are just a few of the problems reported by those who have provided services and reported research findings. Regardless, this is the population that is the target of Parts H and B of the law. It is important to look at the findings of these services.

Shonkoff and his colleagues (1992) published a monograph in which they reported the results of an investigation of developmental change in 190 infants (mean age 10.6 months) and their families following one year of intervention. Infants and families were served in 29 community-based programs in Massachusetts and New Hampshire. Unlike IHDP, there was no single curriculum or service delivery model and the amount of total services rendered to infants and families varied from a few minutes a month to 21 hours, with a mean of 6.9 hours per month. Families of children with developmental delays received higher intensities of services than did families of children with more serious impairments such as Down syndrome and motor impairments. The researchers used the Bayley Scales of Infant Development as the major outcome variable and measured change using residual change scores. The mean mental age of the sample at pretest was 6.9 months and 14.8 months at posttest, with a 7.9 month gain over the 12-month period. When gains were analyzed by disability groups, differences were found. Children with developmental delay made the greatest gains, followed by those with motor impairment, and, lastly, those with Down syndrome. Eleven children displayed no growth or regression, and 35 children changed 12 months or more. Despite these differences, the rates of development were comparable for the three diagnostic groups. In general the best predictor of change at posttest was the severity of the child's psychomotor impairment.

The Shonkoff et al. study performed numerous analyses of family change. Mother-child interaction scores were significantly lower at time one than those reported for the normative sample. These scores sightly increased at time two, particularly in the area of cognitive-growth-promoting behavior. Mean stress scores were found to be significantly lower than those of the normative population. Only modest changes occurred between scores at the two testing periods. Of the three disability groups, mothers of motor-impaired children had the highest stress scores. The researchers also examined mothers' responses to a measure of family adaptation to having a child with a disability and found scores at pretest to be below the means reported by the test authors. Further, they found only slightly more negative effects at time two when compared to time one. Finally, each mother's social support network was assessed at pre- and posttesting. Both the size of the networks and the satisfaction with the network increased from pre- to posttesting. Mothers of children with Down syndrome were significantly more satisfied at both testing periods than mothers of children in the other disability groups.

The Shonkoff study included a number of findings that provide information on program parameters. For children, more positive changes in mental-age scores were found in children whose services were delivered by a single discipline. For families, those who received services primarily from one discipline had significantly less parenting stress than did those whose services came from multiple disciplines. Taken together, these findings call into question some of the practices we have generally accepted as better. Regarding service intensity, this program aspect was not related significantly to child outcomes once they were adjusted for severity. Commenting on the findings of this study, Sameroff (1992) concludes the study found only weak support for program success but strong support for the "powerful thrust of early development" (p. 162). He concluded that the study also demonstrates that different families have different needs and capacities to respond to the interventions we provide for them. Certainly future research must address these issues.

Longitudinal and Multiple Studies of Children with Serious Disabilities. Perhaps the most comprehensive set of studies that can be located are those from the Early Childhood Research Institute (White & Boyce, 1993). This institute was charged with the task of conducting a series of randomized, longitudinal studies to investigate the effects and costs of alternate types of early intervention. It is beyond the scope of this chapter to report all the findings included in the journal issue. Some of the most provocative findings are summarized here.

Two studies examined different dimensions of program intensity. Taylor, White, and Kusmierek (1993) compared results when children were randomly assigned to receive either one or three hours of intervention per week over a five-year period. Measures of child and family outcomes indicated no differences between the two groups. Behl, White, and Escobar (1993) evaluated the immediate and long-term effects of a comprehensive, weekly home intervention for infants and toddlers with visual impairments to a low-intensity treatment using meetings of a parent group once a month over a three-year period. Again, no significant differences were found on child or family measures. Should these findings be replicated at other sites, they could lead to significant policy and program shifts. It could be that what was referred to as an intense model (three hours per week, or one hour of a home visit per week) is simply not intense enough to merit changes in vulnerable children. This could lead to more intense programs. On the other hand, the findings, if supported by subsequent studies, could shift the focus of early intervention to early support services and a decrease in actual hours of direct services for children and families.

A question that is frequently at the forefront of service providers is that of age-at-start. This question was addressed by Boyce, Smith, Immel, Casto, and Escobar (1993) in a study of medically fragile infants. Infants were randomly assigned to begin programs at 3 or 18 months adjusted age. The programs were described as center-based, but home-based if needed, and occurred once a month for about an hour, more often when needed. Parents participated with their children and were given weekly assignments as to their child's developmental needs. The treatment programs were delivered by a physical or occupational therapist. Infants in the early group began at 3 months of age, were tested at 18 months, and then continued in a broader focused and somewhat less intense program with testing at 30 and 42 months. The late start children began the same intervention model at 18 months with sensorimotor intervention continuing as needed by a physical therapist. However, the program of intervention was delivered primarily by a home teacher who focused on child and on family service needs. During the later periods of this study, the groups received between 10 and 14 sessions each year. The earlier start group scored higher on all the covariance-adjusted Battelle Developmental Inventory scores, but these differences were not statistically significant until the last assessment. The group that had the 15-month advantage by starting at the age of three months had scores that were significantly higher than those who started at 18 months of age. No significant group differences were found on family measures. However, at the third assessment period, when the early start group had actually completed 42 months of intervention, their scores were significantly higher than the children whose intervention began 15 months later. No family functioning differences were found at any of the assessment periods. While these findings offer some support for starting early, one must question whether the intensity and breadth of the services delivered were enough to expect real and long-lasting effects. This study appears to be flawed in a number of areas and certainly needs to be replicated before any conclusions can be drawn as to age-at-start.

Finally, Boyce, White, and Kerr (1993) looked at the impact of adding a parent involvement component to an existing center-based program for a period of 15 weeks. The program focused on teaching parents how to implement intervention at home, and additionally, provided them with information and social support. Assessments took place prior to the start of the program, immediately following the intervention, and over the course of a four-year period. No differences were found in child or family functioning at any of the testing times. Again, before concluding that added-on programs for families are not effective, one must ask whether it is reasonable to expect child or family functioning changes in such a short duration. The authors claim the program was as long as those normally provided to parents. Clearly, there is much to learn about what the nature of intervention with parents and the intensity that is needed to produce positive effects.

Conclusions

Early intervention for children with special needs has expanded rapidly over the last 30 years. As a newcomer to the scene, the program has had the advantage of learning from programs for children with special needs that have been in place for many years. Among these many contributions were models for assessment, curricula, and individualized service plans; the need to work with multiple agencies; the importance of working with families, even in their homes, and addressing the needs of the whole family; and the need to be flexible in designing programs. The field emerged at a time when the political, economic, and social climate was ripe to support the movement and send it forward. A gifted group of professionals believed strongly that infants with special needs simply could not wait until the school doors opened for them: They needed to start from the beginning and they needed to start with the support of their families, agencies, and all other providers in their communities. The programs are expanding rapidly, children are arriving at school ready to participate actively, and their families have support systems in place. The field is now ready to turn to the questions that need to be answered to ensure that the best practices are in place and that the services rendered work well and are cost-effective. As we enter the new century, we can anticipate that, for more children, education will begin at birth and with it will come an opportunity to make the most of their abilities.

References

Bagnato, S.J., & Neisworth, J.T. (1991). *System to Plan Early Childhood Services*. Circle Pines, MN: American Guidance Service.

Bailey, D.B., & Simeonsson, R.J. (1984). Critical issues underlying research and intervention with families of young handicapped children. *Journal of the Division of Early Childhood*, 9, 38–48.

Bailey, D.B., & Simeonsson, R.J. (1988). Family Needs Survey. In D.B. Bailey & R.J. Simeonsson (Eds.) *Family assessment in early intervention* (pp.106–109). Columbus, OH: Merrill.

Bayley, N. (1969). *Bayley Scales of Infant Development*. New York: The Psychological Corporation.

Behl, D., White, K.R., & Escobar, C.M. (1993). New Orleans Early Intervention Study of Children with Visual Impairments. *Early Education and Development*, 4, 256–274.

Boyce, G.C., Smith, T.B., Immel, N., Casto, G., & Escobar, C. (1993). Early intervention with medically fragile infants: Investigating the age-at-start question. *Early Education and Development, 4*, 290–305.

Boyce, G.C., White, K.R., & Kerr, B. (1993). The effectiveness of adding a parent involvement component to an existing center-based program for children with disabilities and their families. *Early Education and Development, 4*, 290–305.

Bradley, R.H., Whiteside, L., Mundfrom, D.J., Casey, P.H., Caldwell, B.M., & Barrett, K. (1994). Impact of the Infant Health and Development Program (IHDP) on the home environments of infants born prematurely and with low birthweight. *Journal of Educational Psychology, 85*, 531–541.

Bricker, D.B. (Ed.). (1992). *Assessment, Evalulation and Programming System (AEPS) for infants and children: Vol. 1*. Baltimore, MD: Paul H. Brookes.

Bricker, D., Bailey, E., & Bruder, M. (1984). The efficacy of early intervention and the handicapped infant: A wise or wasted resource? In M. Wolraich & D. Routh (Eds.), *Advances in Developmental and Behavioral Pediatrics, Vol. 5* (pp. 373–423). Greenwich, CT: JAI Press.

Bricker, D., & Sheehan, R. (1981). Effectiveness of an early intervention program as indexed by measures of child change. *Journal of the Division of Early Childhood, 4*, 11–27.

Brooks-Gunn, J., McCarton, C.M., Casey, P.H., McCormick, M.C., Bauer, C.R., Bernbaum, J.C., Tyson, J., Swanson, M., Bennett, F.C., Scott, D.T., Tonascia, J., & Meinert, C.L. (1994). Early intervention in low-birth-weight premature infants. *Journal of the American Medical Association, 272*, 1257–1262.

Bzoch, K.R., & League, R. (1978). *Receptive Expressive Emergent Language Scale*. Austin, TX: Pro-Ed.

Casto, G. (1988). Research and program evaluation in early childhood special education. In S.L. Odom & M.B. Karnes (Eds.), *Early intervention for infants and children with handicaps* (pp. 51–62).

Casto, G., & Mastropieri, M.A. (1986). The efficacy of early intervention programs for handicapped children: A meta-analysis. *Exceptional Children, 52*, 417–424.

Chapter 6A–6.03030. (1993). Special programs for children birth through two years old who have established conditions. *Florida Statute*, Supp. 94–1, 329.

Chapter 6A–6.03031. (1993). Special programs for children birth through two years old who are developmentally delayed. *Florida Statute*, Supp. 94–1, 329.

Cole, K.N., Dale, P.S., Mills, P.E., & Jenkins, J.R. (1993). Interaction between early intervention curricula and student characteristics. *Exceptional Children, 60*, 17–28.

Cole, K.N., Mills, P.E., & Dale, P.S. (1989). A comparison of the effects of academic and cognitive curricula for young handicapped children one and two years post program. *Topics in Early Childhood Special Education, 9*, 110–127.

Crnic, K.A., Friedrich, W.N., & Greenberg, M.T. (1983). Adaptation of families with mentally retarded children: A model of stress, coping, and family ecology. *American Journal of Mental Deficiency, 88*, 125–138.

Dunst, C.J. (1980). *A clinical and educational manual for use with the Uzgiris and Hunt Scale of Infant Psychological Development*. Austin, TX: Pro-Ed.

Dunst, C.J., Jenkins, V., & Trivette, C.M. (1984). Family support scale: Reliability and validity. *Journal of Individual, Community and Family Wellness, 1*(4), 45–52.

Dunst, C.J., McWilliam, R.A., & Trivette, C.M. (1985). Early intervention [Special issue]. *Analysis and Intervention in Developmental Disabilities, 5*(1/2).

Dunst, C.J., & Rheingrover, R.M. (1981). An analysis of the efficacy of infant intervention program with organically handicapped children. *Evaluation and Program Planning, 4*, 287–323.

Early intervention programs for infants and toddlers with disabilities. (1992). *Federal Register, 57*(85), 18987–19012, May 1.

Fewell, R.R. (1991). *Play Assessment Scale.* Unpublished manuscript, Debbie Institute, University of Miami, Miami, FL.

Fewell, R.R., & Glick, P.M. (1993). Observing play: An appropriate process for learning and assessment. *Infants and Young Children, 5*, 35–43.

Folio, M.R., & Fewell, R.R. (1983). *Peabody Developmental Motor Scales and Activity Cards.* Chicago: Riverside Publishing Co.

Gallimore, R., Weisner, T.S., Kaufman, S.Z., & Bernheimer, L.P. (1989). The social construction of ecocultural niches: Family accommodation of developmentally delayed children. *American Journal of Mental Retardation, 94*, 216–230.

Glick, M.P. (1994). *The validity of play assessment: Effects of treatment group and levels of maternal education on play scores.* Unpublished master's thesis, University of Miami, Miami.

Gray, S.W., Ramsey, B.K., & Klaus, R.A. (1982). *From 3 to 20: The early training project.* Baltimore, MD: University Park Press.

Guralnick, M.J. (1988). Efficacy research in early childhood intervention programs. In S.L. Odom & M.B. Karnes (Eds.), *Early intervention for infants and children with handicaps* (pp. 75–88). Baltimore, MD: Paul H. Brookes.

Guralnick, M.J. (1993). Second generation research on the effectiveness of early intervention. *Early Education and Development, 4*, 366–378.

Haley, S.M., Faas, R.M., Coster, W.J., Webster, H., & Gans, B.M. (1989). *Pediatric Evaluation of Disability Inventory: Examiner's manual.* Boston, MA: New England Medical Center.

Hedrick, D.L., Prather, E.M., & Tobin, A.R. (1984). *Sequenced Inventory of Communication Development, revised.* Seattle: University of Washington Press.

Infant Health and Development Program. (1990). Enhancing the outcomes of low birthweight, premature infants: A multisite randomized clinical trial. *Journal of the American Medical Association, 263*, 3035–3042.

Kaufman, A.S., & Kaufman, N.L. (1983). *Kaufman Assessment Battery for children.* Circle Pines, MN: American Guidance Services.

Mahoney, G. (1992). *Maternal Behavior Rating Scale.* Tallmadge, OH: Child Learning Center.

McCarthy, D. (1972). *McCarthy Scales of Children's Abilities.* Chicago: The Psychological Corporation.

Miller, L.J., & Roid, G.H. (1994). *The T.I.M.E.: Toddler and Infant Motor Evaluation.* Tucson, AZ: Communication/Therapy Skill Builders.

Mills, P.E., Dale, P.S., Cole, K.N., & Jenkins, J.R. (1995). Follow-up of children from academic and cognitive preschool curricula at age 9. *Exceptional Children, 61*, 378–393.

Moore, M.G., Fredericks H.D., & Baldwin, V.L. (1981). The long-range effects of early childhood education on a trainable mentally retarded population. *Journal of the Division of Early Childhood, 4*, 94–110.

Newborg, J., Stock, J.R., Wnek, L., Guidubaldi, J., & Svincki, J. (1984). *Battelle Developmental Inventory.* Chicago: Riverside Publishing Co.

Nihira, K., Weisner, T.S., & Bernheimer, L.P. (1994). Ecocultural assessment in families of children with developmental delays: Construct and concurrent validities. *American Journal on Mental Retardation, 98*, 551–566.

Odom, S.L., & Fewell, R.R. (1983). Program evaluation in early childhood special education: A meta-evaluation. *Educational Evaluation and Policy Analysis, 5*, 445–460.

Ramey, C.T., Bryant, D.M., Wasik, B.H., Sparling, J.J., Fendt, K.H., & LaVange, L.M. (1992). Infant Health and Development Program for Low Birth Weight, Premature Infants: Program elements, family participation, and child intelligence. *Pediatrics, 3*, 454–465.

Rosen-Morris, D., & Sitkei, E.G. (1981). Strategies for teaching severely/profoundly handicapped infants and young children. *Journal of the Division of Early Childhood, 4*, 79–93.

Rossetti, L. (1990). *The Rossetti Infant-Toddler Language Scale.* East Moline, IL: LiinguiSystems, Inc.

Sameroff, A.J. (1992). Systems, development and early intervention. *Monographs of the Society for Research in Child Development, 57*(6, Serial No. 230), 154–163.

Schafer, D.S., & Moersch, M.S. (Eds.). (1981). *Early Intervention Developmental Profile (revised edition)*. Ann Arbor: University of Michigan Press.

Schweinhart, L.J., & Weikart, D. (1993). The effects of the Perry Preschool Program on youths through age 15–A summary. In The Consortium for Longitudinal Studies (Ed.), *As the twig is bent: Lasting effects of preschool programs* (pp. 71–101). Hillsdale, NJ: Erlbaum.

Shonkoff, J.P., Hauser-Cram, P., Krauss, M.W., & Upshur, C.C. (1992). Development of infants with disabilities and their families. *Monographs of the Society for Research in Child Development, 57*(6, Serial No. 230).

Simeonsson, R.J., Cooper, D.H., & Scheiner, A.P. (1982). A review and analysis of the effectiveness of early intervention programs. *Pediatrics, 69*, 635–641.

Simmons-Martin, A. (1981). Efficacy report: Early Education Project. *Journal of the Division of Early Childhood, 4*, 5–10.

Sparling, J., Lewis, I., Ramey, C.T., Wasik, B.H., Bryant, D.M., & LaVange, L.M. (1991). Partners: A curriculum to help premature, low birthweight infants get off to a good start. *Topics in Early Childhood Special Education, 11*(1), 36–55.

Taylor, M.J., White, K.R., & Kusmierek, A. (1993). The cost-effectiveness of increasing hours per week of early intervention services for young children with disabilities. *Early Education and Development, 4*, 238–255.

Turnbull, A.P., Summers, J.A., & Brotherson, M.J. (1984). *Working with families with disabled members: A family systems approach*. Lawrence: Kansas University Affiliated Facility.

U.S. Department of Education. (1994). *Sixteenth Annual Report to Congress on the Implementation of the Individuals with Disabilities Education Act*. Washington, DC: Author.

Uzgiris, I., & Hunt, J.M. (1975). *Assessment in infancy: Ordinal scales of psychological development*. Urbana: University of Illinois Press.

Vadasy, P.F., Fewell, R., Meyer, D., & Greenberg, M.T. (1984). Supporting fathers of handicapped young children: Preliminary findings of program effects. *Analysis and Intervention in Developmental Disabilities, 5*, 125–137.

White, K.R., & Boyce, G.C. (Eds.). (1993). Comparative evaluations of early intervention alternatives [Special issue]. *Early Education and Development, 4*(4).

White, K., & Casto, G. (1985). An integrative review of early intervention efficacy studies with at-risk children: Implications for the handicapped. *Analysis and Intervention in Developmental Disabilities, 5*(1/2), 7–31.

Zimmerman, I.L., Steiner, V.G., & Pond, R. (1979). *Preschool Language Scale*. Columbus, OH: Merrill.

3

Elementary-School Programs

ROLLANDA E. O'CONNOR

Department of Instruction and Learning, School of Education, University of Pittsburgh, USA

This chapter addresses service to children with mild disabilities in elementary schools, and includes establishing supportive learning environments, and instruction that is accessible to students with mild disabilities, whether delivered individually or in groups; facilitated by teachers, peers, or computers, centered in pullout or general classroom settings. The scope of this review is broad because the category of "mild disabilities" encompasses a diverse group of students, and also because children with mild disabilities in kindergarten, first, or second grade are rarely formally identified; thus, this review also includes interventions used in compensatory and early intervention programs for children who have not yet been diagnosed for special education services. The studies are organized by type of service (e.g., consultation, school-based, and peer-mediated models) and content area interventions (e.g., reading, writing, mathematics, and cognitive strategies). The implications of increasing service delivery in general education classrooms and of technology in schools are also considered.

Introduction

This chapter will focus on improving learning for students with mild disabilities, and includes establishing supportive learning environments, and designing instruction that is accessible to students with mild disabilities, whether delivered individually or in groups; facilitated by teachers or by peers; centered in pullout or general classroom settings. The scope of this review is broad, first because the category of "mild disabilities" encompasses a diverse group of students, and second, because children with mild disabilities in kindergarten, first, or second grade are rarely formally identified. For young children, this review also includes interventions used in remediation and compensatory programs, and early intervention for children who have not yet been diagnosed for special education services.

The diversity of research described here is a reflection of the decade, in which philosophical conflicts divide the field. The principal of inclusion has gained increasing acceptance, although categorical funding still influences the services children receive in most school districts. Supporters of school-based service delivery models mistrust assumptions about the viability of distinct categories among the broad class "mild disabilities," and of separate treatment for students in compensatory and special education programs. Traditionally, special education has reserved intensive treatment for students with identified disabilities who learn very differently from their peers; however, renewed interest in early intervention—and the hope it offers for curtailing the severity of some forms of mild disabilities—leads some professionals to insist upon immediate and focused assistance for any child who experiences learning difficulties in the early grades.

As of the Fifteenth Annual Report to Congress on the Implementation of IDEA (U.S. Department of Education, 1993), the majority of children identified with mild disabilities received their special services in part-time, pull-out programs, in variations of categorical and cross-categorical configurations. Until evidence converges to document patterns of improved learning for hard-to-teach children, it will be difficult to recommend with certainty the most appropriate grouping and instructional arrangements, and appropriate packages and levels of intervention and support. With individualized planning and instruction the hallmark of special education, a single "best approach" may never be found for struggling learners. This review describes recent research evidence for the effective instruction of very hard to teach children, and is divided into studies that center in general education classrooms, including school-wide models for the provision of services, and studies on the learning of students with mild disabilities conducted in clinical settings, which focus on the instructional conditions that stimulate learning.

Incidence, Overlap, and Distinctions among Categories of Hard-to-teach Children

The U.S. Department of Education (1993) reports that over 4,200,000 students, ages 6-21, were served in 1992 under the categories of learning disabilities (LD), mild mental retardation (MMR), and serious emotional disturbance (SED). In addition to these figures, the American Psychiatric Association estimates that 3-5% of students in schools may have attention deficit disorder (ADD), although some of these students will also have LD, with incidence of co-morbidity variably reported as 26-80% (Holborow & Berry, 1986; Lambert & Sandoval, 1980; Shaywitz, Escobar, Shaywitz, Fletcher & Makush, 1992). These figures exclude other difficult-to-teach students served under Chapter I and other compensatory programs.

Similarities among Hard-to-teach Children

Research into the academic overlap of boundaries among categories of struggling learners (e.g., comparisons of students in programs for learning disabilities and Chapter I) cast doubt on the premise of specific disability. For example, the hump in the lower tail of the normal distribution of reading scores described by Rutter and Yule (1975)—used as evidence for conceptualizing reading disability as discrete from generally poor readers—has not been replicated with other large samples. Shaywitz et al. (1992) proposed a smooth continuum of reading ability/disability, documented by longitudinal data in which no distinguishable cut-off points were found between children with reading disabilities and other children who read poorly in the general population. They also found evidence of instability in diagnoses during the elementary school years, such that children labeled as learning disabled during one school year did not qualify under the same criteria two years later. Their findings suggest that treating poor readers by category (i.e., LD versus remedial) may lack criteria-based support.

The results of two studies in Washington (reported in Jenkins, Pious, & Peterson, 1988) lend support to this suggestion. In a comparison of 7000 students across grade levels and school districts served by special education programs for LD (n = 276), by remedial programs for students without disabilities (n = 955), and by regular education programs (n = 6496), considerable overlap was found in a distribution of reading scores between students with LD and students in remedial programs, although mean performance of the students with LD was lower. Neither low-performing group showed much overlap with the distributions of reading scores of children in the general education program. A second study examined the instructional groupings for children categorized as special education or remedial in elementary schools that had created an integrated program of services. Within each school, teachers formed instructional groups based upon homogeneous reading levels, regardless of funding category. Of the 173 groupings district wide, 67 contained children with mild disabilities along with nonhandicapped students, and only 4 reading groups contained only students with disabilities. This finding raised an important issue about how to consider differences among categorical groups: although mean academic performance of categories of students differs, common academic needs *across* categories may be a useful way to determine instructional groupings within a school. If instructional needs for low-achieving children are similar—regardless of category—does it make sense to devote extensive financial and professional resources on determining the specific categorical label for each child? The answer depends upon whether categorical labels for children, once their academic needs are determined, are likely to influence learning trajectories during instruction.

Differences among Categories of Students Who Are Hard-to-teach

First, several studies reflect significant differences among groups of students in varying categories of mild disabilities and compensatory programs, with little academic overlap. Lombardi (1991) examined scores from a national data base of 22,018 students at risk for learning failure and students with identified disabilities in the 4th, 7th, and 10th grades. Students classified in special education performed significantly lower on standardized measures than their at-risk peers across all categories of disabilities and for all of the sampled grade levels. Jenkins et al. (1988) also reported lower mean scores for the students with mild disabilities, even though school personnel reported substantial overlap in the type of assistance needed. To suggest that instructional needs overlap across categories, however, does not address the issue of rate of learning. In addition to understanding instructional levels, teachers need to organize the conditions of instruction, such as the size and structure of new units of information, and the amount of practice children are likely to require to learn information and processes.

Studies designed to provide evidence for learning differences between categories of children with learning difficulties tend to find them. Innocenti and White (1993) conducted a meta-analysis of studies comparing the gains of young children with disabilities with gains of children "at risk," and documented larger intervention effects (e.g., rate of learning) for students "at risk" than for those with disabilities, suggesting that differences in cognitive performance among populations of children affect the rate of learning when interventions are implemented. Scott and Perou (1994) compared children with LD (IQ > 85) to students with MMR (IQ 50-69) and typically developing children matched for age (six-to-eight-years-old), gender and ethnicity. Measures considered learning and generalization across several types of oddity identification and rhyme production. All groups were able to learn the tasks, and no differences in generalization were found between children with LD and age-matched control children, however, these groups generalized significantly better than the children with MMR. This result replicated an earlier study (Scott & Greenfield, 1992), in which groups of children with LD, MMR, or typical development were compared on taxonomic performance. Again, large, significant performance differences separated the children with mild MR from normally achieving children and from children with LD. When normally achieving children were compared to children with LD, only small differences were observed in their ability to learn new information. Students with LD and MMR diverged in how they used strategic skills, such as organizing material into a permanent knowledge base. The authors suggested that the trend toward grouping children with LD and MMR into learning support classes (e.g., for mild disabilities) may not adequately address the learning differences of the children to be served.

Studies that document the rate of learning between children with disabilities and generally low-achievers also find significant academic differences. Deno, Maruyama, Espin, and Cohen (1990) compared the reading growth of low achieving children with the growth of children with mild disabilities, and found significantly better growth for the low achieving children. Zigmond (1995) reported similar findings in a study of rate of learning over a two year period for elementary-aged children with LD or low achievement, as well as on absolute achievement levels between these groups.

These differences in learning rate and outcomes have implications for class and school-wide instructional groupings. As schools blend funding sources and services, it becomes imperative that teachers and researchers gather data on the performance of children under specific instructional regimes. The issue of learning similarities and differences is likely to become more complicated over the next decade as studies accumulate involving children with attention deficit disorder, alone or in combination with other disabilities. For example, children with ADD *and* LD may have cognitive learning problems which differ from those of children with only LD (Fletcher, Shaywitz & Shaywitz, 1994), and the presence of either disorder is likely to influence how a child performs during instruction (Ackerman, Dykman & Gardner, 1990; O'Neil & Douglas, 1991). The differences in learning rate have prompted some researchers (Berninger & Abbott, 1994; Fletcher & Foorman, 1994) to recommend that rate of learning be substituted for current definitions of learning disabilities, or at the least, amended to the defining criteria for handicapping conditions.

Studies of Students with Mild Disabilities in Regular Program Classes

Policy and practice have shifted since the last edition of the *Handbook* toward greater inclusion of students with mild disabilities in general elementary classrooms. Writers and policy makers comment on fragmentation in services and in the flow of the school day for children who are pulled out of the regular classroom for assistance often unrelated to general education goals and content (Jenkins, Pious, & Peterson, 1988; Reynolds, Wang, & Walberg, 1987; Will, 1986). Education in the mainstream, however, is for many students with mild disabilities the kind of education in which academic failure was first discerned; thus, substantial alteration in that environment— in teacher behavior, the structure and difficulty of tasks, degree of direct assistance, and the like—will be needed to improve the academic outcomes for students with disabilities.

Conditions for Instruction across Settings

Several studies in the last few years have assessed opportunities to learn in various settings for students with mild disabilities. O'Sullivan et al. (1990) compared the opportunities to actively respond during reading instruction for children with three types of mild disabilities (learning disabilities, mild mental retardation, and serious emotional disturbance) and their non-handicapped peers. Children with disabilities (n = 47) were observed during reading instruction in resource room and regular class settings, and children without disabilities (n = 30) observed in regular classrooms. Activities and responses were coded every 10 seconds using the CISSAR technique (Greenwood, Delquadri, Stanley, Terry, & Hall, 1985), which yielded proportion of class time in which children were engaged in academic tasks, and proportion of time in which children were actively responding (e.g., reading words aloud, writing, answering questions, rehearsing, discussing). For the children with disabilities, a greater proportion of time was spent in active responding in resource room settings than in the regular class, however, more of their school day was spent in the regular class, so the number of total minutes actively responding in the two settings did not differ. Moreover, the proportion of academic responding did not differ among the categories of disabilities; Children with all types of mild disabilities responded less in the regular class than their peers without disabilities.

Special education classes which achieved the highest student outcomes were characterized by a high degree of structure, monitoring by teachers, and match between student needs and the instructional content (Christenson, Ysseldyke, & Thurlow, 1989). The authors proposed that even better outcomes could be achieved by allocating more total reading time for children with disabilities, with a combination of reading instruction in the resource room and the regular class to provide extended opportunities for learning. They speculated that taking advantage of regular classroom structures that increase academic responding for all students (for example, peer tutoring) might maximize responding for students with disabilities in the regular class. They also raised the issue of class size, suggesting that the smaller teacher/student ratios in special education classes might contribute to the opportunity to learn for these students. Thurlow (1993) pursued this hypothesis by comparing the outcomes for first through sixth grade students with mild disabilities under varying teacher-student ratios (1:1, 3:1, 6:1, 9:1, 12:1), and significant differences favored lower student-teacher ratios.

A related study of learning conditions conducted by Deno, Maruyama, Espin and Cohen (1990) compared the reading achievement of children in elementary schools with mild disabilities (n = 255) and low achieving children served in integrated and pullout programs. In a comparison of instructional arrangements, they found that special education teachers used more data and spent a higher proportion of the reading time in teacher-directed activities than regular classroom teachers. Nevertheless, students in integrated programs had higher overall gains in reading. A subanalysis revealed that the effects favoring integrated settings were carried by the low achieving children without disabilities. Children with mild disabilities showed poor reading growth in both settings. Like O'Sullivan et al. (1990), Deno et al. found that regular classrooms spent more total time in reading instruction than did pullout programs, making allocated time a serious confound with reading gains. In this large sample, significant differences in initial reading level and annual gains separated the children with disabilities from non-special education low achievers.

Differences have also been found between the quality and quantity of teacher-student interactions for students with disabilities and their non-handicapped peers in mainstream settings. Alves and Gottlieb (1986) used the Brophy-Good interval-based time sampling procedure (Brophy & Good, 1972) to record interactions, and found that children with disabilities had fewer opportunities for academic involvement than the children without disabilities in the same classrooms. Teachers in their sample asked fewer academic questions of children with disabilities, and offered less extended feedback on performance, perhaps because they viewed mainstreaming as more for social than academic purposes.

Taken together, findings from these studies suggest that regular classroom structures will require reorganization if students with mild disabilities are to improve their academic outcomes. Teachers will need to extend learning opportunities to lower-skilled children, and consider ways to increase the academic involvement and responding of students with disabilities. As more children with mild disabilities are served in mainstream classes, researchers have explored methods for providing assistance to regular classroom teachers and to low achieving children in their classes. Some of these methods are designed to help children who have disabilities; others function as prereferral intervention prior to the formal identification of disabilities. Although the trend for placement of students with mild disabilities in mainstream programs may be increasing (Sawyer, McLaughlin, & Wingless (1994), unanswered questions remain about whether these integrated placements are appropriate, "For example, are students with reading disabilities who are moved from resource rooms to general education classes still receiving necessary remedial instruction?" (p. 213)

The studies that follow assess the effects of methods and conditions for providing consultation and advice to regular educators about adjusting instruction for students who are difficult to teach and manage.

Consultation Models

Consultation models evolved from the notion that special educators have particular expertise in instructional design and teaching strategies for students with disabilities. In this view, if regular educators were to modify their instruction, materials, or strategies as recommended by special educators, children with disabilities would be able to succeed in the general education class. The questions then become how should consultation be facilitated among teaching staff, and how much modification to the general education environment is necessary? Models of consultation differ in the degree of directiveness (e.g., the extent to which the regular class teacher is in control), the source of expertise (e.g., from a special educator or same-role colleague), and the desired outcomes (e.g., coping on the part of the teacher or child, increased teaching repertoire, academic progress for students).

Fuchs and colleagues (Fuchs and Fuchs, 1989; Fuchs, Fuchs, Bahr, Fernstrom & Stecker, 1990) took a prescriptive approach to consultation. Over a series of studies aimed at fine-tuning a method for providing teachers with viable prereferral interventions, they designed the Mainstream Assistance Team approach to consultation. Regular teachers identified their most difficult-to-teach students *without* disabilities, and a consultant met with each teacher to define the major problem experienced by the student. The consultant assisted teachers to select the most feasible and effective intervention for the given situation from an array of well-defined potential interventions. In school settings where time is precious and case loads of psychologists and special educators are high, the consultants in these studies adopted consultation as their major role. Fuchs et al. refined the process in a series of experiments to identify critical components of the consultation model, to streamline the procedures to make the process more efficient, and to design interventions that were more feasible and likely to yield generalizable improvements for targeted children. The final list of effective interventions included the details for implementation, record keeping, and a gradual shift from teacher monitoring of behaviors to student self-monitoring and self-evaluation of behaviors. Teacher reports of improved behavior were significant over control students in non-MAT schools, however, observation data for experimental and control children did not differ.

Wesson (1991) designed an experiment to evaluate two conditions of monthly consultation (i.e., expert consultation provided by a special educator, or facilitated problem-solving among small groups of colleagues) crossed with two methods of monitoring the reading/writing progress of students with mild disabilities (i.e., teacher-designed monitoring, or reading fluency graphed through Curriculum-Based Measurement). In the two conditions that used Curriculum-Based Measurement (CBM), reading fluency timings were administered, graphed and compared with an aim line to determine when instructional conditions were adequate (see Deno, Mirkin, & Chiang, 1982, and Fuchs & Deno, 1994 for discussions of procedures, rationale, reliability and validity of CBM). The four conditions were: (1)

teacher-developed monitoring systems, with individual follow-up consultation; (2) CBM to set the goal and inform teachers when modifications would be desirable, with individual follow-up consultation; (3) teacher-developed monitoring with small-group consultation among three regular educators); or (4) CBM, with small-group consultation. Fifty-five teachers of students with mild disabilities, most in the third and fourth grades, were randomly assigned to conditions. Students in both CBM monitoring systems showed significant gains in passage reading over students in the teacher-developed monitoring strategies. Within the CBM conditions, small-group problem-solving was as effective as individual consultation. Wesson concluded that CBM as a monitoring tool lead to better student outcomes, and that the progress graphs helped teachers in the small-group problem solving condition to focus their discussion on the progress of their target student.

Consultation is not viewed by all researchers as the best method for teachers to collaborate on developing interventions to be implemented by regular classroom teachers (Phillips and McCullough, 1990; Villa & Thousand, 1994). Consultation can place the regular educator under the control and advice of a special educator, and generate feelings of inequality, or problems of perceived efficacy. Johnson and Pugach (1990) developed a model of collaboration that depends upon an equal partnership of teachers who operate in similar roles. For example, two regular fourth grade teachers may act as a collaborative team. Their model consists of a structured problem-solving process in which teachers engage repeatedly over time, taking turns as facilitator and problem presenter. The process was designed to assist regular classroom teachers to develop and implement instructional modifications for students with learning difficulties (including students with mild disabilities) in their classes, and includes stages of problem clarification, summarization, developing interventions, predicting probable outcomes, and evaluating the modifications implemented. Forty-four teachers participated in the collaborative problem solving intervention, and 43 comparison teachers acted as controls. Teachers who participated in collaborative problem solving increased their tolerance for the range of students' cognitive abilities in the classrooms, and successfully implemented the interventions they designed, but the effect of the interventions on student outcomes was not assessed, nor was the possibility of differential effectiveness for children with or without disabilities.

As more children with disabilities are placed in general education classrooms for the majority or all of the school day, methods for efficient and effective problem solving will need to be considered. Most schools have limited special education personnel; thus processes that involve teams of regular educators in routine discussions of the progress of individual children merit further exploration, in addition to processes that depend upon the special educator.

School-based Models

Reduction of pullout programs is a goal in many restructuring efforts (Reynolds, 1994; Will, 1986). Reynolds, Zetlin, and Wang (1993) advocated selecting children in the margins—the lowest and highest 20th percentiles of each grade level—to receive special instruction in elementary schools. By examining a simple rank ordering of normed reading scores within a school, they identified the majority of students who had qualified for special education services through extensive testing and professional effort, as well as low-achieving students served by compensatory programs. They argued that the 20/20 approach could save resources by eliminating some of the need for eligibility testing that could be allocated to the provision of more intensive services to more students. Reynolds et al. envisioned the funding from special education, compensatory, and gifted programs combined and concentrated on providing adaptations to the regular educational programs for these students in the margins.

School-based models intrigue researchers and policy makers because of the potential for increasing the capacity of existing resources (e.g., the skills of regular educators and available environments for hard-to-teach children). These models usually affect the structure or services available in the regular classroom, thus affecting not only students with disabilities, but also other low-skilled students. Many school-based research projects have been conducted recently to explore methods for increasing the participation of students with disabilities in regular classes, however, the effects on the academic and social outcomes of students with mild disabilities of focusing adaptations on mainstream classes are not clearly convergent. The unanswered question is whether adaptations to the mainstream will be sufficiently intense for the most resistant learners.

The Cooperative Teaching Project (CTP) in Minnesota (Self, Benning, Marston, & Magnusson, 1992) combined the efforts of regular (n = 14), remedial (n = 2), and special educators (n = 2) in a K-3 school, shifting the focus from identification to prevention of learning difficulties. Their goals were to close the gap in reading between poor and good readers, and to increase the instructional strategy repertoires of the regular teaching staff. Within each classroom, CTP staff provided an extra 25 minutes of daily reading instruction to a small group of poor readers, which included students with mild disabilities and others eligible for compensatory programs. They monitored progress of the children weekly with CBM probes (Deno, Mirkin, & Chiang, 1982), and used these results to make decisions about motivation and instruction in the reading groups. The effectiveness of the model was assessed through individual slopes of reading growth. The district had collected data on rate of fluency increase over time for several years, and the rate of progress in this school grew from .83 words per week before CTP to

2.89 in the second year of implementation. Because the progress was made without additional special education services, the procedures of the CTP represent a cost effective way to increase direct service to young students in mainstream settings.

Studies of school-based models also underscore the difficulty of implementing large-scale change. Zigmond and Baker (1990) developed an elementary school model for accommodating students with learning disabilities in mainstream classrooms (Mainstream Experiences for Learning Disabled Students: MELD) in which regular class teachers received inservice in alternative strategies for teaching reading, and learned to use CBM data for making instructional decisions for students with learning disabilities. In the second year of the project, the 13 children with LD returned full time to the regular classrooms, the roles of the two special educators in the school shifted from direct teaching responsibilities to co-teaching in the regular program half-time, and reviewing CBM data and consulting with teachers for the other half. Although the regular teachers received extensive inservice in the second year (15 hours of district sponsored inservice on teaching literacy, and 2.5 days of inservice on behavior management), teachers did not change their grouping patterns during this year of full integration. Analysis of observation data from Year 1 in resource room settings and Year 2 in the regular program suggested that the children with LD adapted well (e.g., showed no significant increase in behavior or attendance problems), and received more time in teacher-directed reading lessons. "Nevertheless, these students failed to make discernible progress on academic skills as measured by standardized achievement tests, they earned lower grades, and the advantages of the mainstream were not reflected in greater gains on CBM measures." (p. 184)

Jenkins, Jewell, Leicester, and Jenkins (1991) designed a school-based model that included a complex treatment of cooperative learning (at grade six), cross-age peer tutoring for students with mild disabilities and remedial students in the lower grades, and in-class support from special education and compensatory teachers and assistants. Inservice to staff was successful in creating change in the structure of learning arrangements in the regular class, but despite fundamental changes in the delivery of services, the achievement of students with disabilities was not significantly different from those in a matched control school with a traditional pull-out resource room model. Three years later, after cooperative learning had been fully implemented for reading in all of the grade 2-6 classrooms, outcomes for students categorized for special education and participating in the cooperative learning structure still did not differ from controls (Jenkins, Jewell, Leicester, O'Connor, Jenkins, & Troutner, 1994).

These multi-year projects resulted in substantial changes regarding the inservice and classroom support afforded teachers in general education, and in the organizational structures in the classroom, which included a greater emphasis on peer-mediated instruction and opportunities for co-teaching and in-class support. For the children with disabilities, more time was spent in mainstream classes, engaged in the work of the general classroom. The projects included processes for assisting teachers with the academic and behavioral difficulties of struggling learners, increasing the academic engaged time for students, increasing learning opportunities for students with disabilities, frequent progress monitoring procedures, and service to students with disabilities in mainstream settings. Despite large increases in financial and professional support to these schools over a period of several years, and incorporating the major components recommended by advocates of full-time inclusion, "40% of students with learning disabilities who were being educated full-time or primarily in general education settings not only were failing to make average gains, but also were slipping behind at what many would consider a disturbing rate." (Zigmond, Jenkins, Fuchs, Deno, Fuchs, Baker, Jenkins, & Couthino, 1995, p. 539)

Zigmond and Baker (1995) returned to study the MELD model four years later, along with four other models across the United States designed to accommodate students with mild disabilities in mainstream settings, including the model of Jenkins et al. (1991; 1994). These models included cooperative learning, co-teaching, co-planning, and in-class assistance from special education teachers. They shadowed two students with LD (one primary and one intermediate grade student) in each of the five sites for two days, and interviewed the students, their regular and special education teachers, parents, and school administrators. These cases yielded rich and varied educational portraits of students with LD in full time inclusive settings, and also raised questions about the role of the special educator. In each case, the special educator's energy was channeled into accommodation and increasing the participation of students with disabilities in the regular class activities—goals sometimes in opposition to direct teaching designed to close the academic gaps between children with LD and their regular class peers. Children with LD in mainstream reading classes listened to taped stories to prepare them for participating in discussion, or were paired with buddies to finish writing tasks, or were seated close to teachers who could guide them through instructional tasks too difficult to navigate independently. In none of these sites was special education focused on increasing the rate of academic skill development. As these researchers highlight, the fair test—comparing enhanced mainstream instruction with enhanced pullout instruction—has not been conducted.

Peer-mediated Models

Educators believe that learning how to work cooperatively and function as a responsible member of a team is an important educational outcome in its own right (Johnson, Johnson, & Maruyama, 1983; Slavin, 1993). Beside the contribution to the development of social competence, working in peer-mediated teams has been found to improve academic achievement (Johnson & Johnson, 1986; Webb, 1985) and increase interpersonal attraction and friendships (Cohen, Lotan, & Catanzarite, 1990). As reports of beneficial outcomes mount, researchers and practitioners have begun to test peer-mediated approaches with more diverse student subgroups, including students with disabilities. The studies that follow document findings for students with disabilities and low achievement in the two dominant approaches to peer-mediated instruction: peer tutoring and cooperative learning.

Peer tutoring. Classwide peer tutoring affects more than students with disabilities. Because all children in the regular class participate, effects can be assessed across student types (e.g., average learners, students at risk for failure due to economic factors, students with mild disabilities). The Juniper Gardens Children's Project developed a model of Classwide Peer Tutoring (CWPT) to increase the academic success of children at risk for school failure in urban, low SES classrooms (Delquadri, Greenwood, Whorton, Carta, & Hall, 1986). Greenwood (1990) conducted a long-term implementation of CWPT with first grade children, and compared outcomes with those of equivalent at-risk control students, and with children in a non-risk, average to high socioeconomic school. Teachers structured the academic areas of spelling, math and reading into 30-minute blocks for CWPT, in which all children in each class were paired for daily practice sessions, and pairs organized into competitive teams. Children received points for accurate responding and appropriate corrections, which were tallied for team competitions. Among the three comparison groups, differences were found for academic engaged time, which may have contributed to the improved performance of children in the CWPT over their low-SES controls. Balancing the relative gains for CWPT as a model for improving the performance of children at risk was a more disturbing finding: Children in the high-risk schools spent less time doing academic work than children in more advantaged schools. Given the well-documented relation between academic engaged time and achievement (Berliner, 1983; O'Sullivan, Ysseldyke, Christenson, & Thurlow, 1990), interventions designed to increase academic time and academic responding may be critical to alleviate the discrepant outcomes between low and high SES schools.

The Peabody classwide peer tutoring model (Fuchs, Fuchs, & Mathes, 1993) promotes time engaged in reading in classrooms with a broad range of ability. Three times each week, children read aloud in pairs (partner reading), summarize information, and make predictions about story events—activities directed toward promoting reading fluency and comprehension. Points and class teams operate in a similar manner to the Juniper Gardens model. Unlike Greenwood's (1990) study, however, children are paired to include a higher and lower ability student, and for the partner reading, the higher-ability child reads first to provide a model. This pairing of high and low ability students was effective in increasing the reading achievement of the high and low performing students.

CWPT models have a research history documenting effective use in general education classrooms (Greenwood, Delquadri, & Bulgren, 1993; Kohler & Greenwood, 1990), moreover, the effects are positive for children across a spectrum of ability levels, including children with disabilities and low achieving students (Beirne-Smith, 1990; Harper, Mallette, Maheady, & Brennan, 1993; Kamps, Barbetta, Leonard, & Delquadri, 1994). A best-evidence synthesis of peer tutoring research in reading (Mathes, Fuchs, Fuchs, Henley, & Sanders, 1994) documented consistently significant effects for studies that paired high and low-achieving students, in comparison with other whole-class models. The review stressed, however, the CWPT was not more effective than teacher-led one-to-one tutoring or small-group instruction.

Cooperative learning. Cooperative learning (CL) is often recommended as a strategy for increasing the participation of low-skilled children in general education classrooms, and yet the outcomes for children with disabilities are not uniformly superior under this arrangement. Tateyama-Sniezek (1990) conducted a meta-analysis of studies that performed separate analyses for outcomes of children with disabilities. For the 12 studies included in the analysis, Tateyama-Sniezek concluded, "The only firm conclusion is that the opportunity for students to study together does not guarantee gains in academic achievement." (p. 436)

One of the CL models with empirical support for students with mild disabilities is Team Assisted Individualization (TAI), designed by Slavin, Madden and Leavey (1984). In this program, the teacher provides direct instruction to small homogeneous groups on algorithms and concepts, then the students join heterogeneous 4-5 person teams for assignment completion. Teams are encouraged to help one another, and also to perform monitoring tasks such as correcting and recording work. A 24-week study of the approach with 1,371 students in experimental and control groups in 59 third, fourth, and fifth grade classrooms, including 113 students with disabilities (63 in the TAI treatment) showed advantages for students in TAI in computation, but not in concepts and applications. The same pattern was found in a separate analysis of outcomes for the students with disabilities. In computation, the children with disabilities gained an average of .5 grades over their controls, however, the students without disabilities gained significantly more during this time period.

Concerned about the discrepant results reported for children with disabilities in CL arrangements, O'Connor and Jenkins (1996) examined factors related to how students with LD engaged in cooperatively structured reading lessons. They observed pairs of children with LD and average-performing children (grades 3-6) in the same general education classrooms to assess the quality of participation of students with disabilities by noting the amount and kind of help they received (and from whom), and the amount and kind of contributions they made to group efforts. Students with disabilities received more help and offered fewer contributions than non-disabled students, and only 40% of these students were classified as successfully participating in cooperative groups. Analyses disclosed that classroom practices, such as selection of partners, teacher monitoring, the establishment of a cooperative ethic, influenced successful cooperative learning experiences for students with disabilities. Descriptions of the observations suggest that it is a difficult act to balance special education students' need for direct teaching, the group's need to accomplish its work, and the class's need to learn cooperation and tolerance.

Studies on Instructional Conditions and Learning for Students with Mild Disabilities

Reading Approaches

"Mild disabilities" is not a unitary construct—which suggests that the kind of instruction that benefits one child with disabilities will differ from the instruction that benefits another—however, studies of aptitude-treatment interactions have found that matching a student's relative strengths to a particular approach produced no advantage over more eclectic instruction (Arter & Jenkins, 1977; Bateman, 1977). Currently, there is not a dominant "approach" to teaching students with disabilities to read; rather, special educators tend to either shadow instruction in the mainstream by using the same materials in smaller groups, and modifying instruction along gross dimensions such as level of adult support, or pacing of content coverage; or employ alternative instructional materials which focus on a single element of reading, such as decoding or developing sight words or comprehension strategies (Haynes & Jenkins, 1986; McGill-Franzen & Allington, 1991). Neither approach takes advantage of the wealth of knowledge about reading available through studies of the reading progress of typically developing children. The last decade has witnessed advances in the knowledge base about teaching reading well to children in the beginning stages of reading acquisition. Researchers stress the importance of including a range of elements integrated around experiences with story text, such as phonological awareness and decoding, discussion of ideas and concepts central to the stories, reading and rereading connected text, comprehension and vocabulary activities, and writing (Adams, 1990; Anderson, Heibert, Scott, & Wilkinson, 1985; Beck, 1989; Lipson & Wixson, 1986; Slavin, 1990; Stanovich, 1986;

Strickland & Cullinan, 1990). Exposure to print also affects reading acquisition substantially (Adams, 1990; Cipielewski & Stanovich, 1992). Prevailing practices for students with mild disabilities, whether reading is taught in the mainstream or in special education classes, are rarely so comprehensive, although a comprehensive program should be especially important for students experiencing great difficulty learning to read.

There is also a danger, however, in inferring that successful programs for mainstream students will be equally successful for children with disabilities. Most students with mild disabilities are first identified when they *fail* to thrive in traditional mainstream programs; moreover, "an approach" may not be compatible with the notion of individual differences, which is the hallmark of special education. For a reading approach to be effective for children with disabilities, it may need careful adjustments during instruction by reading teachers who understand the elements of effective reading instruction and who also anticipate the difficulties of struggling learners and adjust the instructional blend (e.g., strategies, practice opportunities, scaffolding, review and reinforcement) to suit the learning of an individual child.

The initial acquisition of reading. Evidence suggests that most children who develop severe reading problems have difficulty with phonological processing, and these difficulties can be reliably identified as early as the kindergarten year (Lundberg, Olofsson, & Wall, 1980; Share, Jorm, MacLean, & Matthews, 1984). Phonological processing refers to the ability to isolate and manipulate the individual sounds of spoken words. These skills, such as blending abstract sounds into words or segmenting words into individual sounds, influence a child's grasp of the alphabetic principle, which is necessary for understanding how speech is captured by an alphabetic script. This body of evidence has led some researchers to recommend increasing the phonological understandings of children prior to reading instruction. Experiments which have attempted this task in Denmark, England, Australia, and the United States have yielded superior results over control conditions (Ball & Blachman, 1991; Bradley & Bryant, 1985; Byrne & Fielding-Barnsley, 1993; Lundberg, Frost, & Petersen, 1988; Hatcher, Hulme, & Ellis, 1994; Torgesen, Morgan, & Davis, 1992), however, the subjects of such experiments usually exclude atypically-developing children, such as those repeating kindergarten, or children already identified for special education. Experiments in which children with disabilities just beginning to learn to read were taught phonological blending and segmenting have produced short-term effects (O'Connor, Jenkins, & Slocum, 1995; Williams, 1980), however, the long-term effect on reading development for children with disabilities from this type of intervention has not been tested.

Rack, Snowling, and Olson (1992) reviewed international research aimed at identifying just where in the reading process children with reading disabilities have the greatest difficulties. The weight of evidence derived from reading level matches (older children with LD in reading and younger normal readers) suggests that children with LD have persistent difficulty reading individual words, and particular difficulties with decoding. Instructional approaches that include a code-emphasis are favored in most research findings over approaches that de-emphasize letter/sound correspondences and word analysis (Adams, 1990; Perfetti, 1991; Vellutino, 1991). The theoretical premise for code-based instruction is that reading comprehension is only possible when some degree of fluency has been achieved, and that fluency requires the automatic reading of words, which is generally constructed through a transition from letter/sound decoding to increasingly large structural units of words (Ehri, 1992).

Reading words. Reading often includes decoding instruction in the early grades, but for many students, this type of approach is abandoned as they progress through the grades. Students with mild disabilities are rarely found eligible for special education services until third grade or beyond, a time when basal programs no longer emphasize phonic generalizations. A two-year longitudinal study of dyslexic children conducted by Manis, Custodio and Szeszulski (1993) revealed that children can continue to make decoding progress well up into adolescence, provided that instruction in word analysis continues. Thus it may be advantageous to continue to include decoding and word analysis, even as the emphasis turns toward comprehension for older students. The following studies report findings from a variety of word analysis approaches.

Several studies in the U.S. and the U.K. (e.g., Goswami, 1988; Treiman & Zukowski, 1992) have suggested that typically developing readers during the first year or two of instruction use a level of word analysis in between the letter-by-letter and syllable-by-syllable stages. Evidence for this onset-rime level of processing is found when children choose to break words between the initial consonant or consonant-blend and the first vowel (e.g., sh-ake; m-eet), rather than after the vowel, and to use onset-rime analogies to decode nonsense words. The Benchmark program in Pennsylvania (Gaskins, Downer, Anderson, Cunningham, Gaskins, & Schommer, 1988) uses a reading-by-analogy approach based on onset-rime level units for students with LD in the elementary and middle school years. Preliminary results show significant gains in pseudoword reading, with stronger effects for children who received the intervention for longer duration, but no significant effect on overall standard score gains in total reading.

A series of studies in the Netherlands attempted to tease out the preferences of students with LD (IQ's 90-100) for the size of decoding unit (Van Daal, Reitsma, & Van der Leij, 1994). Students were able to learn words when they were introduced through an individual letter-to-sound approach, through letter clusters, or onset-rime units, but generalization to new words was more effective when children were taught to break words at the phoneme level and blend sounds. As a counterpoint to the recommendations of Goswami, Treiman, and others, Van Daal et al. found that children with reading problems showed no preference for words broken at the onset-rime level, either in learning word lists or in generalizing to new words. Wise (1992), through evidence gathered in experiments with students with LD using computer feedback, found only limited benefits for the onset-rime level of feedback; specifically, this level was useful for one-syllable words of 4 or more letters (see Technology, in this section, for more information on computer-feedback experiments in reading).

Gersten, Keating, and Becker (1988) reported fifth and sixth grade outcomes for children who were part of the Follow Through project and received Direct Instruction (DI) programs during their primary school years. Direct Instruction reading programs are rooted in a synthetic phonics approach with lessons structured on empirically tested design features. Three years later, these children performed significantly better than their matched controls in reading, spelling, and mathematics. All of the children in the Follow Through project were considered at risk for school problems due to economic disadvantages; Gersten et al. did not perform a separate analysis for the children with disabilities. Studies designed to test the effects of DI reading programs for students with mild disabilities are generally favorable. In a meta-analysis of 25 DI experiments with students in special education, including 12 in reading, 53% of the outcomes measured favored DI, and none favored comparison treatments (White, 1988).

Because of the phonological and decoding deficits of children with reading disabilities noted by many researchers, some reading specialists have questioned the advisability of attempting decoding or word analysis methods with these children (Siegel, 1985). Lovett, Warren-Chaplin, Ransby, and Borden (1990) compared the outcomes of whole word and grapheme-phoneme correspondence methods on the learning and transfer to new words. They randomly assigned 54 children (mean age 8.4 years) with reading disabilities to whole word or grapheme-phoneme correspondence conditions, and taught word recognition for 35 one-hour lessons. Children taught through both regimes made greater gains than control children, who received a comparable amount of classroom survival/study skills instruction. The gains, however, did not transfer to untaught words in either condition. These children did not abstract information about letter/sound invariance for application to new words, nor was a letter/sound approach differentially effective in encouraging transfer. The children also failed to use reading-by-analogy strategies documented by Goswami (1988) for younger, normal readers.

Levy, Nicholls, and Kohen (1993) explored the utility of repeated reading for building reading skills of good and poor readers in third, fourth and fifth grade. The task consisted of silently reading three 280-word stories four times each. For each reading, 20 words were misspelled, and the child crossed out the misspellings while reading the story, and then answered 3 comprehension questions. The researchers recorded the total time spent reading, the number of errors detected, and the questions answered correctly. As expected, the children read faster with each repetition, but poor readers also increased their detection of errors across reading, even as their fluency increased. Differential gains in fluency across repeated readings and across the new stories that followed each series favored the poor readers. The speed gains on new stories were significant for the 3rd and 4th grade poor readers, but not for fifth graders. An unexpected bonus to the procedure was an increase in reading comprehension on new, unpracticed stories.

The initial acquisition of reading is the toughest academic challenge faced by young students with disabilities and their teachers. When the evidence so strongly supports the necessity of building a foundation in letter/sound correspondences, word analysis strategies, and automatic word recognition (Adams, 1990; Perfetti, 1991; Stanovich, 1992; Vellutino, 1991), it makes sense to exert effort toward building these skills in the early school years for all children. Many children with disabilities also have difficulty with reading comprehension, and the next studies consider approaches to teaching children with disabilities to understand the text they read.

Cognitive approaches to reading. Although comprehension is the ultimate goal of reading, and comprehension activities are traditionally included in all instructional approaches to encourage children to derive meaning from text, comprehension is rarely directly taught in mainstream classes (Calfee & Drum, 1986). Because children with disabilities often have difficulty comprehending text, several strategies have been developed to help children externalize what people do when they read for comprehension. These strategies include diagrams of story structures, self-questioning techniques, integrating writing with reading activities, summarization strategies, and integrating new with known information, among others.

Griffey, Zigmond, and Leinhardt (1988) tested a metacognitive approach to reading comprehension. Three classes of children with LD (mixed third-fifth grade) were randomly assigned to one of two comprehension treatments or a no-strategy condition, all reading the same basal stories across four sessions. One group was instructed in self-questioning techniques to identify main characters, aims, problems and solutions in stories. One group completed a story-structure frame with the teacher during each lesson. Only the children taught the self-questioning strategy answered significantly more questions correctly following training than the control group. Four sessions was not sufficient for observing improvement on standardized measures, and a longer test of this treatment is clearly warranted, as the authors suggest.

Palincsar and Klenk (1992) suggest that differential instructional practices—from the time a child is first labeled as mildly disabled—preclude opportunities for developing higher level thinking and metacognition, such as the self-questioning strategy of Griffey et al. (1988). Their approach to reading and writing incorporated aspects of emergent literacy and reciprocal teaching into reading programs for primary grade students with LD. Three times each week, teachers combined their story and handwriting time to provide thematic instruction that would integrate activities in the two areas. They introduced invented spellings, encouraged writing about topics from story time, and scaffolded writing by holding words and sentences in memory for the children as they worked with the conventions of getting words and letters on paper. Qualitative analysis of outcomes suggested that children's knowledge of literacy and writing improved, as well as production of written text. The most striking difference they reported was in teacher behavior, which shifted from directive to participatory (Cazden, 1986); no data on student reading outcomes were reported.

Gaultney (1995) examined the effect of prior knowledge on students' application and retention of a strategy for deriving the main idea. Fourth and fifth grade boys with poor reading comprehension but good baseball knowledge (mean IQ = 90) were taught to ask "why" questions during six 30-minute lessons. Half of the boys read stories about baseball (high prior knowledge), and the other half read stories about sports other than baseball, all with similar readability. The boys who read the baseball stories demonstrated better use of the strategies and better recall of new story information following training, and on a follow-up test three weeks later. The experiment demonstrated the effect of domain specific knowledge on the acquisition of new reading strategies, and suggested that a strong knowledge base might ease this acquisition. They recommend beginning strategy instruction in areas where student expertise is strong, then generalizing new skills to less familiar areas. Krupski, Gaultney, Malcolm, and Bjorklund (1993) also documented the effect of prior knowledge on memory and comprehension, particularly for students with disabilities.

Englert, Tarrant, Mariage, and Oxer (1994) used preservice special education teacher interns to teach 109 children with LD in grades 1-8 one of two reading comprehension strategies: the POSSE (*p*redict, *o*rganize, *s*earch, *s*ummarize, *e*valuate) strategy which emphasized cognitive apprenticeship and the social construction of knowledge through elaborated graphic display of reading content, or the K-W-L (*k*now, *w*ant to find out, and have *l*earned following reading) strategy in which teachers directed children to make lists pertinent to the three categories. Six weeks later, children were tested on reading comprehension with measures of written retelling of stories, and passage recall, following group practice using one of the learned strategies. Significant results favored the POSSE strategy, however, children in this condition also spent more time on the comprehension of each story.

Children taught to comprehend passages through each of these strategies demonstrated gains; moreover, all of the researchers stressed the importance of frequent practice, and of providing opportunities for children to generalize the use of strategies to a range of appropriate text.

Early Intervention in Reading

Early intervention offers hope for reducing the intensity of support needed later in school. The arguments supporting early intervention resonate among researchers and practitioners concerned about children whose disabilities who often go undetected until serious and prolonged school failure marshals the necessary professional attention. Ideally, intensive instruction would be delivered as soon as the need is apparent, to prepare these primary children to be brought into the inclusive classroom ready to participate and learn. The timing of instruction may be particularly crucial, however, providing services to struggling learners prior to labeling for special education requires tapping funding sources that are not typically available until after formal identification—which occurs during the middle elementary years for most students with mild disabilities—too late for the proposed interventions to be useful. Funding early intervention for non-labeled students that is intensive enough to decrease the likelihood of needing special services later remains a challenge for school, district, and state administration.

Phonological awareness in kindergarten. Most children who have difficulty learning to read also have difficulty with tasks requiring phonological manipulation skills, such as auditory blending, segmenting, and rhyming. For typically developing children, training in these skills either before or alongside initial reading instruction improves reading acquisition in the early years. Programs have been developed to stimulate phonological awareness prior to the beginning of formal reading instruction and the onset of reading difficulties (Byrne & Fielding-Barnsley, 1991; Lindamood & Lindamood, 1984; Torgesen & Bryant, 1994), but the long term effects of early training for children with disabilities has not been determined. The focus on experimentation with phonological skill development is likely to continue over the next several years, or until issues concerning instructional content, methods for resistant learners, support during initial reading acquisition, and long term effects on reading progress are resolved.

Tutoring approaches. First grade tutoring approaches have received considerable attention in reading research. Most of these approaches have not included children with known disabilities, however, their intent is to forestall the need for more intensive special education assistance in the later school years.

Wasik and Slavin (1993) compared the effects of five one-to-one remedial tutoring programs in reading for first-grade students achieving below grade level expectations: Reading Recovery (Clay, 1985), Success for All (Slavin et al., 1992), Prevention of Learning Disabilities (Silver & Hagin, 1990), the Wallach Tutoring Program (Wallach & Wallach, 1976), and Programmed Tutorial Reading (Ellson, Baber, Engle, & Kampwerth, 1965). All of the programs demonstrated positive effects for tutoring, but several features of programs led to particularly strong results. The programs that achieved the strongest effects were those with the most comprehensive models of reading (e.g., including perceptual analysis, error correction strategies, decoding, comprehension, and reading strategies), delivered by certified teachers rather than paraprofessionals, and coordinated with regular classroom instruction.

Several early intervention programs have, in fact, developed a mix of elements in a successful intervention program (Clay, 1979; Hiebert et al., 1992; Pinnell, 1989; Slavin et al., 1992). The Reading Recovery (RR) Program, for example, targets low-achieving first-grade students; they are pulled from their classroom and instructed on an individual basis by a trained teacher for 30 minutes each day for 60 lessons or until the student can handle the average level of text reading for her/his class, monitor his/her own reading, and correct self-detected errors. But Reading Recovery is not the solution for all struggling readers: Although subjects who receive instruction in Reading Recovery perform better, overall, than subjects who do not receive one-to-one tutoring, 27% of Reading Recovery subjects do not "catch up" to their classmates within the 60 lessons the program allows (DeFord, Pinnell, Lyons, & Young, 1988). This percentage translates into 5.4% of the first-grade sample, a proportion similar to that of the population reading disabilities. Reading Recovery does not appear to be intensive enough for the children about whom we are most concerned.

Although many school districts around the country are channeling Chapter 1 funds and/or special education funds into Reading Recovery, the long term outcomes for children in these programs is far from settled (Hiebert, 1994). The longitudinal reports are clouded by missing data: First, the number of children who are dropped from the program during the first two weeks because they are judged unsuited for this type of intervention is rarely recorded; second, a percentage of children never do catch up to their first grade peers in reading ability, and for several years, these children were not included in analyses of the long term effects of the program. Even accepting these experimental flaws, by the 4th grade, the mean difference in overall reading level between at-risk children who received or did not receive the intervention (n = 212) was not significant (grade levels of 3.0 and 2.9, respectively, on the Woodcock Reading Mastery Test-revised).

Two studies tested student outcomes for the Reading Recovery program against variations on RR lessons. Pinnell, Lyons, DeFord, Bryk, and Selzer (1994) used a split plots design, in which each district's existing Reading Recovery program was compared with 3 other models, all consisting of 30 minute lessons taught by certified teachers: (1) Reading Success (RS), a one-to-one tutoring program using RR materials and similar lesson formats. These teachers received a condensed training cycle—70 hours of staff development training over a two-week period (compared to year-long training for the RR teachers). (2) Direct Instruction Skills Plan (DISP), an individual tutorial linked to classroom instruction, but focusing on deficits identified through a skills survey (e.g., sight vocabulary, word analysis, comprehension, language development, and study skills), and usually focused on learning words. The description provided by Pinnell et al. suggests that children seldom read books in this condition, and that teachers often read aloud to the students. (3) Reading and Writing Group (RWG) followed a format similar to RR, and was taught by RR teachers, however, children met in small groups with the teacher, instead of individually.

All of the first grade children (n = 403) fell below the 37th percentile on district-administered reading tests. All four program variations were also compared to a control group of children receiving the district-provided Chapter I assistance. A comparison of the time spent on components of reading showed that children in RR and Reading Success lessons spent about 60% of their time reading, compared to less than 30% in the DISP, RWG, and control group. The number of books read by the children varied from 84-94 in RR and RS, to 4 in the DISP. Short term effects favored the RR group in level of text reading and dictation, which is not surprising given the amount of reading in that condition. Three months following the treatment, however, there were no statistical differences between the experimental conditions on the Gates-MacGinitie, although small RR effects on dictation persisted through the beginning of second grade. An interesting comparison in this study is that between RR conducted individually, as designed, and RR conducted by similarly trained teachers, with the same lesson format and materials, in small groups. At the end of treatment, children in the individually conducted lessons performed significantly higher than those taught in groups on dictation, text reading, and the Gates-MacGinitie. In a climate which encourages children to receive the majority of their instruction in the mainstream, it appears that when lesson components are equal, one-to-one instruction is still the most effective format for changing the reading behavior of children.

In another Reading Recovery variation, Iversen and Tunmer (1993) tested the effect of increasing the amount of phonological recoding instruction in Reading Recovery lessons. From low-skilled first graders drawn from 30 elementary schools, they matched children on letter identification and dictation scores, and formed three groups of 32 children each. One group received standard Reading Recovery lessons, one group modified lessons, and one the traditional remedial reading instruction provided by their districts. The modifications began around lesson 15, and focused on making children aware that many words with common sounds also share common letter patterns. Children manipulated plastic letters to make, break and build new words with similar sound patterns. The rest of the lesson followed the Reading Recovery format. Children in both RR variations were discontinued as they reached the mean of their classroom performance, as specified in the standard RR program. Both variations of Reading Recovery performed better than children receiving district-supplied remediation programs, and the authors suggest, as did Pinnell et al. (1994), that one-to-one tutoring (as opposed to grouped instruction) is a likely factor in these results. Although the outcomes for the two variations of RR did not differ, children in the modified treatment reached the average levels of their class performance sooner than those in the standard RR program (42 lessons for the modified RR children, compared to 57 lessons for the standard RR).

Neither of these experiments involved children with documented disabilities, and Reading Recovery was not designed as a treatment for children in special education. Considering, however, that most children with LD are not identified until after their first grade year, and that increasing numbers of schools nationwide are forestalling identification of children for special education services in an attempt to encourage inclusion, the findings from these studies have bearing on the discussion of services for students with mild disabilities.

Juel (1994) took a less structured approach to tutoring, and used college students who were at risk for failure due to poor literacy as tutors for young children with reading acquisition difficulties, including children in special education. Lessons conducted twice weekly included books written by the college students for their tutees, predictable children's books, phonemic awareness, writing in journals, and letter/sound activities. The tutoring produced significant gains for the children; Furthermore, through observations of tutoring sessions Juel extracted the elements that appeared to contribute to improved outcomes. Her observations suggested that children learned best when tutors personalized stories, provided visual and auditory support as children read, and broke word recognition and spelling down into small steps. She also emphasized the importance of the relations between the tutor and tutee, particularly by establishing a social context that is relevant, meaningful, powerful, and caring.

Hiebert, Colt, Catto, and Gury (1992) restructured a typical Chapter I remediation format for first graders which significantly increased the power of the intervention. At the end of first grade, 77% of the children in the restructured program reached a level of primer or higher, compared to 18% of children in the typical program. The components of the restructured program that differed included extensive reading practice, a little writing, word level strategy instruction that included segmentation and letter/sound instruction, and the integration of words used for study with the text to be read by students.

Taylor, Strait, and Medo (1994) designed extra reading instruction for the lowest readers in first grade. Teachers provided an additional 15-20 minutes of daily reading time for their lowest five-seven readers, focusing on word attack and phonological awareness, on sounding out, blending, and writing words and sentences, and monitoring comprehension by asking questions such as "Does that word make sense?" Children engaged in repeated reading of short texts, which were rewritten versions of the longer stories read during regular, whole-class instruction. Although outcomes were not uniformly successful, end of first- and second-grade follow-up measures revealed that children with low entering reading skills who received early intervention targeted on reading could be raised from the lowest quartile, in which Juel's (1988) classic study predicts low achievers will remain.

Slavin (1991) recommended an intensive preventative approach he termed "neverstreaming", in which students would be identified very early in their school careers and receive immediate and continued intervention as necessary to prevent large achievement lags from developing in the early years. Success for All (Slavin et al., 1992), modeled on that philosophy, again documents the power of early intervention. Tutorial support for 20 minutes daily was offered immediately to children who made less than average progress in the first years of school. The regular class reading lessons included cooperative learning for 90 minutes daily, comprehension lessons taught at the level of children's receptive language, shared reading with teachers, and school-wide regrouping across grades so that children read materials at appropriate reading levels. Slavin et al. reported decreased incidence of referrals to special education, however, children who were identified for special education services needed additional support beyond that offered by Success for All.

These methods of individual and small-group tutoring, although varying in content and focus, adopt a comprehensive approach to reading acquisition. All of these models have found positive short-term effects over traditional approaches. The hope of early interventionists is that—if implemented promptly—early intervention may decrease the incidence of reading failure and the need for prolonged special services for children whose primary disability is in reading. The long-term effects of comprehensive and continuing early intervention on the incidence and severity of reading disability have not been determined.

Writing and Spelling

Writing strategies. The cognitive approaches for reading comprehension, described earlier, have also been applied effectively to writing instruction for students with mild disabilities. In these approaches, students learn a strategy for designing and executing compositions, with practice in applying strategies across a range of writing situations. Graham, Harris, and colleagues (Graham & Harris, 1989; 1992; Graham, MacArthur, Schwartz, & Page-Voth, 1992) developed a line of research that combined writing development with cognitive behavior modification. Their procedures address the problems documented for children who have particular difficulty producing appropriate pieces of writing: planning, revising, production of text, and monitoring its quality. Outcomes of these procedures for children with LD in elementary classes indicated that students spent more time planning, more time writing, and improved the quality as well as the quantity of work. For example, in a study of 4 fifth-grade children with LD, the writing strategy was structured around a goal setting process which began with structuring argumentative essays (Graham, MacArthur, Schwartz, & Page-Voth, 1992). Teachers modeled using strategies, students memorized steps and practiced. Children received eight 40-minute lessons in which they learned about the facets of good essays, how to use word processing software, and the general strategy (planning, writing, and testing goals). Teaching continued until students wrote essays which included all of the components of good essays they had learned. Maintenance probes documented generalization from essays to stories, and increases in the length and quality of the essays 15 weeks after instruction. MacArthur, Schwartz, and Graham (1991) added reciprocal peer revision to the process, in which pairs of 4th, 5th and 6th grade students in special education classes helped each other improve their compositions. The editing procedure included (1) listening to the author read the paper; (2) telling what it was about, and the best part; (3) rereading the paper, and making notes about revision; and (4) discussing suggestions with the author. Children were given two questions to consider during step three: Is there any part which is not clear? Where could more information or details be added? As well as learning the process for editing papers, children decreased errors of spelling and punctuation in their final products and improved the overall quality of compositions, as reflected in holistic ratings.

Key in this kind of instruction is developing methods for generalizing learned strategies to new types of writing tasks, and for maintaining gains over time. For students who did not initially transfer learned strategies to new tasks, a booster session (practice in applying the strategy to new writing situations) improved subsequent performance (Graham et al., 1992). Reinforcement and practice in logical extensions of new learning assisted with generalization. Most of these studies were conducted in small-group, laboratory settings, and applications need to be extended to regular schools and classrooms. The outcomes stimulated through these interventions are consistent with the goals of regular class teachers in upper elementary grades. If regular education teachers can successfully implement this type of instruction, it may be a cost effective method for including children with mild disabilities in regular class writing instruction.

Spelling. Fuchs and colleagues (Fuchs, Allinder, Hamlett, & Fuchs, 1990; Fuchs, Fuchs, Hamlett, & Allinder, 1991) have developed and tested computer programs to provide teachers with data to illuminate the types of instructional modifications likely to improve the spelling performance of their children. Teachers who measured the spelling performance of their children with mild disabilities weekly using curriculum-based measurement (CBM) received three types of computer feedback: graphed performance of each child; graphed performance with lists of words spelled incorrectly, indicating how children actually spelled the words; or graphed performance with skills analysis that documented problematic word patterns for individual children (e.g., specific suffixes or vowel error types). Children whose teachers received computer-generated specific feedback (either detailing the errors made, or clustering the most common errors) spelled significantly better 15 weeks later than children whose teachers received only graphed performance, or whose spelling routine did not include CBM. These results suggest that inspection of the errors children make can lead to improved instruction and improved outcomes for students with mild disabilities.

O'Connor & Jenkins (1995) taught spelling through the manipulation of letter tiles to 6-year-old children with mild disabilities who were already receiving code-based reading instruction. They formed matched pairs of children with mild developmental delays based on their progress on reading lessons in Distar *Reading Mastery I*, and randomly assigned one of each pair to an experimental treatment of 20 ten-minute spelling lessons on words drawn from their curriculum, or to a reading control group that practiced reading the same words. Children in the spelling treatment significantly improved their spelling and reading of practiced and untaught words over their matched controls. These results suggest that the children who practiced forming letter representations of spoken words developed more complete generalizations of their current knowledge, which facilitated learning to read, as well as spell, words.

Gordon, Vaughn, and Schumm (1993) reviewed spelling studies that investigated effects for students with LD, and derived a set of empirically-based findings for instruction. No advantages were found for practice media, whether paper and pencil, letter tiles, or computer keyboarding (Vaughn & Schumm, 1992), although students with disabilities preferred computer practice, and CAI improved attitudes toward spelling practice (Watkins, 1989). As suggested by the CBM studies (Fuchs et al., 1993), including specific feedback on types of errors (Gerber, 1986) improved students' overall spelling ability. Because practice is key to improvement, strategies which increased persistence on spelling tasks and academic responding, such as CAI or peer mediated structures (Harper, Mallette, & Moore, 1991) were the most effective. Although most of the reviewed studies were conducted in controlled settings, these recommendations could be feasibly tested in regular classrooms that include students with disabilities.

Mathematics. Teachers in general education classrooms have increased their use of manipulatives for concept development and problem solving, however, few studies have tested this approach for students with disabilities. Peterson, Mercer, and O'Shea (1988) considered the utility of manipulatives for teaching the concept of place value to 24 students with LD, ages 8–13 by randomly assigning children to treatments that either used manipulatives, or only numbers to teach place value for ones and tens. Both groups received nine days of instruction using the same set of problems. In the manipulatives treatment, the first three days of instruction used concrete representations, such as unifix cubes, the next three days used pictorial representations, such as bundles of sticks, and the last three days used only printed numbers. The abstract condition taught the same items, but used only numbers for all nine days of instruction. A 2 (treatment) X 3 (time) MANOVA demonstrated a strong treatment effect favoring the use of manipulatives, and a follow-up test 3 weeks later again favored the concrete-to-abstract condition on retention, as well as on skill acquisition. Children in *neither* treatment, however, demonstrated generalization to 3–4 digit numbers. As in other academic disciplines, it is likely that children with disabilities will need to be taught to generalize by increasing the range of potential applications for new learning.

Students with disabilities that affect mathematics performance are likely to need support beyond that offered in traditional regular classrooms. Geary (1990) suggested that memory for math facts and using appropriate strategies are both factors in the problem solving success of children with disabilities. In a later study, Geary, Bow-Thomas, and Yao (1992) compared the relations between counting and addition computation for typical first grade children and first graders with math disabilities. Children with math disabilities made significantly more errors in counting (e.g., counting some objects twice, difficulty detecting their errors), suggesting that immature counting behaviors may also contribute to the problems these children exhibit with beginning computations. Since counting small numbers is rehearsed much more in kindergarten than in first grade classrooms, where these difficulties were discerned, additional experience with prerequisite skills assumed by most regular class teachers may need to be provided for children with disabilities. The identification of areas on which to concentrate support, and methods for providing it in general education settings have received little research attention in recent years.

Problem solving in mathematics. Methods for teaching mathematics strategies to students with disabilities in small group research settings have been successful. Case, Harris, and Graham (1992) taught four fifth and sixth grade boys with LD self-regulated strategy procedures for solving word problems, which included cue word identification (e.g., how many less, how many more, how much more), problem solving steps (e.g., read problem, circle cue words, draw pictures, write a math sentence, write the answer), self-evaluation and self-reinforcement. Procedures included activities to stimulate generalization, such as taking math folders into the regular class, and discussing instances when students successfully applied the strategies. A multiple-baseline design across subjects demonstrated positive effects for their procedures, and effects maintained for three of the four students 10-12 weeks following instruction. The authors attribute the success of the procedure to the combination of learning new skills along with a strategy for how to apply the skills in word problems.

Wilson and Sindelar (1990) used a Direct Instruction approach to teach a variety of problem solving procedures to children with LD in grades 2-5. Sixty-two children were randomly assigned to: strategy only (e.g., if the big number is given, subtract); strategy plus sequence (in addition to strategy training, problem types were grouped so that children had only one problem type daily); or sequence only (only one problem type daily, along with questioning, such as "What are we supposed to find?"). Providing the problem solving strategy led to superior results over children who received the traditional approach (sequence only), but there was no reliable difference between the two approaches that included the strategy. The authors concluded that children with learning disabilities in math can learn to solve word problems, but that explicit instruction and practice was necessary for them to do so.

As in all academic tasks, children with disabilities consistently need more explicit teaching, more opportunities to practice learned skills, and encouragement to use new learning across a range of applications. A growing body of research suggests that for students who have difficulties with comprehension, memory, problem solving, and other multi-step academic procedures, strategy instruction to develop and internalize these processes may be useful.

Cognitive Strategy Instruction and Maintenance of Learned Behavior

Several studies have suggested that children with mild disabilities lack knowledge or use of appropriate learning strategies, however, children with disabilities use some strategies for learning, even without special training (Papetti, Cornoldi, Pettavino, Mazzoni, & Borkowski, 1992). For example, a study of second and fifth grade good readers and children with reading disabilities in Italy suggests that the children with reading disabilities accurately distinguish between items that will be easy or difficult to learn, and differentially study more difficult items. Still, their outcomes on memory tasks tend to be significantly lower than that of good readers. Recall is higher for older students, and for better readers at both grades, but higher recall was not related to longer study times (e.g., children who read poorly spent a longer time studying, and received lower scores), rather, it appears that good readers study more efficiently.

Specific strategy training improves the use of appropriate and efficient strategies for learning tasks. Cornoldi and Vianello (1992) taught students with mild mental retardation strategies for memorization tasks, and found positive effects on tasks similar to the training tasks, but poor generalization. In a meta-analysis of 268 studies of the memory problems of students with MR, Kavale and Forness (1992) concluded that these children have deficits compared to age-matched students without MR, and that strategy training did not alleviate the discrepancies. In comparisons of effects with non-treated control children with MR, training had a considerable effect on improved learning and memory (effect sizes averaged .7). Outcomes affected in these studies included number of errors, recall, identification (e.g., discrimination or recognition), and number of learning trials. The least likely areas to be affected were incidental learning, transfer, and generalization. Kavale and Forness concluded, as in other meta-analyses of treatment effects, that interventions of longer duration were more effective.

The aim of most instructional packages in learning strategies is independent use of appropriate strategies—a downfall for students with disabilities (Ellis & Friend, 1991; Torgesen, 1980; Wong, 1991). Researchers have tried varying the types of feedback children receive on the utility of learned strategies (Schunk & Rice, 1987), and fading techniques for strategy rehearsal (Schunk & Rice, 1993), however, these studies were not conducted with students with disabilities.

Attitudes toward learning. Other researchers have suggested that the attitudes children have toward learning gains are powerful mediators of future learning, and have set out to study this phenomenon. "Motivational processes and the self-system are intimately related to the development of self-regulation, the component of metacognition essential for generalized strategy use." (p.1) (Borkowski, Day, Saenz, Dietmeyer, Estrada, & Groteluschen, 1992) For long-term performance gains, strategy training must be accompanied by self-regulation strategies, which include feedback about the importance of effort (Lucangeli, Galderis, & Cesare, 1995). In a study with LD and hyperactive children, Reid and Borkowski (1987) discussed task performance with children, and helped them to verbalize the relation between their performance and the use of the taught strategy. They emphasize that educators should not assume that children with disabilities will find increases in learning or scores pleasurable, and that pleasure will sustain effort in the future on difficult tasks. Rather, they suggest teachers explore the notion of causality with their students (i.e., success or failure on learning tasks) by explicitly forging the connections between knowledge, effort, and achievement, and by linking success or failure on tasks to effort and application of known strategies. The authors attribute the maintenance of strategy use ten months later to children's understanding of this connection.

In a similar study in Finland, Vaurus, Lehtinen, Kinnunen and Salonen (1992) taught text-processing skills aimed toward promoting long-term learning and generalization to 48 fourth grade children with LD. In one condition, children learned a text-processing strategy and how and when to apply the strategy. In a second condition, children learned methods for coping with difficult tasks (e.g., approaching and beginning tasks, dealing with obstacles, the meaning of errors). A third intervention encompassed both the other two: teaching the initial acquisition of knowledge, metacognitive skills (application of strategy), and socioemotional skills (coping). In all conditions, the researchers gradually faded scaffolding to promote generalization. As expected, the more complex treatment yielded the strongest outcomes, however, the outcomes of training were also dependent upon ego defensiveness of children prior to training (e.g., avoidance tendencies, loss of personal control), and on the learning environment following training. In classrooms that fostered the application of cognitive strategies, children were more likely to continue to use them. Vaurus et al. noted, however, that in Finland most classrooms were not structured to foster this kind of internalized persistence. The researchers needed to reorganize classrooms to promote retention of learned behaviors. "Only when regular classroom practices have been organized to support the laboratory intervention—that is, to foster self-directed use of cognitive strategies, task involvement, less ego-related coping with failure, and the sense of control—can long-term and lasting progression be achieved with severely LD children." (pp. 183–4)

Most strategy interventions for struggling learners include the development of preskills necessary for performance of the strategy, modeling and practice of the strategy to be used, discussion and demonstration of applications for the strategy, internalization of the strategy (verbal to silent rehearsal, self-recording or monitoring), and independent application of the strategy across relevant conditions. Several general aspects of strategy instruction discriminate more and less successful interventions for students with mild disabilities, particularly in maintaining and generalizing appropriate application over time. Instruction is more effective when (1) students plan when to use strategies; (2) a self-monitoring component is taught and rehearsed; (3) students practice across a range of useful applications; (4) teachers probe usage over time; and

(5) teachers promote application opportunities outside of special education settings, including applications outside of school (Harris and Pressley, 1991).

Technology and Monitoring Program Effects

Teachers find it difficult to alter instructional approaches and content, and to choose more effective teaching routines when children make less than optimal growth (Fuchs, Fuchs, & Bishop, 1992; Jenkins & Leicester, 1992). Computer applications for monitoring hard-to-teach children by special education and regular class teachers have been developed by Deno, Fuchs, and colleagues (Deno, 1985; Fuchs, Fuchs, Hamlett, & Ferguson, 1992), who advocate Curriculum-Based Measurement (CBM)—frequent short samples of reading, mathematics, spelling, or writing behavior over the course of a school year—as a means judging the adequacy of progress of children with mild disabilities or at risk for school failure. The expert feedback systems developed by Fuchs and colleagues assist teachers in making better instructional matches to address the kinds of errors produced by particular children. Expert systems have also been used to advise teachers about problems or adjustments needed in classwide peer tutoring (Greenwood, Finney, Terry, Arreaga-Mayer, Carta, Delquadri, Walker, Innocenti, Lignugaris-Kraft, Harper, & Clifton, 1993). The feedback provided by expert computer systems may assist teachers to make these difficult changes, and to adjust instruction in more diverse ways to achieve better outcomes for their students.

Technology and Children with Mild Disabilities

Ellsworth (1993) surveyed 68 teachers of students with LD in New York City on their use of technology for instruction, and on the availability of appropriate software in schools. Teachers reported using video recording and playback equipment, and older computers with limited software. Although the teachers surveyed were most interested in using interactive software, they said that newer equipment tended to be reserved for students in regular education classrooms. According to Nicklin (1992), an average of only one computer for each 30 students is available in most schools, and the most commonly used computers are obsolete versions of the Apple II (Cooper, 1993). Nevertheless, several research teams are investigating ways to improve academic outcomes for students with mild disabilities through computer assisted instruction (CAI).

CAI and reading. McGregor, Drossner and Axelrod (1990) compared the effect of adding synthesized speech to computer presentation of prereading practice for 6 and 7 year old children with LD or MR. The children knew the letters and letter sounds; the computer program was designed to provide practice in matching pictures and letter/sounds, pictures and words, and numbers with words. The children received each method of computer presentation for 25 days of instruction. The children were more motivated to continue practice when the on-screen tasks were accompanied by synthesized speech that "spoke" the text on the screen, and reduced their error rates for the voice plus text condition. Nevertheless, synthesized speech adds considerable expense to the equipment needed to run this type of software.

Wise, Olson and colleagues experimented with different levels of visual and synthesized speech computer feedback and the effect on word learning for students with reading disabilities. Early experiments (e.g., Olson & Wise, 1991) found that poor readers who read with computerized assistance made much stronger gains in word recognition and decoding than children who received traditional instruction. In their most sophisticated computerized reading instruction, children read connected text presented on the screen, and used a touch window to highlight unknown words. The computer, equipped with digitized speech, "read" the word for the student. They have investigated the relative utility of several forms of feedback (Wise, 1992; Wise & Olson, 1994), such as pronouncing whole words, words divided into syllables, onset-rime divisions of short words, subsyllables for longer words, or phoneme-by phoneme units. Their results have yielded several useful insights. First, presenting phoneme-by-phoneme feedback for words four or more phonemes in length was ineffective. It is possible that the memory difficulties associated with learning disabilities (Brady, 1992) may make blending of many separate phonemes difficult. Next, an onset-rime format appeared effective for short words, but not as subsyllables for longer words. The most effective presentation for longer words was syllable-level presentation. A main effect was also found for the severity of the reading disability, such that less severely impaired readers were able to read more pseudowords, and profited more from breaking words apart at the level of onset-rime than students with more severe reading difficulties. The poorest of the poor readers read more words correctly on outcome measures when the practice condition presented words in larger units.

In the Netherlands, Van den Bosch, van Bon, and Schreuder (1995) used computer technology to speed up the presentation of words to be decoded by children with LD, aged 7.8-12.8 years. Students with unlimited time to decode new words made less progress than children who studied with very limited exposure conditions. The authors concluded that stimulating children to decode faster increased their level of automaticity, which contributed to reading efficiency. The computer allowed control over exposure duration to be optimally effective for each child. Transfer effects to unstudied words were also found, suggesting that the decoding advantage underpinned improved reading generally.

For reading comprehension practice, however, Swanson and Trahan (1992) found no advantage to computer-mediated text for 60 students with LD in the 4th-6th grades. They recommended that teachers carefully evaluate the effects of computer practice, because in many cases it will produce results no different from other media, and for students with LD, may take much longer than text and paper tasks, and be considerably more expensive.

The effectiveness of any computer program is likely to depend on the design of the program and the opportunities it affords students who have severe learning difficulties (Torgesen, 1986). *Who* delivers instruction (e.g., teacher, peer tutor, or computer) is less important than the *design* of the instruction, and computers may be able to provide some advantages during practice stages of learning if the programs take into account sound learning principles. For example, CAI designers need to consider how much information to present to poor readers at one time (Gleason, Carnine, & Vala, 1990), and how to insure sufficient practice opportunities for students with disabilities.

Although the technology applications developed by Hasselbring and Woodward and their colleagues have been used primarily with secondary students, it is likely that extensions for younger students will soon be available. Persistence with difficult tasks and application of information are the focus of the work of Hasselbring (Hasselbring, Going, & Wissick, 1989). In a series of computer applications involving hypermedia, factual information is linked to video to provide simulations which demonstrate critical and motivating applications of the information to be learned. The Genisys Program (Woodward & Carnine, 1989) also makes use of videodisk technology in the content areas to teach problem solving in applied situations. Their program includes curriculum design features with demonstrated effects (e.g., modeling, cumulative review, a broad range of appropriate examples to encourage generalization).

The use of computers for instruction with hard-to-teach children is certain to expand over the next decade, but the issues raised in studies with children with disabilities—the need for extensive practice, careful introduction of new material, the utility of different methods of feedback, opportunities for a range of application of learned skills or knowledge—could require modification of commercial programs in the same ways that teachers modify other instructional materials for these children. We have learned from the large body of research on teaching and learning that children with disabilities need extensive practice to master new information, and more opportunities to apply new learning where it is most appropriate. For effective instruction, the amount of practice may need to be greater than that offered in most commercial learning materials for computers.

Summary

In many studies, students with more severe impairments (e.g., lower IQ's, lower entering reading levels, lower memory scores) realize smaller outcomes following treatments than students who are less impaired. But some studies also show an interaction between levels of impairment and the effectiveness of particular treatments (e.g., Wise & Olson, 1994). The notion that severity of disability or other attributes of particular students facilitates or inhibits the effectiveness of particular treatments, techniques, or strategies suggests that special education may have an important role in selecting and adjusting educational treatments as the science of instruction becomes more clearly understood. The shift toward more instructional time in general education classes for students with mild disabilities—begun with the Least Restrictive Environment clause of PL94-142 and escalating through the last decade—will also affect learning opportunities in complex ways. We need a solid research base that defines how regular class environments can best stimulate learning for students with disabilities. Promising models have been considered, but as yet we have no clearly converging trends.

Even after "best practices" in the mainstream have been thoroughly researched, the best arrangement overall is likely to fail for some students. New statistical options, such as hierarchical linear modeling and latent trait methods may eventually make it possible to determine, in the earliest stages of instruction, the educational conditions most likely to be effective for a given individual. This level of educational science will depend upon studies with careful descriptions of the target population, including the variance within the population, the treatments delivered, and the effects—not just on experimental and control groups, but on individual learners.

Looking Ahead

Keogh (1990), in her discussion of policy and practice, comments that the growing body of research on academic interventions confirms that "some intervention is better than no intervention." (p. 187) The rich variety of effective components and intervention packages for students with disabilities suggests that the field is unlikely to settle on a single "best approach" for instruction and learning support for students with mild disabilities. Given the range of response among students with disabilities to any particular method of instruction, generating an array of empirically validated approaches is critical for assuring that options will be available for adjusting instruction to facilitate optimal learning. Coupled with effective instructional approaches, better methods for monitoring the progress of children will allow teachers to respond quickly to poor academic growth in general or special education settings.

Despite the wealth of new research with difficult-to-teach children, we still lack research evidence in a number of areas:

1. *Selecting the targets for early intervention.* Most children with mild disabilities are not identified for assistance programs early enough in their

school careers to circumvent the negative side effects of school failure. Those who are identified as mild MR, LD or SED by the first grade are often excluded from one-to-one tutoring projects funded through compensatory education systems. Research is beginning to identify powerful early intervention processes which could help to avert some of the emotional and cognitive consequences of poor academic competence, but we need to identify children who need such assistance early, and provide access to the most powerful interventions available.

2. *Long-term effects for instructional approaches.* Longitudinal research for the most promising interventions is critical. These projects should involve large enough subject samples to vary the length of intervention, to continue approaches longer than a school year, and to provide a range of follow-up support over time, so that early gains can be maintained.

3. *Clear descriptions of children who participate in research experiments.* Studies that track the progress of individual students, as well as groups, usually note a small percentage of failures, along with the successes. The knowledge base should include descriptions of subjects that might allow better predictions of who is most likely to profit from a particular intervention.

4. *Frequent measures taken over the course of treatments.* This recommendation accompanies the request for clear subject descriptions above. Research on rates of growth among groups of students with different mild disabilities and in compensatory programs suggests that for cross-categorical groupings, such as grouping all low-achieving children in a classroom for a particular treatment, or including children with mild disabilities in regular classroom instruction, very careful monitoring of academic progress could assist teachers to make the kinds of adjustments in instructional conditions that will lead to improved outcomes. For researchers, determining the likely patterns of growth could help practitioners know sooner whether a particular intervention is effective for an individual student, thus avoiding a long-term educational mistake that could have been averted.

5. *Careful descriptions of complex treatments in general education programs.* Research on recommended general class practices (e.g., cooperative learning, collaborative teaching) has produced mixed findings for children with disabilities. Research that identifies the conditions under which these practices produce reliable effects is needed.

References

Ackerman, P., Dykman, R., & Gardner, M. (1990). ADD student with and without dyslexia differ in sensitivity to rhyme and alliteration. *Journal of Learning Disabilities, 23,* 279–283.

Adams, M. (1990). *Beginning to read: Thinking and learning about print.* Cambridge, MA: MIT Press.

Alves, A.J. & Gottlieb, J. (1986). Teacher interactions with mainstreamed handicapped students and their non-handicapped peers. *Learning Disability Quarterly, 8*, 77–83.

Anderson, R., Heibert, E., Scott, J., & Wilkinson, I. (1985). *Becoming a nation of readers: The report of the commission on reading.* Champaign, IL: Center for the Study of Reading.

Arter, J. & Jenkins, J. (1977). Examining the benefits and prevalence of modality considerations in special education. *Journal of Special Education, 11*, 281–298.

Bateman, B. (1977). Teaching reading to learning disabled children. In L. Resnick & P. Weaver (Eds.), *Theory and practice of early reading.* Hillsdale, NJ: Lawrence Erlbaum Associates.

Ball, E., & Blachman, B. (1991). Does phoneme awareness training in Kindergarten make a difference in early word recognition and developmental spelling? *Reading Research Quarterly, 26*, 49–66.

Barkley, R., DuPaul, G., & McMurray, M. (1990). A comprehensive evaluation of attention deficit disorder with and without hyperactivity as defined by research criteria. *Journal of Consulting and Clinical Psychology, 58*, 775–789.

Barkley, R. Grodzinsky, G. & DuPaul, G. (1992). Frontal lobe functions in attention deficit disorder with and without hyperactivity. *Journal of Abnormal Child Psychology, 20*, 163–188.

Beck, I. (1989). Improving practice through understanding reading. In Resnick & Klopfer (Eds.), *Toward the thinking curriculum: Current cognitive research.* Yearbook: Association of Supervision and Curriculum Development (pp. 40–58). Washington, DC: ASCD.

Beirne-Smith, M. (1990). Peer tutoring in arithmetic for children with learning disabilities. *Exceptional Children, 57*, 330–337.

Berliner, D. (1983). Developing conceptions of classroom environments: Some light on the T in ATI. *Educational Psychologist, 18*, 1–13.

Berninger, V. & Abbott, R. (1994). Redefining learning disabilities: Moving beyond aptitude-achievement discrepancies to failure to respond to validated treatment protocols. In G. R. Lyon (Ed.), *Frames of reference for the assessment of learning disabilities: New views on measurement issues* (pp. 163–183). Baltimore: Paul Brookes Publ.

Borkowski, J.G., Day, J.D., Saenz, D., Dietmeyer, D., Estrada, T.M., & Groteluschen, A. (1992). Expanding the boundaries of cognitive interventions. In B.Y.L. Wong (Ed.), *Contemporary Intervention Research in LD: An international perspective* (pp. 1–21). NY: Springer-Verlag.

Bradley, L., & Bryant, P. (1985). *Rhyme and reason in reading and spelling.* Ann Arbor: The University of Michigan Press.

Brady, S.A. (1991). The role of working memory in reading disability. In S. Brady & D. Shankweiler (Eds.), *Phonological processes in literacy* (pp. 129–152). Hillsdale, NJ: Erlbaum.

Brophy, J. & Good, T. (1972). Brophy-Good System (Teacher-Child Dyadic interaction). In A. Simon & E. Bayer (Eds.), *Mirrors for behavior: An anthology of observation instruments continued* (Vol. A., pp. 191–197). Philadelphia: Research for Better Schools.

Bursuck, W. (1989). A comparison of students with learning disabilities to low achieving and higher achieving students on three dimensions of social competence. *Journal of Learning Disabilities, 22*, 188–194.

Byrne, B., & Fielding-Barnsley, R. (1991). *Sound foundations.* Artarmon, New South Wales, Australia: Leden Educational Publishers.

Byrne, B., & Fielding-Barnsley, R. (1993). Evaluation of a program to teach phonemic awareness to young children: A 1-year follow-up. *Journal of Educational Psychology, 85*, 104–111.

Calfee, R. & Drum, P. (1986). Research on teaching reading. In M. Wittrock (Ed.), *Handbook of research on teaching.* (pp. 804–849) New York: Macmillan.

Case, L.P., Harris, K.R., & Graham, S. (1992). Improving the mathematical problem-solving skills of students with learning disabilities: self-regulated strategy development. *Journal of Special Education, 26*, 1–19.

Cazden, C. B. (1986). Classroom discourse. In M. Wittrock (Ed.), *Handbook of research on teaching*. (pps. 432–463) New York: Macmillan.

Christenson, S.L., Ysseldyke, J.E., & Thurlow, M.L. (1989). Critical instructional factors for students with mild handicaps: An integrative review. *Remedial and Special Education, 10*, 5, 21–31.

Cipielewski, J. & Stanovich, K.E. (1992). Predicting growth in reading ability from children's exposure to print. *Journal of Exceptional Child Psychology, 54*, 74–89.

Clay, M.M. (1985). *The early detection of reading difficulties*. Exeter, NH: Heinemann.

Cohen, E., Lotan, R., Catanzarite, L. (1990). Treating status problems in cooperative classrooms. In S. Sharan (Ed.), *Cooperative learning: Theory and research* (pp. 203–230). NY: Longman.

Coleman, J.M. & Minnett, A.M. (1992). Learning disabilities and social competence: A social ecological perspective. *Exceptional Children, 59*, 3, 234–246.

Cooper, R.J. (1993). Ask R.J. *Closing the Gap, 12*, 8.

Cornoldi, C. & Vianello, R. (1992). Metacognitive knowledge, learning disorders, and mental retardation. In T. Scruggs & M. Mastropieri (Eds.), *Advances in learning and behavioral disabilities* (pp. 107–134.) Greenwich, CT: JAI Press.

Delquadri, J., Greenwood, C.R., Whorton, D., Carta, J.J., and Hall, R.V. (1986). Classwide peer tutoring. *Exceptional Children, 52*, 535–542.

Deno, S. (1985). Curriculum–Based Measurement: The emerging alternative. *Exceptional Children, 53*, 5–22.

Deno, S., Maryama, G., Espin, C., & Cohen, C. (1990). Educating students with mild disabilities in general education classrooms: Minnesota alternatives. *Exceptional Children, 57*, 150–161.

Deno, S. L., Mirkin, P. K., & Chiang, B. (1982). Identifying valid measures of reading. *Exceptional Children, 49*, 36–45.

Ehri, L. C. (1992). Reconceptualizing the development of sight word reading and its relationship to recoding. In P. Gough, L. Ehri & R. Treiman (Eds.), *Reading acquisition* (pp. 107–144). Hillsdale, NJ: Erlbaum.

Ellis, E. & Friend, P. (1991). Adolescents with learning disabilities. In B. Wong (Ed.), *Learning about learning disabilities* (p. 506–563). San Diego: Academic Press.

Ellson, D.G., Baber, L., Engle, T.L., & Kampwerth, L. (1965). Programmed tutoring: A teaching and a research tool. *Reading Research Quarterly, 1*, 77–127.

Ellsworth, N.J. (1993). Technology and education: Applications for students with learning disabilities. *Learning Disabilities: A Multidisciplinary Journal, 4*, 45–52.

Englert, C.S., Tarrant, K.L., Mariage, T.V., & Oxer, T. (1994). Lesson talk as the work of reading groups: The effectiveness of two interventions. *Journal of Learning Disabilities, 27*, 165–85.

Espin, C., Deno, S., Maruyama, G., & Cohen, C. (1989). *The Basic Academic Skills Samples (BASS): An instrument for the screening and identification of children at risk for failure in regular education classrooms*. Paper presented at the annual meeting of the American Educational Research Association, San Francisco.

Felton, R. H. (1993). Effects of instruction on the decoding skills of children with phonological-processing problems. *Journal of Learning Disabilities, 26*, 583–589.

Fletcher, J.M. & Foorman, B.R. (1994). Issues in definition and measurement of learning disabilitiies: The need for early intervention. In G. R. Lyon (Ed.), *Frames of reference for the assessment of learning disabilities: New views on measurement issues*. (pp. 185–200) Baltimore: Paul Brookes Publ.

Fletcher, J.M., Shawitz, B., & Shawitz, S. (1994). Attention as a process and as a disorder. In G. R. Lyon (Ed.), *Frames of reference for the assessment of learning disabilities: New views on measurement issues*. (pp. 103–116) Baltimore: Paul Brookes Publ.

Fuchs, D. & Fuchs, L.S. (1989). Exploring effective and efficient prereferral interventions: A component analysis of behavioral consultation. *School Psychology Review, 18*, 260–281.

Fuchs, D., Fuchs, L.S., Bahr, M.W., Fernstrom, P., & Stecker, P.M. (1990). Prereferral intervention: a prescriptive approach. *Exceptional Children, 56*, 493–513.

Fuchs, L.S., Fuchs, D., & Bishop, N. (1992). Teacher planning for students with learning disabilities: differences between general and special educators. *Learning Disabilities Research and Practice, 7*, 120–128.

Fuchs, D., Fuchs, L.S., & Mathes, P.G. (1993, April). *Peer-mediated learning strategies: Effects on learners at different points on the reading continuum*. Paper presented at the annual meeting of the American Educational Research Association, Atlanta.

Fuchs, L.S., Fuchs, D., Hamlett, C.L., & Allinder, R.M. (1991). The contribution of skills analysis to curriculum-based measurement in spelling. *Exceptional Children, 57*, 443–452.

Fuchs, L.S., Fuchs, D., Hamlett, C.L., & Ferguson, C. (1991). Effects of expert system consultation within Curriculum-Based Measurement. using a reading maze task. *Exceptional Children, 58*, 436–450.

Gaskins, I.W., Downer, M.A., Anderson, R.C., Cunningham, P., Gaskins, R.W., & Schommer, M. (1988). A metacognitive approach to phonics: Using what you know to decode what you don't know. *Remedial and Special Education, 9*, 36–41.

Gaultney, J.F. (1995). The effect of prior knowledge and metacognition on the acquisition of a reading comprehension strategy. *Journal of Experimental Child Psychology, 59*, 142–163.

Geary, D.C. (1990). A componential analysis of an early learning deficit in mathematics. *Journal of Experimental Child Psychology, 49*, 363–383.

Geary, D.C., Bow-Thomas, C.C., & Yao, Y. (1992). Counting knowledge and skill in cognitive addition: A comparison of normal and mathematically disabled children. *Journal of Experimental Child Psychology, 54*, 372–391.

Gerber, M.M. (1986). Generalization of spelling strategies by learning disabled students as a result of contingent imitation/modeling and mastery criteria. *Journal of Leaning Disabilities, 19*, 530–537.

Gersten, R., Keating, T. & Becker, W. (1988). The continued impact of the Direct Instruction model: Longitudinal studies of Follow Through students. Special Issue: Direct Instruction. A general case for teaching the general case. *Education and Treatment of Children, 11*, 318–327.

Gleason, M., Carnine, D., & Vala, N. (1990). Cumulative versus rapid introduction of new information. *Exceptional Children, 57*, 353–358.

Gordon, J., Vaughn, S., and Schumm, J.S. (1993) Spelling interventions: A review of literature and implications for instruction for students with learning disabilities. *Learning Disabilities Research and Practice, 8*, 175–181.

Goswami, U. (1988).Orthographic analogies and reading development. *Quarterly Journal of Experimental Psychology, 40*, 239–268.

Graham, S. & Harris, K.R. (1989). Improving learning disabled students' skills at composing essays: Self-instructional strategy training. *Exceptional Children, 56*, 201–214.

Graham, S. & Harris, K.R. (1992). Self-regulated strategy development: Programmatic research in writing. In B.Y.L. Wong (Ed.), *Contemporary Intervention Research in LD: An international perspective* (pp. 47–64). NY: Springer-Verlag.

Graham, S., MacArthur, C., Schwartz, S. & Page-Voth, V. (1992). Improving the compositions of students with learning disabilities using a strategy involving product and process goal setting. *Exceptional Children, 58*, 322–334.

Greenwood, C.R. (1990). Longitudinal analysis of time, engagement, and achievement in at risk versus non-risk students. *Exceptional Children, 57*, 521–535.

Greenwood, C.R., Carta, J.J., Arreaga-Mayer, C. & Rager, A. (1991). The behavior analysis consulting model: Identifying and validating naturally effective instructional models. *Journal of Behavioral Education, 1*, 165–191.

Greenwood, C.R., Delquadri, J.C., & Bulgren, J. (1993). Current challenges to behavioral technology in the reform of schooling: Large-scale, high-quality implementation and sustained use of effective education practices. *Education and Treatment of Children, 16,* 401–440.

Greenwood, C.R., Delquadri, J.C., Stanley, S.O., Terry, B., & Hall, R.V. (1985). Assessment of ecobehavioral interaction in school settings. *Behavioral Assessment, 7,* 331–347.

Greenwood, C.R., Finney, R., Terry, B., Arreaga-Mayer, C., Carta, J., Delquadri, J., Walker, D., Innocenti, M., Lignugaris-Kraft, J., Harper, G.F., & Clifton, R. (1993). Monitoring, improving, and maintaining quality implementation of the classwide peer tutoring program using behavioral and computer technology. *Education and Treatment of Children, 16,* 19–47.

Griffey, Q.L., Zigmond, N., & Leinhardt, G. (1988). The effects of self-questioning and story structure training on the reading comprehension of poor readers. *Learning Disabilities Research, 4,* 45–51.

Harper, G.F., Mallette, B., Maheady, L., & Brennan, G. (1993). Classwide students tutoring teams and Direct Instruction as a combined instructional program to teach generalizable strategies for mathematics word problems. *Education and Treatment of Children, 16,* 115–134.

Harper, G.F., Mallette, B. & Moore, J. (1991). Peer-mediated instruction: Teaching spelling to primary school children with mild disabilities. *Reading, Writing and Learning Disabilities, 7,* 137–151.

Harris, K. R. & Pressley, M. (1991). The nature of cognitive strategy instruction: Interactive strategy construction. *Exceptional Children, 57,* 392–404.

Harter, S. (1985). *Self-Perception: Profiles for Children* (Revision of the Perceived Competence Scale for Children). Denver, CO: University of Denver.

Hasselbring, T.S., Goin, L.I., & Wissick, C. (1989). Making knowledge meaningful: Applications of hypermedia to create integrated, anchored instructional environments. *Journal of Special Education Technology, 10,* 61–72.

Hatcher, P., Hulme, C., & Ellis, A. (1994). Ameliorating early reading failure by integrating the teaching of reading and phonological skills: The phonological linkage hypothesis. *Child Development, 65,* 41–57.

Haynes, M. & Jenkins, J.R. (1986). Reading instruction in special education resource rooms. *American Educational Research Journal, 23,* 161–190.

Hiebert, E.H. (1994). Reading Recovery in the United States: What difference does it make to an age cohort? *Educational Researcher, 23, 9,* 4–14.

Hiebert, E.H., Colt, J.M., Catto, S.L., & Gury, E.M. (1992). Reading and writing of Grade 1 students in a restructured Chapter 1 program. *American Education Research Journal, 29,* 545–572.

Hieronymous, A.N., Hoover, H.D., & Lindquist, E.F. (1982). *Iowa Tests of Basic Skills.* Chicago: Riverside.

Holborow.P. & Berry, P. (1986). A multi-national, cross-cultural perspective on hyperactivity. *American Journal of Orthopsychiatry, 56,* 320–322.

Innocenti, M.R. & White, K.R. (1993). Are more intensive early intervention programs more effective? A literature review. *Exceptionality, 4,* 31–50.

Iversen, S. & Tunmer, W.E. (1993). Phonological processing skills and the Reading Recovery program. *Journal of Educational Psychology, 85,* 112–126.

Jenkins, J.R., Jewell, M., Leicester, N. & Jenkins, L. (1991). Development of a school building model for educating students with handicaps and at-risk students in general education classrooms. *Journal of Learning Disabilities, 24,* 311–320.

Jenkins, J.R., Jewell, M., Leicester, N., O'Connor, R.E., Jenkins, L. & Troutner, N. (1994). Accommodations for individual differences without classroom ability groups: An experiment in school restructuring. *Exceptional Children, 60,* 344–358.

Jenkins, J.R. & Leicester, N. (1992). Specialized instruction within general education: A case study of one elementary school. *Exceptional Children, 58,* 555–563.

Jenkins, J.R., Pious, C.G., & Peterson, D.L. (1988). Categorical programs for remedial and handicapped students: Issues of validity. *Exceptional Children, 55*, 147–158.

Johnson, D. W., & Johnson, R. T. (1986). *Learning Together and Alone* (2nd ed.). Englewood Cliffs, NJ: Prentice-Hall.

Johnson, D., Johnson, R., & Maruyama, G. (1983). Interdependence and interpersonal attraction among heterogeneous and homogeneous individuals: A theoretical formulation and a meta-analysis of the research. *Review of Educational Research, 53*, 5–54.

Johnson, L.D. & Pugach, M.C. (1990). Peer collaboration: Accommodating students with mild learning and behavior problems. *Exceptional Children, 57*, 454–461.

Juel, C. (1988). Learning to read and write: A longitudinal study of 54 children from first through fourth grades. *Journal of Educational Psychology, 80*, 437–447.

Juel, C. (1994). Learning to learn from effective tutors. In L. Schauble & R. Glaser (Eds.), *The contributions of instructional innovation to understanding learning.*

Kamps, D., Barbetta, P.M., Leonard, B.R., & Delquadri, J. (1994). Classwide peer tutoring: An integration strategy to improve reading skills and promote interaction among students with autism and general education peer. *Journal of Applied Behavior Analysis, 27*, 49–61.

Kavale, S.A. & Forness, S.K. (1992) Learning difficulties and memory problems in mental retardation: A meta-analysis of theoretical perspectives. In T. Scruggs & M. Mastropieri (Eds.), *Advances in learning and behavioral disabilities* (pp. 177–219). Greenwich, CT: JAI Press.

Keogh, B.K. (1990) Narrowing the gap between policy and practice. *Exceptional Children, 57*, 186–190.

Keogh, B. K., & Weisner, T. (1993). An ecocultural perspective on risk and protective factors in children's development: Implications for learning disabilities. *Learning Disabilities Research & Practice, 8*, 3–10.

Kohler, F.W. & Greenwood, C.R. (1990). Effects of collateral peer supportive behaviors within the classwide peer tutoring program. *Journal of Applied Behavior Analysis, 23*, 307–322.

Krupski, A., Gaultney, J.F., Malcolm, G., & Bjorklund, D.F. (1993). Learning disabled and nondisabled children's performance on serial recall tasks: The facilitating effects of knowledge. *Learning and Individual Differences, 5*, 199–210.

Lambert, N. & Sandoval, J. (1980). The prevalence of learning disabilities in a smaple of children considered hyperactive. *Journal of Abnormal Child Psychology, 8*, 33–50.

Levy, B.A., Nicholls, A. & Kohen, D. (1993). Repeated readings: Process benefits for good and poor readers. *Journal of Experimental Child Psychology, 56*, 303–327.

Lindamood, C.H. & Lindamood, P.C. (1984). *Auditory Discrimination in Depth.* Allen, TX: DLM/Teaching Resources.

Lipson, M. & Wixson, K. (1986). Reading disability research: An interactionist perspective. *Review of Educational Research, 56*, 111–136.

Lombardi, T.P. (1991). Special education and students at risk: Findings from a national study. *Remedial and Special Education, 12*, 56–62.

Lovett, M.W., Warren-Chapin, P.M., Ransby, M.J. & Borden, S.L. (1990). Training the word recognition skills of reading disabled children: Treatment and transfer effect. *Journal of Educational Psychology, 82*, 769–780.

Lucangeli, D., Galderisi, D. & Cornoldi, C. (1995). Specific and general transfer effects following metamemory training. *Learning Disabilities Research and Practice, 10*, 11–21.

Lundberg, I., Frost, J., & Petersen, O. (1988). Effects of an extensive program for stimulating phonological awareness in preschool children. *Reading Research Quarterly, 23*, 263–284.

Lundberg, I., Olofsson, A., & Wall, S. (1980). Reading and spelling skills in the first school years predicted from phonemic awareness skills in Kindergarten. *Scandinavian Journal of Psychology, 21*, 159–173.

MacArthur, C.A., Schwartz, S.S., & Graham, S. (1991). Effects of a reciprocal peer revision strategy in special education classrooms. *Learning Disabilities Research and Practice, 6*, 201–210.

Manis, F.R., Custodio, R, & Szeszulski, P.A. (1993). Development of phonological and orthographic skill: A two-year longitudinal study of dyslexic children. *Journal of Experimental Child Psychology, 56*, 64–86.

Mathes, P.G., Fuchs, D., Fuchs, L.S., Henley, A.M., & Sanders, A. (1994). Increasing strategic practice with Peabody Classwide Peer Tutoring. *Learning Disabilities Research and Practice, 9*, 44–48.

Mattingly, J.C. & Bott, D.A. (1990). Teaching multiplication facts to students with learning problems. *Exceptional Children, 56*, 438–449.

McGill-Franzen, A. & Allington, R. (1991). The gridlock of low reading achievement: Perspectives on practice and policy. *Remedial and Special Education, 12*, 20–30.

McGregor, G., Drossner, D., & Axelrod, S. (1990). Increasing instructional efficiency: A comparison of voice plus text versus test alone on the error rate of students with mild disabilities during CAI. *Journal of Special Education Technology, 10*, 192–97.

Mudre, L.H., & McCormick, S. (1989). Effects of meaning-focused cues on underachieving readers' context use, self-corrections, and literal comprehension. *Reading Research Quarterly, 24*, 89–113.

Nicklin, J.L. (1992, July 1). Information technology: Teachers' use of computers stressed by education college. *Chronicle of Higher Education*, pp. 15–17.

O'Connor, R.E. & Jenkins, J.R. (1996). Cooperative learning: A closer look. *Exceptionality*.

O'Connor, R.E., & Jenkins, J.R. (1995). Improving the generalization of sound/symbol knowledge: Teaching spelling to kindergarten children with disabilities. *Journal of Special Education, 29*, 255–275.

O'Connor, R.E., Jenkins, J.R., & Slocum, T.A. (1995). Transfer among phonological tasks in kindergarten: Essential instructional content. *Journal of Educational Psychology, 87*, 202–217.

Olson, R.K. & Wise, B.W. (1992). Reading on the computer with orthographic and speech feedback: An overview of the Colorado remedial reading project. *Reading and Writing: An Interdisciplinary Journal, 4*, 107–144.

O'Neil, M. & Douglas, V. (1991). Study strategies and story recall in attention deficit disorder and reading disability. *Journal of Abnormal Child Psychology, 17*, 671–692

O'Sullivan, P.J., Ysseldyke, J.E., Christenson, S.L., & Thurlow, M. (1990). Mildly handicapped elementary students' opportunity to learn during reading instruction in mainstream and special education settings. *Reading Research Quarterly, 25*, 131–146.

Palincsar, A.S. & Klenk, L. (1992). Fostering literacy learning in supportive contexts. *Journal of Learning Disabilities, 25*, 211–225.

Papetti, O., Cornoldi, C., Pettavino, A., Mazzoni, G., & Borkowski, J. (1992). Memory judgments and allocation of study times in good and poor comprehenders. In T. Scruggs & M. Mastropieri (Eds.), *Advances in learning and behavioral disabilities* (pp. 3–33) Greenwich, CT: JAI Press.

Perfetti, C. A. (1991). On the value of simple ideas in reading instruction. In S. A. Brady & D. P. Shankweiler (Eds.), *Phonological processes in literacy* (pp. 211–218). Hillsdale, NJ: Erlbaum.

Perfetti, C., Beck, I., Bell, L., & Hughes, C. (1987). Phonemic knowledge and learning to read are reciprocal: A longitudinal study of first grade children. *Merrill-Palmer Quarterly, 33*, 283–319.

Peterson, S.K., Mercer, C., & O'Shea (1988). Teaching learning disabled students place value using concrete to abstract sequence. *Learning Disabilities Research, 4*, 52–60.

Pinnell, G.S., Lyons, C.A., DeFord, D.E., Bryk, A.S., & Seltzer, M. (1994). Comparing instructional models for the literacy education of high-risk first graders. *Reading Research Quarterly, 29*, 8–39.

Rack, J.P., Snowling, M.J. & Olson, R.K. (1992). The nonword reading deficit in developmental dyslexia: A review. *Reading Research Quarterly, 27*, 29–53.

Reid, M.K. & Borkowski, J.G. (1987). Causal attributions of hyperactive children: Implications for training strategies and self-control. *Journal of Educational Psychology, 79*, 296–307.

Reynolds, M.C. (1994). Counterpoint: 20/20 analysis. *Exceptional Children, 60,* 278–9.

Reynolds, M.C., Wang, M. C., & Walberg, H.J. (1987). Restructuring "special" school programs: A position paper. *Policy Studies Review, 2,* 189–212.

Reynolds, M.C., Zetlin, A.G., & Wang, M.C. (1993). 20/20 analysis: Taking a close look at the margins. *Exceptional Children, 59,* 294–300.

Rutter, M & and Yule, W. (1975). The concept of specific reading retardation. *Journal of Child Psychology and Psychiatry, 16,* 181–197.

Sawyer, R., McLaughlin, M., & Winglee, M. (1994). Is integration of students with disabilities happening? *Remedial and Special Education, 15,* 204–215.

Schunk, D.H. & Rice, J.M. (1987). Enhancing comprehension skill and self-efficacy with strategy value information. *Journal of Reading Behavior, 19,* 285–302.

Schunk, D.H. & Rice, J.M. (1993). Strategy fading and progress feedback: Effects on self-efficacy and comprehension among students receiving remedial reading services. *Journal of Special Education, 27,* 257–276.

Scott, M.S. & Greenfield, D.B. (1992). A comparison of normally achieving, learning disabled, and mildly mentally retarded students on a taxonomic information task. *Learning Disabilities Research and Practice, 7,* 59–67.

Scott, M.S. & Perou, R. (1994). Some observations on the impact of learning disabilities and mild mental retardation on the cognitive abilities of young grade school children. In T.E. Scruggs & M.A. Mastropieri (Eds.), *Advances in learning and behavioral disabilities, Vol. 8* (pp. 215–233). Greenwich, CT: JAI Press.

Self, H., Benning, A., Marston, D., & Magnusson, D. (1991). Cooperative Teaching Project: A model for students at risk. *Exceptional Children, 58,* 26–34.

Siegel, L.S. (1985). Psycholinguistic aspects of reading disabilities. In L.S. Siegel & F.J. Morrison (Eds.), *Cognitive development in atypical children.* New York: Springer-Verlag.

Share, D., Jorm, A., MacLean, R., & Matthews, R. (1984). Sources of individual differences in reading acquisition. *Journal of Educational Psychology, 76,* 1309–1324.

Shaywitz, S.E., Escobar, M.D., Shaywitz, B.A., Fletcher, J.M., & Makuch, R. (1992). Evidence that dyslexia may represent the lower tail of a normal distribution of reading ability. *The New England Journal of Medicine, 326,* 145–150.

Shaywitz, S.E., Shaywitz, B.A., Fletcher, J.M., & Escobar, M.D. (1990). Prevalence of reading disability in boys and girls: Results of the Connecticut longitudinal study. *Journal of the American Medical Association, 264,* 8, 998–1002.

Silver, A.A. & Hagin, R.A. (1990). *Disorders of learning in childhood.* NY: John Wiley & Sons.

Slavin, R.E. (1991). Neverstreaming: Prevention and early intervention as an alternative to special education. *Journal of Learning Disabilities, 24,* 373–378.

Slavin, R. (1983). *Cooperative learning.* New York: Longman.

Slavin, R.E., Madden, N.A., Karweit, N.L., Dolan, L. & Wasik, B.A. (1992). *Success for All: A relentless approach to prevention and early intervention in elementary schools.* Arlington, VA: Educational Research Service.

Slavin, R.E., Madden, N.A., & Leavey, M. (1984). Effects of Team Assisted Individualization on the mathematics achievement of academically handicapped and nonhandicapped students. *Journal of Educational Psychology, 76,* 813–819.

Speckman, N. J., Goldberg, R. J., & Herman, K. L. (1993). An exploration of risk and resilience in the lives of individuals with learning disabilities. *Learning Disabilities Research & Practice, 8,* 11–18.

Stanovich, K. (1986). Matthew effects in reading: Some consequences of individual differences in the acquisition of literacy. *Reading Research Quarterly, 21,* 360–406.

Stanovich, K.E. (1992). Speculations on the causes and consequences of individual differences in early reading acquisition. In P. Gough, L. Ehri & R. Treiman (Eds.), *Reading acquisition* (pp. 307–342). Hillsdale, NJ: Erlbaum.

Strickland, D. & Cullinan, B. (1990). Afterword, in M.J. Adams, *Beginning to Read,* (pps. 426–433). Cambridge, MA: The MIT Press.

Sullivan, G.S. & Mastropieri, M.A. (1994). Social competence of individuals with learning disabilities. In T.E. Scruggs & M.A. Mastropieri (Eds.), *Advances in learning and behavioral disabilities, Vol. 8* (pp. 177–213). Greenwich, CT: JAI Press.

Swanson, H.L., Trahan, M.F. (1992). Learning disabled readers' comprehension of computer mediated text: The influence of working memory, metacognition, and attribution. *Learning Disabilities Research and Practice, 7,* 74–86.

Tateyama-Sniezek, K.M. (1990). Cooperative learning: Does it improve the academic achievement of students with handicaps? *Exceptional Children, 56,* 426–437.

Taylor, B.M., Strait, J., & Medo, M.A. (1994). Early intervention in reading: Supplemental instruction for groups of low-achieving students provided by first-grade teachers. In E. H. Hiebert & B.M. Taylor (Eds.), *Getting reading right from the start.* Boston: Allyn & Bacon.

Thurlow, M. (1993). Instruction in special education classrooms under varying student-teacher ratios. *Elementary School Journal, 93, 3,* 305–320.

Torgesen, J. K. (1980). Memory processes in exceptional children. In B. Keogh (Ed.), *Basic constructs and theoretical orientation: A research annual.* Greenwich, CT: JAI Press.

Torgesen, J. K. (1986). Using computers to help learning disabled children practice reading: A research-based perspective. *Learning Disabilities Focus, 2,* 72–81.

Torgesen, J. K. & Bryant, B.(1994). *Phonological awareness training for reading.* Austin, TX: Pro-Ed.

Torgesen, J., Morgan, S., & Davis, C. (1992). Effects of two types of phonological awareness training on word learning in Kindergarten children. *Journal of Educational Psychology, 84,* 364–370.

Uhry, J.K. & Shepherd, M.J. (1993). Segementation/ spelling instruction as part of a first-grade reading program: Effects on several measures of reading. *Reading Research Quarterly, 28,* 219–233.

United States Department of Education, Office of Special Education Programs (1993). *Fifteenth annual report to Congress on the implementation of the Individuals with Disabilities Education Act.* Washington, DC: Author.

Van Daal, H.P., Reitsma, P, & Van der Leij, A. (1994). Processing units in word reading by disabled readers. *Journal of Experimental Child Psychology, 57,* 180–210.

Van den Bosch, K., van Bon, W., Schreuder, R. (1995). Poor readers' decoding skills: Effects of training with limited exposure duration. *Reading Research Quarterly, 30,* 110–125.

Vaughn, S., Schumm, J.S., & Gordon, J. (1992). Early spelling acquisition: Does writing really beat the computer? *Learning Disabilities Quarterly, 15,* 223–228.

Vaurus, M., Lehtinen, E., Kinnunen, R., & Salonen, P. (1992). Socioemotional coping and cognitive processes in training learning disabled students. In B.Y.L. Wong (Ed.), *Contemporary Intervention Research in LD: An international perspective* (pp. 163–189). NY: Springer-Verlag.

Vellutino, F.R. (1991). Introduction to three studies on reading acquisition: Convergent findings on theoretical foundations of code-oriented versus whole-language approaches to reading instruction. *Journal of Educational Psychology, 83,* 437–443.

Villa & Thousand, (1994). One divided by two or more: Redefining the role of a cooperative education team. In J.S. Thousand, R.A. Villa, and A.I Nevin (Eds.), *Creativity and collaborative learning.* Baltimore, MD: Paul Brooks.

Wallach, M.A. & Wallach, L. (1976). *Teaching all children to read.* Chicago: University of Chicago Press.

Wasik, B. & Slavin, R. (1993). Preventing early reading failure with one-to-one tutoring: a review of five programs. *Reading Research Quarterly, 28,* 179–200.

Watkins, M.W. (1989). Computerized drill and practice and academic attitudes of learning disabled students. *Journal of Special Education Technology, 9,* 168–172.

Webb, N. (1985). Student interaction and learning in small groups: A research summary. In R. E. Slavin, S. Sharan, S. Kagan, R. Hertz-Lazarowitz, C. Webb, and R. Schmuck (eds.), *Learning in cooperate, cooperating to learn* (pp. 147–172). New York: Plenum.

Wesson, C.L. (1991). Curriculum-Based Measurement and two models of follow-up consultation. *Exceptional Children, 57,* 246–256.

White, W.A.T. (1988). A meat-analysis of the effects of Direct Instruction in special education. *Education and Treatment of Children, 11,* 364–374.

Will, M. (1986). *Educating students with learning problems: A shared responsibility.* Washington, D.C.: Office of Special Education and Rehabilitation Services, U.S. Department of Education.

Williams, J. (1980). Teaching decoding with an emphasis on phoneme analysis and blending. *Journal of Educational Psychology, 72,* 1–15.

Wilson, C.L. & Sindelar, P.T. (1990) Direct instruction in math word problems: Students with learning disabilities. *Exceptional Children, 57,* 512–519.

Wise, B. (1992). Whole words and decoding for short-term learning: comparisons on a "Talking Computer" system. *Journal of Exceptional Child Psychology, 54,* 147–167.

Wise, B. & Olson, R. (1991). Remediating reading difficulties. In J. Obrzut & G. Hynd (Eds.), *Neuropsychological foundations in learning disabilities: A handbook of issues, methods, and practice* (pp. 631–652). San Diego: Academic Press.

Wise, B. & Olson, R. (1994). Computer speech and the remediation of reading and spelling problems. *Journal of Special Education Technology, 12,* 207–220.

Wong, B. (1991). The relevance of metacognition to learning disabilities. In B. Wong (Ed.), *Learning about learning disabilities* (p. 232–261). San Diego: Academic Press.

Woodward, J.P. & Carnine, D.W. (1989). The Genisys program: Linking content area knowledge to problem-solving through technology-based instruction *Journal of Special Education Technology, 10,* 99–

Zigmond, N. & Baker, J. (1990). Mainstream experiences for learning disabled students (Project MELD): Preliminary report. *Exceptional Children, 57,* 176–185.

Zigmond, N. & Baker, J. (Eds.) (1995). An exploration of the meaning and practice of special education in the context of full inclusion of students with learning disabilities. *Journal of Special Education,* special issue.

Zigmond, N., Jenkins, J.R., Fuchs, L.S., Deno, S., Fuchs, D., Baker, J.N., Jenkins, L., & Couthino, M. (1995). Special education in restructured school: Findings from three multi-year studies. *Phi Delta Kappan, 76,* 531–540.

4

Secondary Educational Programs and Transition Perspectives[1]

PAULA D. KOHLER and FRANK R. RUSCH
Transition Research Institute, University of Illinois at Urbana-Champaign, USA

Within five years of leaving high school, a majority of youths with disabilities do not achieve full time employment, do not live independently, and are not self-sufficient. As a result of these findings, legislation has been enacted recently that addresses the educational programs and targeted post-school outcomes of youths in secondary education. These legislative acts provide the framework for an emerging paradigm that shifts educational perspectives from a focus on disability categorization to a focus on the abilities and needs of individual students. This transition perspective is supported by educational practices that link post-school goals with specific instructional activities and educational experiences through an individual planning process. To facilitate implementation of a transition perspective, secondary education restructuring should address current practice across five areas: (a) student development, (b) student-focused planning, (c) interagency and interdisciplinary collaboration, (d) family involvement, and (e) program structures and attributes.

Preparing Youths for Tomorrow's Challenges

An effective high school experience is a cornerstone of lifelong success. Thus, it is essential that we provide meaningful high school experiences to our young people with disabilities. Our failure to do so results in their inability to achieve valued adult outcomes and to participate fully in our society. Currently, we must ask serious questions about whether we are providing an education that prepares youths with disabilities for the demands of adult life, that is, working, living independently, and enjoying the recreational and civic opportunities that are available in our communities.

The difficulties faced by individuals with disabilities as young adults are evidenced by an unemployment rate that exceeds 50% within a two-year period after leaving high school (Wagner, 1989). Further, as many as 40% of all students with disabilities have been reported to drop out of high school (Bruin-inks, Thurlow, Lewis, & Larson, 1988; Gajar, Goodman, & McAfee, 1993). Although some youths who drop out leave school for employment, many do not find jobs. In fact, Mithaug, Horiuchi, and Fanning (1985) reported that youths with disabilities who leave school early have less than a 35% chance of finding work. Youths with disabilities also live dependently; over 80% live with their parents after high school (Wagner, D'Amico et al., 1992).

Tens of thousands of youths with disabilities leave high schools without the skills or the support needed to survive independently in their communities. High schools appear primarily to be a training ground for the nation's most promising students; those who will graduate and pursue a college education. Students who do not aspire to a postsecondary education, by contrast, appear to be left isolated from any unified system that addresses their needs; theirs is an uncertain future.

This chapter serves several purposes. First and foremost, we present knowl-edge that has emerged from research conducted over a relatively short time period (1985-present). Much of this research has focused on students with disabilities. With this in mind, we have reviewed the research in search of generalizations that may be made to all students who are not college bound and who form a larger network of youths who aspire to graduate and get on with their lives. Consequently, the chapter focuses upon the bigger picture— *preparing youths for tomorrow's challenges.* We begin with a review of student outcomes. We follow this with an overview of legislation in support of transition-related activities, and then introduce evidence of an emerging paradigm, which represents a *transition perspective* of education. Finally, we present and discuss transition-related research as it relates to organizing secondary education.

Review of Student Outcomes

We begin this chapter with a brief overview of the outcomes that have been associated with youths with disabilities. Such a review is important because it helps to construct a picture of the extent to which we need to adopt practices that can redirect current practice, and ultimately, our expectations for youths with disabilities.

The outcomes of integrated, competitive employment, residential indepen-dence, and a formation of social and interpersonal networks, as a result of one's high school experience, have improved marginally for students with disabilities over the past 10 years (Hasazi, Gordon, & Roe, 1985; Mithaug, et al., 1985; Wagner, D'Amico et al., 1992). According to the *Twelfth Annual Report to Congress* (U. S. Department of Education, 1990), employment figures for youths with disabilities lag behind those of their peers without disabilities, without exception. Overall, 20% of the entire population of youths with disabilities are never employed at any time in their lives, another 20% are unemployed and looking for a job, 6% enter sheltered employment, and 14% are employed part-time. Forty-three percent are employed full time, the majority of them earning less than $6.00 per hour. Fifty-seven percent of youths with disabilities are employed at some point three to five years after they leave high school, compared to 70% of the general population. Not surprisingly, the majority of the youths with disabilities who become employed are young people with mild disabilities, including learning disabilities (70.8%) and speech impairments (65.4%), whereas employment among youths who are deaf, hard of hearing, or blind is lower. There also are major employment differences based on gender. About half (53.4%) of males with disabilities who leave school have part or full-time employment within two years; only 29.6% of females with disabilities who leave school are employed half or full time during that same period (D'Amico, 1991).

Only 1 in 10 youths with disabilities lives independently two years after they have exited high school, a figure that increases to almost 4 in 10 three to five years after leaving school (37.4%). By comparison, approximately 60.4% of youths in the general population live independently three years after leaving school.

After departing high school, youths typically establish patterns of community involvement that include such roles as finding a mate and parenting. However, the rates at which individuals assume these responsibilities vary according to gender and among individuals with and without disabilities. For example, within four years of departing high school, females with (30.4%) and without (37.7%) disabilities are married at higher percentages than men with (14.8%) and without (21.7%) disabilities. Wagner (1992) found that a much higher percentage of females with disabilities are becoming parents within five years of their leaving high school (40.6%) compared to young women without disabilities (27.8%).

The arrest rate among youths with disabilities increases as this population of youths get older. For example, Wagner (1993) reported that almost 30% of all youths with disabilities are arrested three to five years out of school, with youths with emotional disturbances accounting for more than half (57.6%) of those arrested.

The dropout rates for youths with disabilities have been reported to exceed 40% (Seidel & Vaughn, 1991), while the rates for the general population vary from 25% to 30% (Marder & D'Amico, 1992). The *Twelfth Annual Report to Congress* reported that 65,395 youths with disabilities dropped out of school during the 1987-88 school year. Gajar et al. (1993) reported that 92% of these youths were students with mild disabilities, including learning disabilities, emotional disturbances, and "educable" mental retardation. Further, dropout rates are greater for students who are African American or Hispanic (Harnisch, 1995).

The benefits of a special education come under particular scrutiny when the outcomes of special education graduates are compared to those of students who dropped out. In a study of semi-rural students with learning disabilities, deBettencourt, Zigmond, and Thornton (1989) found that employment rates of students who graduated did not differ significantly from those of students who dropped out. Sitlington, Carson, and Frank (1992) found similar results when comparing employment outcomes for youths with behavior disorders. Even more disturbing, Sitlington et al. found that, on the average, youths who had dropped out were earning higher wages in their present jobs than students who had graduated. Investigating students with mild mental retardation, learning disabilities, emotional disturbances, and speech impairments, Kranstover, Thurlow, and Bruininks (1989) found that graduates did not differ significantly from nongraduates with respect to employment outcomes, wages, and a majority of independence and social outcomes.

In summary, in the period immediately following school-leaving, the majority of America's youths with disabilities are not employed, are not living on their own, do not take advantage of their communities, and are not satisfied with their lives. If we are to consider the outcomes of youths with disabilities an important benchmark in our ability to provide an effective education—we have failed. We have begun, however, to make significant changes in the way we educate our young people with disabilities. Many of these changes have been brought about by changes in education policy as reflected in recent legislation.

Overview of Legislation Establishing Parameters of Special Education and Transition

Although typically viewed as separate systems over the past 20 years, special education and regular education have been more similar than different, with the differences becoming increasingly blurred. Special education was once considered a field that focused almost exclusively upon students with disabilities. Today, however, the focus of special education parallels major thrusts in regular education, including the inclusion of students with disabilities in regular education classrooms. The primary differences are value based, and these values are now beginning to define regular education. Legislation passed in 1994, referred to as Goals 2000: Educate America Act (PL 103-227), was the outgrowth of values and goals being articulated by the governors of each state and the President of the United States. These new goals include higher school completion rates, promotion of adult literacy and lifelong learning, and participation of parents in promoting the social, emotional, and academic growth of their children.

Since the passage of the inaugural Education of All Handicapped Children Act in 1975 (PL 94-142), the way special education is to be delivered in our nation's schools has changed incrementally. The Individuals with Disabilities Education Act (IDEA) of 1990 (PL 101-476) provided important, new definitions to help guide the delivery of special education services, including a range of outcomes expected as a result of obtaining an effective education. By defining *transition services* and by requiring a statement of such in a student's individual education program, IDEA did more than any of the previous special education amendments to promote the involvement of students with disabilities in the development of educational programs focused on their postschool goals. Over the course of the remainder of this decade, further substantial changes in special education legislation can be expected. Foremost, we can expect additional complementary legislative actions that embrace the necessary cooperation between related fields to more effectively serve persons with disabilities.

Other changes affecting secondary education are a result of the School-to-Work Opportunities Act of 1994. This bill, in part, involves a broad cross-section of students, integrates vocational and academic instruction, and combines classroom and on-the-job instruction. Importantly, the School-to-Work Opportunities Act gives priority to students who have been neglected by previous school reforms. Equally important is the extent to which proposed activities will be shaped by the values shared by the regular, vocational, and special education communities. The School-to-Work Opportunities Act is jointly administered by the Departments of Education and Labor. As a result of co-sponsorship, certain core elements must appear in every state's activities related to education and training, including (a) work-based learning, (b) school-based learning, and (c) connecting activities. Briefly, *work-based activities* include job training experiences in a broad range of occupational areas, as well as workplace mentoring. *School-based activities* include high school and postsecondary education tied to occupational and academic standards that are consistent with Goals 2000. Finally, *connecting activities* refer to efforts that relate to administering work-based and school-based learning. Such activities may include collecting follow-up data on students after they have received training, matching students to jobs related to their occupational interests, and providing technical assistance in designing work-based and school-based learning.

Specifically, the School-to-Work Opportunities Act will result in every student (a) gaining access to a relevant education; (b) receiving an education that combines classroom, community, and work-related experiences; and (c) realizing an education that is individualized and based upon the student's needs, interests, and abilities. All students will receive a high school diploma, a certificate or diploma that recognizes one or two years of postsecondary education, and a certificate that is recognized by business and industry. The fundamental components of the School-to-Work Opportunities Act parallel those that emerged in PL 94-142 in 1975: a free, appropriate public education; zero exclusion; least restrictive environments; and non-discriminatory testing.

Other significant recent legislation prohibits discrimination against people with disabilities in the areas of private employment, public accommodations and services, transportation, and telecommunications. Referred to as civil rights legislation for individuals with disabilities, the Americans with Disabilities Act (ADA) was signed into law by President Bush on July 26, 1990.

The Rehabilitation Act Amendments of 1992 (PL 102-569) also extended the concept of legislative acts working in concert to promote improved outcomes for individuals with disabilities. These Amendments are consistent with the Individuals with Disabilities Education Act of 1990 (IDEA) and include a number of important provisions related to transition. Specifically, these provisions define *transition services* the same as IDEA and emphasize that there can be no gaps in the provision of services provided by rehabilitation and education. The Rehabilitation Act Amendments of 1992 require states to describe how they intend to promote services that demonstrate such collaboration and coordination of the two systems.

The federal disability legislation of the 1990s—Americans with Disabilities Act of 1990 (ADA), the Individuals with Disabilities Education Act of 1990 (IDEA), and the Rehabilitation Act Amendments of 1992—represent a major shift in public policy with regard to persons with disabilities and their families—away from what people can *not* do, to what people *can* do. Thus, these separate legislative acts put in place a consistent set of public legislation which recognizes that a disability in no way diminishes the right of individuals to live independently, make choices, and pursue meaningful careers. Key concepts in these separate legislative actions include inclusion, integration, full participation, living and working independently, meaningful and informed choices, involvement of students and families, and natural supports in the workplace. Of particular interest in the context of this discussion, the umbrella civil rights legislation—the ADA—works in concert with IDEA and the Rehabilitation Amendments of 1992 to facilitate the transition of individuals with disabilities from school to non-discriminatory post-school environments.

Changing Educational and Service Delivery Paradigm

Traditionally, both education and adult service delivery systems have operated from the notion that individuals must "fit" into systems (Clark & Kolstoe, 1995; Szymanski, Hanley-Maxwell, & Parker, 1990). Thus an examination of educational planning procedures and typical service development practices reflects the idea that it is the individual who must adjust to the system rather than the system adjusting to the individual. In reality, therefore, "individual educational programs" and "individual habilitation or rehabilitation programs" have been individual in name only. The organization of educational services around a person's disability label reflects this practice. For example, rather than focusing on the individual needs of students, educational programs have tended to focus on the categorical characteristics of the disability. Teacher certification requirements, curricula, educational opportunities, and funding for educational programs have all been organized according to disability classifications. For example, most states certify special education teachers by disability category for grades K-12, with the result being that these individuals are "endorsed" to teach students with a particular disability any subject across both elementary and secondary educational settings. In contrast, "regular" education teachers are typically certified for either elementary or secondary education and, within the secondary endorsement, for particular subjects.

Because of the focus on disability categorization, Individualized Education Programs (IEPs) and curricula at the secondary level have created, in essence, a "Catch 22" situation for students. Typically, curricula have been developed to address students' "needs," learning styles, and deficits. As a result, in many schools separate blocks of curriculum are developed for students with learning disabilities, students with behavior disorders, students with mild mental retardation, students with moderate and severe disabilities, and so on. Thus, when an IEP is developed for a student, his or her program consists of assignment to courses within a designated disability block, resulting in many instances, in few opportunities for elective coursework, classes that address individual needs, preferences, or post-school goals, or education in classrooms with students without disabilities. For example, a student with cerebral palsy whose post-school goal is postsecondary education may be required, because of the existence of a disability, to participate in a work training program during her senior year, which eliminates the opportunity for academic coursework required for college admission.

The concept of attending to deficits has affected students with disabilities in fundamental ways. Traditionally, the education of students with learning disabilities and mild mental retardation has focused on remediation of identified deficits in knowledge rather than on the development of alternative knowledge or skills. The amount of time spent on remediation reduces the time available for instruction in other areas. Often, the result for students is poor academic performance, lack of basic skills, low self-esteem, and increased feelings of frustration and failure—student characteristics often associated with dropping out (Bearden, Spencer, & Moracco, 1989; Ekstrom, Goertz, Pollack, & Rock, 1986; Wolman, Bruininks, & Thurlow, 1989).

Assessment practices also have focused on deficit identification. Historically, assessment has been conducted primarily for making classification and placement decisions, to predict academic or vocational success (failure), and to determine program eligibility (exclusion) (Gajar et al., 1993; Menchetti & Flynn, 1990). Thus, traditional methods have relied on formalized testing procedures and standardized testing instruments. Data generated from such assessments have helped fuel the system wherein students' disability characteristics and deficits are identified and students are placed into "appropriate" educational programs taught by disability specialists. In contrast, assessment focused on instructional planning is classroom-oriented, ongoing, and informal, and relies on the teacher as the primary diagnostician (Gajar et al., 1993). Rather than focus on establishing eligibility and determining disability, the purpose of assessment for instructional planning is to promote student performance and monitor student progress in several domains (Gajar et al., 1993; Menchetti & Flynn, 1990).

An Emerging Paradigm

The recent changes in policy, introduced above, have helped begin to shift the focus from systems to individuals (Szymanski et al., 1990; Wehman, 1992). As a result, a new educational and service delivery paradigm appears to be emerging that is based upon abilities, options, and self-determination (Szymanski et al., 1990). Both IDEA and the Rehabilitation Amendments of 1992 demonstrate significant shifts in philosophy and values concerning individuals with disabilities. Both legislative acts support self-determination and the importance of incorporating individual preferences and interests into service delivery. For example, IDEA requires student participation in the development of transition-focused IEPs and stipulates that educational programs be developed on the basis of post-school goals reflecting student needs, preferences, and interests. Similarly, the Rehabilitation Act Amendments of 1992 focus on holistic services rather than employment alone, placing a greater emphasis on the role of the consumer in developing an individualized educational program.

In the light of these policy developments, the concepts of zero exclusion, least restrictive environment, nondiscriminatory testing, and appropriate and individualized education programs appear to be taking on new meaning. This new meaning is reflected in innovative practices emerging across the United States that are serving to define and give substance to a "consumer-oriented paradigm." These practices include instructional-focused assessment, inclusion of students with disabilities in regular education classrooms, development of individual education programs based upon post-school goals, and active student involvement in program planning.

In support of these innovations, curriculum-based assessment (Salvia & Hughes, 1990), curriculum-based vocational assessment (Ianacone & Leconte, 1986; Stodden, Ianacone, Boone, & Bisconer, 1987), situational assessment (Clark & Kolstoe, 1995), and ecological vocational evaluation (Menchetti & Flynn, 1990) focus on the abilities, aptitudes, and behaviors of students in multiple and natural environments. Data generated through these models are useful for identifying target goals for students, planning instructional processes, and assessing student progress (Gajar et al., 1993). The emergence of these assessment models and their application in secondary schools has direct implications for developing appropriate individualized education programs based on the post-school goals, needs, and preferences of students, and are particularly relevant for assessing students' progress toward their curricular goals within regular education classrooms.

Whereas the concept of mainstreaming exemplified the notion of "fixing" students, the more contemporary *regular education initiative* (REI) supports the emerging focus upon *individuals.* Mainstreaming typically involved educating students in regular classrooms with support services provided by a "resource" teacher. Support services were focused on bringing the performance of the mainstreamed student up to the level expected of all students in the classroom. Thus, mainstreaming tended to address the deficits of students in their attempts to perform in a typical academic curriculum. In contrast, inclusive education focuses on *individual* goals and abilities of students with disabilities within regular education classrooms. With this focus on individuals rather than on subject matter, course content and associated performance standards may vary for particular students. Although inclusive education remains embroiled in controversy, more and more communities are attempting to include their students with disabilities in regular education classrooms. Research on student performance in inclusive classrooms is in its beginning stages; however, to date, data indicate that teachers who have participated in inclusion have positive perceptions and see benefits to their students, both with and without disabilities (Giangreco, Dennis, Cloninger, Edelman, & Schattman, 1993; Janney, Snell, Beers, & Raynes, 1995).

Prior to the passage of IDEA, local education agencies often debated whether transition-related goals and objectives should be included in a student's IEP. The fear was that such a focus would hold the school district legally responsible for providing the transition services. IDEA ended this debate by requiring that once a student reaches age 16, the IEP must specifically address transition services, and further, that whenever transition services are addressed, the student must be involved in all aspects of the meeting. In response, the field is now investigating ways to actively include students in their program development, including the identification and development of student skills relative to quality participation and the identification and expansion of opportunities to participate. Thus, for the first time, those most directly affected by IEP content are gaining a voice in IEP development. Further, as PL 101-476 required an outcome-oriented process, specific post-school goals for students, based on needs, interests, and preferences must be identified. This approach is directly opposite of the traditional approach whereby students' needs and educational programs were determined by disability category.

The traditional paradigm which focused on student deficits rather than abilities is not effective for preparing students for productive and fulfilling lives in today's adult world. By comparison, the emerging paradigm, discussed here, which is based upon abilities, options, and self-determination, holds greater promise for the development of educational programs that prepare students to assume a productive and valued place in society. This emerging paradigm reflects a transition perspective, useful for planning educational programs and services for all students.

A Transition Perspective

Many researchers, educators, and service providers perceive transition planning and services as a narrow band of activities pertaining to identifying specific services needed by students as they transition from school to adult life. However, a number of authors (e.g., Clark & Kolstoe, 1995; Gajar et al., 1993; Kohler, 1995; Rusch, DeStefano, Chadsey-Rusch, Phelps, & Szymanski, 1992) have suggested the need for a much broader interpretation of transition planning and services to support the emerging paradigm. In fact, it is this broader interpretation of transition planning and services that serves to define the emerging paradigm. This definition includes the notion that all educational programs and instructional activities should be (a) based on the post-school goals of students and (b) developed on the basis of individual needs, interests, and preferences. For example, a typical college-bound student enrolled in college preparation curricula during high school takes two years of foreign language instruction and academic coursework required by most four-year universities; registers for and takes the SAT or ACT exams, also required for admission; identifies and applies to appropriate colleges and universities; identifies and applies for appropriate scholarships or financial aid; and participates in extracurricular activities which develop personal, social, and leadership skills that enhance one's ability to succeed. Many individuals within the school and community setting work with this college-bound student to see that these various tasks are accomplished, including teachers, guidance counselors, coaches and other club sponsors, administrators, parents, and even employers. Importantly, this student is actively involved in planning his or her schedule each year, choosing electives, identifying careers and colleges of interest, and choosing the clubs and sports in which to participate.

The new broader interpretation of these interrelated activities considers these events and this process as *transition planning*. Thus, we define transition services as they are defined by IDEA, but interpret the "coordinated set of activities" (34 C.F.R. Section 300.18) to mean *all* the educational activities and programs in which a student participates. This transition perspective does not view "transition planning" as an add-on activity for students with disabilities once they reach age 16, but rather a fundamental basis of education which guides the development of all educational programs.

With respect to specific postschool goals, transition planning occurs for students who plan and go on to postsecondary education: their educational programs and instructional activities are designed to help them attain their post-school training goal related to employment–a college education. Our educational systems are set up to facilitate this transition planning process for students who are college bound, albeit informally. Generally, the system has been effective for this population of students (Berliner, cited in Edgar & Polloway, 1994). However, the statistics reviewed earlier suggest that for students who are not college bound—both students with and without disabilities—the system has not been very effective.

To summarize, the transition perspective views the educational planning process as consisting of the following steps: (a) post-school goals are identified based upon student abilities, needs, interests, and preferences; (b) instructional activities and educational experiences are developed to prepare students for their post-school goals; and (c) a variety of individuals, including the student, work together to identify and develop the goals and activities. The various practices discussed previously with respect to the emerging paradigm—instructional-focused assessment, inclusion, outcome-based planning, and student involvement—are consistent with this transition perspective and provide specific examples for its application. This transition perspective reinforces an emerging education paradigm that provides structure for educational planning that is outcome-oriented. Our challenge is to build educational systems and structures which facilitate transition planning for all students, regardless of their abilities, and are mindful of their post-school goals.

Organization of Schools and Instruction

In order to promote student-oriented outcome planning, special education will require significant modification. Specifically, secondary education programs must be transformed from deficit-based, disability-driven programs to outcome-based, ability-driven programs (Clark & Kolstoe, 1995; Gajar et al., 1993; Rusch et al., 1992). The IDEA Amendments of 1990, the Rehabilitation Act Amendments of 1992, the Carl D. Perkins Vocational Education and Applied Technology Act of 1990, and the School-to-Work Opportunities Act of 1994 provide a policy framework for structuring secondary education. Further, research on effective transition practices provides the substance with which to build these programs.

Kohler and her colleagues (e.g., Kohler, 1993a, 1995; Kohler, DeStefano, Wermuth, Grayson, & McGinty, 1994; Rusch, Kohler, & Hughes, 1992) developed a taxonomy for transition programming, which presents a comprehensive, conceptual organization of practices that represent the transition perspective of secondary education. Education practices that define the taxonomy are organized into five categories that are relevant for organizing schools and instruction to facilitate transition: student development, student-focused planning, interagency and interdisciplinary collaboration, family involvement, and program structure and attributes. A discussion of each of these areas follows.

Student Development

During their secondary education experience, students need to gain skills that enable them to live and work as independently as possible in their adult lives, including academic and employment skills. To teach these skills, schools must provide academic and vocational instruction through a number of curricular options, identify and provide the necessary supports that facilitate learning by all students, conduct and utilize assessment related to instructional planning, and, where appropriate, provide structured work experience (Clark & Kolstoe, 1995; Gajar et al., 1993; Kohler, 1995; Rusch et al., 1992).

Heal and his colleagues (Heal, Copher, DeStefano, & Rusch, 1989; Heal, Gonzalez, Rusch, Copher, & DeStefano, 1990; Heal & Rusch, 1995) found that work quality, attitude, social skills, absence of asocial behavior, living skills, and academic skills are related to post-school employment. Similarly, D'Amico and Marder (1991) and Goldberg, McLean, LaVigne, Fratillo, and Sullivan (1990) noted that on-the-job training (OJT) enhanced employment rates. Further, Kohler (1994a) found that students participating in an organized OJT curriculum that included work-based and school-based learning showed significant increases in employment-related skills and behaviors. Kohler (1993b) also found that student self-esteem and self-advocacy were considered most important to the success of students with disabilities in postsecondary education programs. It is essential, therefore, that secondary education programs provide curricula that not only focus on academic skill development, but enhance the vocational, social, and personal development of students.

However, if state credit requirements for high school graduation remain the same, course and program content must be re-evaluated so that competency development in the areas described here can be adequately addressed within the framework of the graduation requirements. For example, students must have the option to earn specific academic credits related to math, science, and writing through career and vocational curricula, as provided for in the School-to-Work Opportunities Act of 1994. Self-determination, social, and living skill development must be included in curricula related to health and social studies. In other words, instructional or educational maps must be drawn, indicating those competencies identified in research as being important for students' post-school success along with opportunities for students to develop such competencies within their schools. Based on individual needs and abilities, students will gain these competencies through differing curricular options and experiences. Therefore, rather than organizing instructional opportunities rigidly within schools by disability category, educational offerings should be flexible and adaptive to individual students. Individual needs and student choices should be a paramount consideration.

Student-Focused Planning

Individualized planning is the key to effectively matching a student's educational program and school experiences to his or her post-school goals. Theoretically, an individualized education program (IEP) was intended to serve as a means of adapting education to meet needs of students with disabilities. In essence, "the IEP has been the cornerstone of special education policy" (Martin, Marshall, & Maxson, 1993, p. 53). As discussed previously, however, over the years individual planning has been characterized by disability-based planning with little student involvement.

The IDEA Amendments of 1990 require that a statement of needed transition services be included in a student's IEP beginning at age 16, and at age 14 when considered appropriate. In defining transition services, the legislation focuses on outcomes, activities, students' preferences and interests, and student, parent, and service provider involvement. The planning vehicle is the individual education program. A comprehensive approach to developing outcome-focused educational programs must address the IEP document and process, student and family participation, and accommodations and planning strategies (Kohler, 1994b). Student participation in this process is essential, and self-determination skills may be fundamental for participation.

Self-determination. Ward (1988) defined self-determination as "the attitudes which lead people to define goals for themselves and the ability to take the initiative to achieve those goals" (p. 2). According to Wehmeyer (1992), the construct of self-determination includes "the attitudes and abilities required to act as the primary causal agent in one's life and to make choices regarding one's actions free from undue external influence or interference" (p. 305). In arguing for including self-determination in special education curricula, Schloss, Alper, and Jayne (1994) defined it as "the ability to consider options and make appropriate choices regarding residential life, work, and leisure time" (p. 215). Stowitschek (1992), in turn, saw self-determination as representing "a composite of traits which may vary according to the person and the level of application" (p. 3), and comprised of five major categories of skills: goal-setting, self-advocacy, assertive pattern of responding, decision making, and interpersonal problem solving. Common elements of these various definitions of self-determination include attitudes and skills, goals, and choices relevant to decisions that affect one's future—all aspects related to a student's participation in the development of outcome-based educational programs.

Wehmeyer (1992) proposed that one reason for poor quality of life after exiting high school is that students in special education lack self-determination skills and infrequently have opportunities to experience self-determination. Educators, therefore, must begin to take responsibility for giving students opportunities to experience self-determination. Van Reusen and Bos (1990) warned, "If special educators plan and carry out instructional activities without involving or considering the adolescent's perceptions and priorities, they may be minimizing the student's self-determination" (p. 30).

Ward and Kohler (in press) reviewed 20 model demonstration projects that had developed curricula and expanded opportunities for students to practice self-determination. These curricula focused on teaching students "to evaluate their skills, recognize their limits, set goals, identify options, accept responsibility, communicate their preferences and needs, and monitor and evaluate their progress" (Ward & Kohler, in press). Curricular activities were conducted in school-based and community-based settings and included modeling by teachers, parents, and mentors. To promote the acceptance by others of students' self-determined behavior, projects trained teachers, parents, and other significant adults in students' lives. Field and Hoffman (1994) and The Arc (1994) found that students showed improvements in skills associated with self-determination after completing the project curricula. The University of Colorado at Colorado Springs (1991) developed curricula through which students learned and then directed their individualized education program staffing. Initial findings indicated that students in the program participated in IEP activities at much greater rates and identified more IEP goals than peers in control groups (J. Martin, personal communication, April, 1994). Similarly, Van Reusen and Bos (1994) developed a strategy to train students and their parents to actively participate as partners in the IEP process. The IEP participation strategy (IPARS) consisted of five steps: (a) Inventory, (b) Provide inventory during IEP conference, (c) Ask questions, (d) Respond to questions, and (e) Summarize the IEP goals. Strategy instruction was conducted across multiple stages (orientation, model and prepare, verbal rehearsal, strategy practice and feedback, and generalization) during three training sessions. Van Reusen and Bos (1994) reported that strategy-instructed students identified more goals and communicated more effectively during the conferences than did students in the contrast group.

Thus, issues related to student participation in transition planning include the concept of self-determination. Related to self-determination, students must develop skills associated with self-evaluation, problem solving, reviewing choices, and making decisions. Further, students must have opportunities to practice and apply these skills in relation to decisions about their future, particularly their post-school goals and educational objectives.

Goals, objectives, and outcomes. The underlying purpose of the IEP is to specify the goals and objectives of a student's educational program and the mechanisms for achieving and evaluating progress. The IEP document should reflect activities and services relevant to achieving the post-school goals, as well as the persons or agencies responsible for conducting the activities and providing the services. Further, the IEP should reflect student needs and interests and be based upon assessment information, which reflects the student's current level of functioning. As a result, there is a fundamental relationship between the IEP content as reflected in the document, assessment data on student abilities and interests, the educational activities in which a student participates, and student outcomes. However, research indicates that all too often, one or more of these variables are missing in the IEP document.

Trach (1995) reviewed 486 IEPs of transition-aged students, including 258 written for the 1992–93 school year and 228 for the 1993–94 school year. Findings indicate that, in general, specific activities associated with transition-related goals were not identified in IEP documents. Stodden, Meehan, Bisconer, and Hodell (1989) obtained similar results after reviewing the educational records of 127 secondary education students. Students included in the study met three criteria: they had participated in a work-study or equivalent vocational program, vocational assessment information was available for review, and an IEP was available for review for the period prior to and after the vocational assessment activities (Stodden et al., 1989). Regardless of students' specified disability level (i.e., mild, moderate, or severe), there was no significant difference in the number of IEP vocational goals and objectives written before and after vocational assessment was conducted (Stodden et al., 1989). These authors also noted that student IEPs included few vocational objectives, and those that existed were vague. Further, in each school investigated, "nearly every student, regardless of handicapping condition or level of programming need, had identical IEP vocational goals and objectives" (Stodden et al., 1989, p. 35).

In a follow-up study of students with disabilities in Oregon and students with and without disabilities in Nevada, Benz and Halpern (1993) identified significant discrepancies between identified student needs and services actually provided to students with disabilities prior to their leaving school. Specifically, their findings indicated that (a) students with mild mental retardation were reported to have the greatest number of needs related to transition planning; (b) 25% - 50% of all identified needs for students with disabilities were not addressed at all during the transition planning process; (c) one third to one half of students' unmet needs occurred in the areas of remedial academics, social skills, vocational training, postsecondary education, and independent living skills; and (d) students with learning disabilities and emotional disabilities were most likely to present unmet transition needs when they left school (Benz & Halpern, 1993).

In summary, if we expect to improve the adult outcomes of individuals with disabilities, it is essential that we improve the IEP process and IEP content (Benz & Halpern, 1993; Edgar & Polloway, 1994; Martin et al., 1993; Stodden et al., 1989). Educators must begin early to assist and guide students in developing appropriate education programs based on individual transition goals (Newman & Cameto, 1993). As required in the legislation, educational program planning must become outcome- rather than disability-focused (Edgar & Polloway, 1994; Wehman, 1992). As a student's IEP is the primary vehicle for identifying educational objectives, activities, services, and service providers, educators must reform the IEP process to include student involvement and so that it includes the development of relevant assessment information and identification of valued and attainable post-school goals.

Interagency and Interdisciplinary Collaboration

As illustrated in previous discussion, a number of individuals participate in helping non-disabled college-bound students achieve their long- and short-term goals. For example, teachers, counselors, coaches, parents, and peers typically work in concert with the student to ensure that he or she engages in curricular and extracurricular activities associated with the goal of postsecondary education.

By contrast, special education teachers have traditionally taken on the primary role of educational planner for students with disabilities, often with little cooperation from or collaboration with general or vocational education teachers, school administrators, or community service providers (Wagner, 1993; Wagner, Blackorby, Cameto, Hebbeler, & Newman, 1993). The Individuals with Disabilities Education Act of 1990, the School-to-Work Opportunities Act, and the Rehabilitation Act Amendments of 1992 all require collaboration on both the individual planning level and the community level. Collaboration focused on individual students will address post-school goals, educational programs, and related services, whereas community-level collaboration will focus on programs, systems, and service delivery (Everson & McNulty, 1992; Kohler, 1994b).

Kohler et al. (1994) found that interagency and interdisciplinary collaboration were present in 60% of the exemplary transition programs identified in their analysis. In a review of transition literature, Kohler (1993a) noted that 50% of the documents reviewed cited interagency collaboration and service delivery as an important component of transition planning. In contrast, Rusch, Kohler, and Hughes (1992) found that employment-focused projects cited personnel issues and a lack of collaboration or cooperation as primary barriers to achieving anticipated outcomes. Thus, it appears that collaboration and cooperation are present in programs considered effective, and that the lack of collaboration and cooperation can serve as a barrier.

Gajar et al. (1993) described a model for consultation and collaboration that is based upon the work of several authors pertaining to collaborative consultation (e.g., Conoley & Conoley, 1988; Idol, Paolucci-Whitcomb, & Nevin, 1986; West & Idol, 1990; West, Idol, & Cannon, 1988; cited in Gajar et al., 1993). The model, which concentrates on the relationships among professionals, the student, and significant others in the life of the student, includes the following stages: (a) identification of members, responsibilities, roles, objectives, and expectations; (b) identification of the problem; (c) development of intervention strategies; (d) implementation of intervention strategies; (e) evaluation of interventions; and (f) reconsideration, change, or maintenance of specific strategies (Gajar et al., 1993). Through such a model, educators, students, parents, and service providers can work together to identify appropriate post-school goals and subsequently develop relevant educational programs and services.

Individual members of a student's collaborative team vary across time and from student to student, depending on the student's needs as well as age and grade level (Clark & Kolstoe, 1995; Gajar et al., 1993; Stodden et al., 1987). Members might include, but are not limited to, (a) special, vocational, and regular education teachers; (b) speech, occupational, or physical therapists; (c) adult service providers, including rehabilitation or independent living counselors; (d) educational program support staff and guidance counselors; and (e) employers or postsecondary education representatives.

At the community level, collaboration focuses on eliminating service gaps, avoiding service duplication, increasing efficient use of scarce resources; it also reduces professional territoriality and increases holistic planning and service delivery (Everson & Moon, 1990). Halpern, Benz, and Lindstrom (1992) presented a systems-change model designed to support improvements in program capacity to address service needs identified through individual-focused planning. The model, Community Transition Team Model (CTTM), is based on the concepts of local control, a developmental perspective of change, improvement of program capacity, an intervention-centered approach, and networking among teams. Through this model, local communities address questions associated with their present levels of services and programs and plan for where they want to go in the future. The purpose of the CTTM is to improve secondary special education and transition services in local communities, thus improving and expanding the options available to individual students. Findings indicate that the CTTM has resulted in the development of new instructional programs and improved communication and collaboration among service providers (Halpern et al., 1992).

Family Involvement

Involvement of family in the education of students with disabilities has been mandated since 1975 with the implementation of PL 94–142. In addition, the passage of IDEA in 1990 addressed the mandate for family involvement in the development of transition-related IEPs. Research indicates that students with higher levels of family involvement are more successful in school than students with little or no family involvement. Specifically, students whose parents are very involved in their education miss significantly fewer days of school and are significantly less likely to fail a class than their peers whose parents are not at all involved in their schooling (Wagner, Blackorby, & Hebbeler, 1993). Significant positive relationships have also been found between high parent expectations and high parent involvement and residential independence and full community participation (Wagner, Blackorby, Cameto et al., 1993).

In addition, strong positive relationships have been shown between parent involvement, parent expectations, and student enrollment in postsecondary education. Specifically, Newman and Cameto (1993) found that of those students who went on to postsecondary vocational and postsecondary academic programs, 90% and 82% of their parents, respectively, participated in their school conferences, as compared to only 75% of the parents of students who did not attend postsecondary education. Newman and Cameto found also that youths in both academic and vocational postsecondary programs were significantly more likely than those who did not attend to be expected to do so by their parents. With respect to employment outcomes, Rusch, Kohler, and Hughes (1992) found that parent and family resistance often serves as a significant barrier to projects that attempt to obtain employment for individuals with disabilities.

An additional benefit of continuing parental involvement in the school program relates to the problem of unrealistic parental expectations. Cummings and Maddux (1987, cited in Gajar et al., 1993) found that parents' expectations for their sons or daughters with a disability ranged from being much too low to, more often, unrealistically high. In conjunction with their expectations, parents often advocate for an educational program that does not support realistic post-school goals; as a result, adequate time to provide training and experiences associated with more realistic goals may not be available. Continued participation of parents in the schooling of their children provides for ongoing opportunities to assess student abilities and progress toward specified post-school goals (Gajar et al., 1993).

With respect to their participation in educational program planning, parents can participate in numerous roles, including that of advocate and case manager (Gajar et al., 1993). Clark and Kolstoe (1995) identified specific responsibilities for parents who participate in transition-oriented secondary school programs:

1. Encourage student self-determination and independence at home.
2. Encourage and facilitate setting goals.
3. Teach, and assist in teaching, daily living and personal-social goals.
4. Encourage the student to work at home and at a neighborhood or community job.
5. Reinforce work-related behaviors at home (work habits, hygiene, grooming, etc.).
6. Explore and promote community resources for transition.
7. Assist in the student assessment process.
8. Assist the student in developing personal and social values, self-confidence, and self-esteem.
9. Work with legal and financial experts, as appropriate, to plan financial, legal, and residential alternatives. (pp. 99–100)

Parents can also provide assistance with homework and facilitate student involvement in extracurricular clubs and sports, for example, by providing transportation. In addition, Kohler (1994b) found that family involvement can extend beyond the individual student to the areas of policy development, program evaluation, staff development, strategic planning, and resource allocation.

Program Structure and Attributes

As indicated in the review of student outcome data earlier in this chapter, evidence suggests that our school programs have not been effective at preparing students to achieve positive post-school outcomes. This evidence is particularly disturbing when comparing the outcomes of high school graduates with disabilities with those of dropouts. These data indicate that for many, participation in special education did not enhance their prospects for the future. To improve these prospects, schools must become more responsive to the needs of our young people—needs associated with becoming participating and productive members of society.

In order to achieve student development, conduct individual planning, and facilitate collaboration and family involvement, schools and programs must be organized in a way that promotes these activities. In line with the transition perspective presented previously, fundamental change must occur in two areas: (a) educational programs (i.e., curricular decisions) for all students must be based upon post-school goals, and relevant to these goals, (b) a variety of curricular options must be made available to students (Edgar & Polloway, 1994; Gajar et al., 1993; Stodden & Leake, 1994). Stodden and Leake noted that past attempts to include transition planning and services in educational programs have met with resistance and have achieved limited success because they have been "hampered by a pervasive tendency to add programs to the core of the education system, rather than infusing essential changes into the core itself" (p. 65). They suggested that by infusing a transition perspective into our educational system,

> We would not need more add-on programs with new personnel to assess, plan, and teach, because instruction in the science and English classrooms and the vocational shops would be focused on post-school outcomes through an integrated continuum of steps. Transition values would guide the decision-making of teachers regarding, most importantly, *why* they teach what they do: to prepare students for the day when they leave the school system, whether that is one year or twelve years down the road. Once the *why* of teaching is established, it guides *what*, *when*, *where*, and *how* to assess, plan, and teach, and for exceptional students this implies an individually tailored continuum. (p. 69)

Program structures and attributes associated with outcome-based education and expanded curricular options include community-level strategic planning, cultural and ethnic sensitivity, a clearly articulated mission and values, qualified staff, and sufficient allocation of resources (Kohler, 1994b). Transition-oriented schools must focus also on systematic community involvement in the development of educational options, community-based learning opportunities, systematic inclusion of students in the social life of the school, and increased expectations related to skills, values, and outcomes for all students (Edgar & Polloway, 1994). Polloway, Patton, Epstein, and Smith (1989) identified specific attributes of programs that focus on post-school outcomes and presented a continuum of curricular options that responded to student needs at specific points in time and across the lifespan. Through such an approach, postsecondary education is not overlooked as an option for students with disabilities. A focus on individual abilities, rather than eligibility labels or diagnostic categories, facilitates identification of relevant post-school goals, and focuses on the possible utility of postsecondary education.

Summary

Data generated by the National Longitudinal Transition Study and other transition-related research have clear implications for the way we conduct and organize secondary education. In general, high school graduates experience significantly greater employment outcomes, particularly with respect to wages and salary, than do youths who do not graduate from high school (U.S. Department of Education, 1994). Unfortunately, as we have shown, this is not always the case for young people who graduate from special education programs. Emerging research, however, indicates that particular aspects of one's high school experience do appear related to improved chances for post-school success.

Participation in extracurricular groups and activities in secondary school is significantly negatively associated with the number of days students are absent from school and the rate of course failure; it is positively associated with attendance in postsecondary education (Newman, 1991). Paid work experience in high school is negatively associated with the average number of days absent and course failure (Wagner, Blackorby, & Hebbeler, 1993), and positively related with post-school employment (Hasazi, Gordon, & Roe, 1989; Hudson, Schwartz, Sealander, Campbell, & Hensel, 1988; Mithaug, et al., 1985; Sitlington, Frank, & Carson, 1992.). Further, parents' involvement in their child's education is negatively related to the number of days absent and course failure, and positively related to attendance in post-secondary education.

Our challenge is to build secondary education systems that facilitate identification of post-school goals for individual students and then to provide opportunities for students to choose appropriate curricular activities. These systems must *engage* students, parents, teachers, and communities in the educational process and in ongoing evaluation of that process.

References

The Arc (1993). *Final report: Self-determination curriculum project.* Arlington, TX: The Arc, National Headquarters.

Bearden, L. J., Spencer, W. A., & Moracco, J. C. (1989). A study of high school dropouts. *The School Counselor, 37,* 113–120.

Benz, M. R., & Halpern, A. S. (1993). Vocational and transition services needed and received by students with disabilities during their last year of high school. *Career Development for Exceptional Individuals, 16,* 197–211.

Bruininks, R. H., Thurlow, M. L., Lewis, D. R., & Larson, N. W. (1988). Post-school outcomes for students in special education and other students one to eight years after high school. In R. H. Bruininks, D. R. Lewis, & M. L. Thurlow (Eds.), *Assessing outcomes, costs and benefits of special education programs* (pp. 9–111). Minneapolis: University of Minnesota, University Affiliated Programs.

Clark, G. M., & Kolstoe, O. P. (1995). *Career development and transition education for adolescents with disabilities* (2nd ed.). Boston: Allyn and Bacon.

D'Amico, R., & Marder, C. (1991). *The early work experience of youth with disabilities: Trends in employment rates and job characteristics. A report from the National Longitudinal Transition Study of Special Education Students.* Menlo Park, CA: SRI International.

deBettencourt, L. U., Zigmond, N., & Thornton, H. (1989). Follow-up of postsecondary-age rural learning disabled graduates and dropouts. *Exceptional Children, 56,* 40–49.

Edgar, E., & Polloway, E. A. (1994). Education for adolescents with disabilities: Curriculum and placement issues. *The Journal of Special Education, 27,* 438–452.

Ekstrom, R. B., Goertz, M. E., Pollack, J. M., & Rock, D. A. (1986). Who drops out of high school and why? Findings from a national study. *Teachers College Record, 87* (3), 356–373.

Everson, J. M., & McNulty, K. (1992). Interagency teams: Building local transition programs through parental and professional partnerships. In F. R. Rusch, L. DeStefano, J. Chadsey-Rusch, L. A. Phelps., & E. Szymanski (Eds.), *Transition from school to adult life: Models, linkages, and policy* (pp. 341–352). Sycamore, IL: Sycamore.

Everson, J. E., & Moon, M. S. (1990). Developing community program planning and service delivery teams. In F. R. Rusch (Ed.), *Supported employment: Models, methods, and issues* (pp. 381–394). Sycamore, IL: Sycamore.

Field, S., Hoffman, A., Sawilowsky, S., & St. Peter, S. (1994). *Skills and knowledge for self-determination: Final report.* Detroit: Wayne State University.

Gajar, A., Goodman, L., & McAfee, J. (1993). *Secondary schools and beyond: Transition of individuals with mild disabilities.* New York: Merrill.

Giangreco, M. F., Dennis, R., Cloninger, C., Edelman, S., & Schattman, R. (1993). "I've counted Jon": Transformational experiences of teachers educating students with disabilities. *Exceptional Children, 59,* 359–372.

Goldberg, R. T., McLean, M. M., LaVigne, R., Fratolillo, J., & Sullivan, F. T. (1990). Transition of persons with developmental disability from extended sheltered employment to competitive employment. *Mental Retardation, 18,* 199–304.

Halpern, A. S., Benz, M. R., & Lindstrom, L. E. (1992). A systems change approach to improving secondary special education and transition programs at the community level. *Career Development for Exceptional Individuals, 15,* 109–120.

Harnisch, D. L., & Gierl, M. J. (1995, April). *Factors associated with dropping out for youths with disabilities.* Paper presented at the Council for Exceptional Children Annual Meeting, Indianapolis, IN.

Hasazi, S. B., Gordon, L. R., & Roe, C. A. (1985). Factors associated with the employment status of handicapped youth exiting high school from 1979 to 1983. *Exceptional Children, 51,* 455–469.

Heal, L. W., Copher, J. I., DeStefano, L., & Rusch, F. R. (1989). A comparison of successful and unsuccessful placements of secondary students with mental handicaps into competitive employment. *Career Development for Exceptional Individuals, 12,* 167–177.

Heal, L. W., Gonzalez, P., Rusch, F. R., Copher, J. I., & DeStefano, L. (1990). A comparison of successful and unsuccessful placements of youths with mental handicaps into competitive employment. *Exceptionality, 1,* 181–195.

Heal, L. W., & Rusch, F. R. (1995). Predicting employment for students who leave special education high school programs. *Exceptional Children, 61*(5), 472–487.

Hudson, P. J., Schwartz, S. E., Sealander, K. A., Campbell, P., & Hensel, J. W. (1988). Successfully employed adults with handicaps. *Career Development for Exceptional Individuals, 11,* 7–14.

Ianacone, R. N., & Leconte, P. J. (1986). Curriculum-based vocational assessment: A viable response to a school-based service delivery issue. *Career Development for Exceptional Individuals, 9,* 113–120.

Individuals with Disabilities Education Act of 1990, 20 U.S.C. § 1400 ff. (1990).

Janney, R. E., Snell, M. E., Beers, M. K., & Raynes, M. (1995). Integrating students with moderate and severe disabilities into general education classes. *Exceptional Children, 61,* 425–439.

Kohler, P. D. (1993a). Best practices in transition: Substantiated or implied? *Career Development for Exceptional Individuals, 16,* 107–121.

Kohler, P. D. (1993b). *Serving students with disabilities in postsecondary education settings: A conceptual model of program outcomes.* Unpublished doctoral dissertation, University of Illinois at Urbana-Champaign.

Kohler, P. D. (1994a). On-the-job training: A curricular approach to employment. *Career Development for Exceptional Individuals, 17,* 187–202. 29–40.

Kohler, P. D. (1994b). *A taxonomy for transition programming.* Champaign: University of Illinois at Urbana-Champaign, Transition Research Institute,

Kohler, P. D. (1995). *Preparing youths with disabilities for future challenges: A taxonomy for transition programming.* Manuscript submitted for publication.

Kohler, P. D., DeStefano, L., Wermuth, T., Grayson, T., & McGinty, S. (1994). An analysis of exemplary transition programs: How and why are they selected? *Career Development for Exceptional Individuals, 17,* 187–202.

Kranstover, L. L., Thurlow, M. L., & Bruininks, R. H. (1989). Special education graduates versus non-graduates: A longitudinal study of outcomes. *Career Development for Exceptional Individuals, 12,* 153–166.

Marder, C., & D'Amico, R. (1992). *How well are youth with disabilities really doing? A comparison of youth with disabilities and youth in general.* Menlo Park, CA: SRI International.

Martin, J. E., Marshall, L. H., & Maxson, L. L. (1993). Transition policy: Infusing self-determination and self-advocacy into transition programs. *Career Development for Exceptional Individuals, 16,* 53–61.

Menchetti, B. M., & Flynn, C. C. (1990). Vocational evaluation. In F. R. Rusch (Ed.), *Supported employment: Models, methods, and issues* (pp. 111–130). Sycamore, IL: Sycamore.

Mithaug, D. E., Horiuchi, C. N., & Fanning, P. N. (1985). A report on the Colorado statewide follow-up survey of special education students. *Exceptional Children, 51,* 397–404.

Newman, L. (1991). *The relationship between social activities and school performance for secondary students with learning disabilities.* Menlo Park, CA: SRI International.

Newman, L., & Cameto, R. (1993). *What makes a difference? Factors related to postsecondary school attendance for young people with disabilities.* Menlo Park, CA: SRI International.

Polloway, E. A., Patton, J. R., Epstein, M. H., & Smith, T. E. C. (1989). Comprehensive curriculum for students with mild handicaps. *Focus on Exceptional Children, 21*(8), 1–12.

Rusch, F. R., DeStefano, L., Chadsey-Rusch, J., Phelps, L. A., & Szymanski, E. (1992). *Transition from school to adult life: Models, linkages, and policy.* Sycamore, IL: Sycamore.

Rusch, F. R., Kohler, P. D., & Hughes, C. (1992). An analysis of OSERS-sponsored secondary special education and transitional services research. *Career Development for Exceptional Individuals, 15,* 121–143.

Salvia, J., & Hughes, C. (1990), *Curriculum based assessment.* New York: Macmillan.

Schloss, P. J., Alper, S., & Jayne, D. (1994). Self-determination for persons with disabilities: Choice, risk, and dignity. *Exceptional Children, 60,* 215–225.

Seidel, J. F., & Vaughn, S. (1991). Social alienation and the learning disabled school dropout. *Learning Disabilities Research and Practice, 6,* 152–157.

Sitlington, P. L., Carson, R., & Frank, A. R. (1992). *Iowa statewide follow-up study: Adult adjustment of individuals with behavioral disorders.* Des Moines: State of Iowa Department of Education.

Sitlington, P. L., Frank, A. R., & Carson, R. (1992). Adult adjustment among high school graduates with mild disabilities. *Exceptional Children, 59*(3), 221–233.

Stodden, R. A., Ianacone, R. N., Boone, R. M., & Bisconer, S. W. (1987). *Curriculum-based vocational assessment: A guide for addressing youth with special needs.* Honolulu: Centre Publications.

Stodden, R. A., & Leake, D. W. (1994). Getting to the core of transition: A re-assessment of old wine in new bottles. *Career Development for Exceptional Individuals, 17,* 65–76.

Stodden, R. A., Meehan, K. A., Bisconer, S. W., & Hodell, S. L. (1989). The impact of vocational assessment information and the individualized education planning process. *The Journal for Vocational Special Needs Education, 12,* 31–36.

Stowitschek, J. (1992). *Development of a model self-determination program and taxonomy for youth with moderate and severe disabilities (Project Proposal).* Seattle: University of Washington, Experimental Education Unit WJ-10.

Szymanski, E. M., Hanley-Maxwell, C., & Parker, R. M. (1990). Transdisciplinary service delivery. In F. R. Rusch (Ed.), *Supported employment: Models, methods, and issues* (pp. 199–214). Sycamore, IL: Sycamore.

Trach, J. S. (1995). *Impact of curriculum on student post-school outcomes.* Champaign: University of Illinois, Transition Research Institute. (Manuscript in preparation.)

U.S. Department of Education. (1990). *Twelfth annual report to congress.* Washington, DC: U.S. Government Printing Office.

U.S. Department of Education. (1994). *Sixteenth annual report to congress.* Washington, DC: U.S. Government Printing Office.

University of Colorado at Colorado Springs. (1991). *Choice makers.* Colorado Springs: Author.

Van Reusen, A. K., & Bos, C. S. (1990). I Plan: Helping students communicate in planning conferences. *Teaching Exceptional Children, 22(4)*, 30–32.

Wagner, M. (1993). *The secondary school programs of students with disabilities.* Menlo Park, CA: SRI International.

Wagner, M., Blackorby, J., & Hebbeler, K. (1993). *Beyond the report card: The multiple dimensions of secondary school performance of students with disabilities.* Menlo Park, CA: SRI International.

Wagner, M., Blackorby, J., Cameto, R., Hebbeler, K., & Newman, L. (1993). *The transition experiences of young people with disabilities.* Menlo Park, CA: SRI International.

Wagner, M., D'Amico, R., Marder, C., Newman, L., & Blackorby, J. (1992). *What happens next? Trends in postschool outcomes of youth with disabilities. The second comprehensive report from the National Longitudinal Transition Study of Special Education Students.* Menlo Park, CA: SRI International.

Wagner, M. (1989, May). *Youth with disabilities during transition: An overview of descriptive findings from the National Longitudinal Transition Study.* Menlo Park CA: SRI International.

Ward, M. J. (1988). The many facets of self-determination. *Transition Summary, 5*, 2–3. (A publication of the National Information Center for Children and Youth with Handicaps.)

Ward, M. J., & Kohler, P. D. (in press). Teaching self-determination: Content and process. In L. Power (Ed.), *Making our way: Building self-competence among youth with disabilities.* Baltimore: Brooks.

Wehman, P. (1992). Transition for young people with disabilities: Challenges for the 1990's. *Education and Training in Mental Retardation, 27*, 112–118.

Wehmeyer, M. L. (1992). Self-determination and the education of students with mental retardation. *Education and Training in Mental Retardation, 27*, 302–314.

Wolman, C., Bruininks, R., & Thurlow, M. L. (1989). Dropouts and dropout programs: Implications for special education. *Remedial and Special Education, 10(5)*, 6–20.

Note

1. This research was sponsored in part by the Office of Special Education and Rehabilitative Services (OSERS), US Department of Education, under a cooperative agreement (H158-T-00-1) with the University of Illinois. Opinions expressed here do not necessarily reflect those of OSERS.

5

Gifted and Talented Students

SALLY M. REIS and JOSEPH S. RENZULLI
University of Connecticut, USA

In this chapter, psychological conceptions of giftedness are reviewed as are operational definitions used by states and school districts: the current status of the field of education of the gifted is discussed, as are current issues in the field. Promising alternate strategies for the future and new directions are mentioned, including the extension of services often reserved for gifted students to develop talents in all students.

Introduction

Rocio, who lives in a large urban city in the Northeast, reads at a fourth-grade level and often finishes a 50-page paperback book in the hour she reads to herself before she falls asleep each night. The next day in school she is bored and sleepy; on several occasions she has begged her parents to let her stay home and read. It's only September, and she is becoming distracted in school, rarely finishes her assignments, and on two occasions, Rocio's parents have been called by Rocio's teacher. The problem with Rocio is that she is six years old and in first grade. Only one other student in her classroom reads and knows what to do with her. While her peers learn beginning sounds, Rocio is left on her own and spends most of her time in school finishing her assigned work in a fraction of the time it takes other students to finish their work and learning to daydream and pass the time. There is a gifted program in Rocio's school, but it provides services for identified students who are in fourth grade. The teacher in the program spends only a day and a half at Rocio's school, which is one of the three schools that she travels to each week. No mandate exists to provide services to gifted students in the state in which Rocio lives, and there is no certification for teachers who work in gifted programs. The teacher who provides services to students in Rocio's school has had no formal training or course work, but she has attended a few staff development sessions and one conference on gifted students. Rocio's parents watch as their child, who was excited about learning and motivated to begin first grade, gradually begins to dislike school and cry each morning as she must get ready to leave their apartment. As Rocio finishes first grade, she has become bored by school and disinterested in learning.

Rocio's problems in school are not uncommon for students who learn at an advanced level. These young people do not often receive instruction in school that enables them to grow and learn at an appropriate level and rate because they are often ahead of their chronological aged peers in mastering of basic skills and able to learn the skills they haven't mastered more rapidly than peers. Yet many classroom teachers are not prepared to adjust curriculum or modify instruction for high achieving students.

Recent research indicates that 64% of third- and fourth-grade classroom teachers who responded to a large national survey indicated that they had never had any training in staff development on meeting the needs of gifted students (Archambault et al., 1993). In this chapter, an overview of the current policies and practices related to gifted education is discussed as are current definitions of giftedness. Current research indicating that the needs of gifted students are not being addressed in many classrooms is presented as are a number of service delivery systems designed for this population. The chapter concludes with a discussion of current issues such as special populations of gifted students and future trends that should be considered for this group of young people.

In the recently released federal report on the status of education for our nation's most talented students entitled *National Excellence, A Case for Developing America's Talent* (O'Connell-Ross, 1993), a quiet crisis is described in the education of talented students in the United States. The report clearly indicates the absence of attention paid to this population: "Despite sporadic attention over the years to the needs of bright students, most of them continue to spend time in school working well below their capabilities. The belief espoused in school reform that children from all economic and cultural backgrounds must reach their full potential has not been extended to America's most talented students. They are underchallenged and therefore underachieve" (p. 5). The report further indicates that our nation's talented students are offered a less rigorous curriculum, read fewer demanding books, and are less prepared for work or postsecondary education than top students in many other industrialized countries. Talented children who come from economically disadvantaged homes or are members of minority groups are especially neglected, the report also indicates, and many of them will not realize their potential without some type of intervention. What types of talents and gifts do these children have and in what ways have educators and psychologists defined talent and proposed to identify potential in various populations?

Psychological Perspectives of Giftedness

For many years, psychologists defined people with extraordinary abilities through the use of intelligence quotients for measuring intellectual abilities. In 1916, Lewis M. Terman revised an instrument developed by French psychologist Alfred Binet and entitled the instrument the Stanford-Binet Intelligence Scale. The wide use of this instrument contributed to the belief that a single score was equivalent to one's intelligence. Since that time, other researchers have argued that intellect cannot be expressed in such a unitary manner (Cattell, 1971; Guilford, 1959), and more recently, many more psychologists and researchers have agreed that intelligence can be expressed in a variety of ways. Sternberg and Davidson (1986) reassessed what is meant by giftedness when they edited *Conceptions of Giftedness*, which includes 17 different conceptions of giftedness that, although distinct, seem to be interrelated in certain ways. Sternberg and Davidson distinguish between implicit and explicit theories of giftedness and further subdivide explicit theoretical approaches into cognitive theory, developmental theory, and domain specific theory. Implicit theories consist of definitions that cannot be empirically tested, whereas explicit theories presuppose definitions, "seeking to interrelate such definitions to a network of psychological or educational theory" (p. 3). Explicit theories are testable by usual empirical means, yet the evaluation of these theories is often dependent on the usefulness of the underlying conception of giftedness that has generated the theory. Sternberg and Davidson believe that giftedness is a concept that we invent, not something we discover, and that it is what one society or another wants it to be, making it, therefore, subject to change according to time and place (p. 4).

Implicit theoretical conceptions are embodied in the work of Tannenbaum (1986), who proposed a psychosocial approach to defining giftedness by classifying talents of four types: scarcity (those always in short supply that make life easier, safer, healthier, and more intelligible); surplus (those able to elevate people's sensibilities and sensitivities to new heights through the production of great art, literature, music, and philosophy); quota (specialized, high-level skills needed to provide goods and services for which the market is limited, such as those of physicians, lawyers, and teachers); and anomalous (reflecting prodigious feats that show how far the powers of the human mind and body can be stretched and yet not be recognized for excellence, e.g., speed-reading or gourmet cooking). Renzulli's three-ring conception of giftedness (Renzulli, 1986a) is also considered an implicit theory, focusing on the interaction of above-average ability, creativity, and task commitment. According to Renzulli, all of these traits can be developed in young people, and he has advocated providing opportunities for self-selected study in which students can learn the appropriate methodology and creative outlets of practicing professionals (Renzulli & Reis, 1985). Common themes identified by the implicit theorists include the need to identify the domain that serves as the basis of one's definition, whether individual or societal; the essential role that cognitive abilities and motivation play in giftedness; the importance of the developmental course of one's talents for whether or how they are expressed; and the inevitability of how one's abilities come together or coalesce as affected by societal forces (Sternberg & Davidson, 1986, pp. 6–7).

Sternberg's (1986) explicit theoretic approach emphasizes three aspects of intellectual giftedness: the superiority of mental processes, including metacomponents relating intelligence to the internal world of the individual; superiority in dealing with relative novelty and in automating information processing, an experiential aspect relating cognition to one's level of experience in applying cognitive processes in particular tasks or situations; and superiority in applying the processes of intellectual functioning, as mediated by experience, to functioning in real-world contexts, a contextual aspect. Sternberg and Davidson believe that "the outward manifestation of giftedness is in superior adaptation to, shaping of and selection of environments" (1986, p. 9) and would agree with Renzulli and Tannenbaum that it can be attained in a number of ways, differing from one person to another. Recurrent themes among the explicit theorists include questioning the cognitive bases of giftedness—asking "what is it that a person can do well to be identified by this term" (Sternberg & Davidson, 1986, p. 10)—and emphasizing the importance of theory-driven empirical research as the primary means for advancing our understanding of giftedness (pp. 10–11).

Explicit psychological approaches to giftedness based on developmental theory are embodied in the work of Gruber; Csikszentmihalyi and Robinson; Feldman; Walters and Gardner; and Albert and Runco. Gruber (1986) emphasizes the study of child and adult development. To gain an understanding of the development of the extraordinary individual, he believes in the intense study of a small number of these persons. Gruber is interested in the kind of gift "that can be transformed by its possessor into effective creative work for the aesthetic enrichment of human experience, for the improvement of our understanding of the world, or for the betterment of the human condition and of our prospects for survival as a species" (p. 247). He further believes that the individual's interests and activities are essential to the development of an extraordinary individual, and that every personal resource is needed to surmount difficulties. Gruber recognizes the importance of the social and historical circumstances of the times, as do Csikszentmihalyi and Robinson, who, like Renzulli, recognize that we should not "view giftedness as an absolute concept—something that exists in and of itself, without relation to anything else" (Renzulli, 1980, p. 4). Csikszentmihalyi and Robinson (1986) view giftedness as emerging throughout one's lifetime.

Feldman (1986), like Csikszentmihalyi and Robinson, views development as a movement through a sequence of stages. Feldman, however, believes that the development of giftedness is domain specific, observing that the movement through the levels of a domain is not mastered by all individuals includes three forms: the rate at which one moves to the level of mastery, the number of levels one achieves, and the domain one selects. According to Feldman, giftedness "is the outcome of a sustained coordination among sets of intersecting forces, including historical and cultural as well as social and individual qualities and characteristics" (p. 303). Walters and Gardner (1986) add the concept of crystallizing experiences that is derived from Gardner's theory of multiple intelligences. According to Gardner (1983), all normal individuals are capable of seven forms of intellectual accomplishment: linguistic, musical, logical-mathematical, spatial, bodily-kinesthetic, inter-personal, and intrapersonal. These multiple intelligences manifest themselves early in life as abilities to process information in certain ways. During crystallized experiences, latent skills of underutilized intelligence may be activated, and an individual's major life activities may change as a result of such an experience.

Albert and Runco (1986) believe that giftedness is both biological and experiential in nature and focus on family emphasis and history. Their longitudinal study of exceptionally gifted boys and their families may provide insight into the development of talent. At the halfway mark of their longitudinal research, it was already evident that those who may cultivate remarkable careers "do it with hard work, faith in the future and a healthy degree of openness to freshly appearing opportunities" (p. 354).

Bloom and his associates at the University of Chicago also engaged in a study of the development of talent in children, examining the processes by which young people who reached the highest levels of accomplishment developed their capabilities. Groups studied included concert pianists, sculptors, research mathematicians, research neurologists, Olympic swimmers, and tennis champions who attained these high levels of accomplishment before the age of 35. According to Bloom and his associates, the following factors play a role in the development of talent: the home environment, which develops the work ethic and the importance of doing one's best at all times; the encouragement of parents in a highly approved talent field; the involvement of families and teachers; and the presence of achievement and progress, which are necessary to maintain a commitment to talent over a decade of increasingly difficult learning (Bloom, 1985, pp. 508-509).

The importance of development throughout the lifespan of the individual is reinforced by each of these developmental theorists, as is the domain-specific nature of giftedness. Gifted individuals are seen as those who can excel usually in one domain, providing that the environmental factors allow this excellence to manifest itself. These developmental psychologists also emphasize the insights gained from intensive case study research and qualitative or naturalistic methodology.

The domain-specific explicit theorists represented in *Conceptions of Giftedness* are Stanley and Benbow, and Bamberger. Stanley and Benbow (1986) deal with mathematically precocious youths who are identified early by high scores on the mathematics section of the College Board Scholastic Aptitude Test. Early identification and subsequent programming is essential for the enhancement of mathematical talent. Bamberger (1986), reporting on her studies of musically gifted children, believes in a transition period when a young musician moves from early prodigiousness to adult artistry and focuses on the internal representations of musical structure. Bamberger's analysis of the transition made by adolescents who must find a way to coordinate a network of representations that is different than it was for younger performers seems to include a combination of developmental and cognitive perspectives.

Definitions of Giftedness in Schools

Definitions of gifts and talents utilized by educators generally follow those proposed in federal reports. For example, the Marland Report (Marland, 1972) commissioned by the Congress included the following federal definition of giftedness, which has been widely adopted or adapted by many states:

> Gifted and talented children are those, identified by professionally qualified persons, who by virtue of outstanding abilities are capable of high performance. These are children who require differentiated educational programs and/or services beyond those normally provided by the regular school program in order to realize their contribution to self and society.
>
> Children capable of high performance include those with demonstrated high achievement and/or potential ability in any of the following areas, singly or in combination: general intellectual ability, specific academic aptitude, creative or productive thinking, leadership ability, visual and performing arts, and/or psychomotor ability. (p. 2)

The selection of a definition of giftedness has been and continues to be the major policy decision made at state and local levels. It is interesting to note that policy decisions are often either unrelated or marginally related to actual procedures or to research findings about a definition of giftedness or identification of the gifted, a fact well documented by the many ineffective, incorrect, and downright ridiculous methods of identification used to find students who meet the criteria in the federal definition (Alvino, McDonnel, & Richert, 1981; Jenkins-Friedman, 1982; Renzulli & Delcourt, 1986).

This gap between policy and practice may be caused by many variables. Unfortunately, although the federal definition was written to be inclusive, it is, instead, rather vague, and problems caused by this definition have been recognized by experts in the field (Renzulli, 1978). Despite these problems, 24 states had adopted the federal definition by 1977 (Karnes & Collins, 1978), and the numbers have increased since then (Council of State Directors of Programs for the Gifted, 1987; Gallagher, Weiss, Oglesby, & Thomas, 1983). In the most recent federal report on the status of gifted and talented programs entitled *National Excellence* (O'Connell-Ross, 1993), a new federal definition is proposed based on new insights provided by neuroscience and cognitive psychology. Arguing that the term "gifted" connotes a mature power rather than a developing ability and, therefore, is antithetic to recent research findings about children, the new definition "reflects today's knowledge and thinking" (p. 27) by emphasizing talent development:

Children and youth with outstanding talent perform or show the potential for performing at remarkably high levels of accomplishment when compared with others of their age, experience, or environment. These children and youth exhibit high performance capability in intellectual, creative, and/or artistic areas, possess an unusual leadership capacity, or excel in specific academic fields. They require services or activities not ordinarily provided by the schools. Outstanding talents are present in children and youth from all cultural groups, across all economic strata, and in all areas of human endeavor. (p. 27)

The report further recommends systems of identification that: seek variety, use multiple assessment measures, are free of bias, are fluid in that they use assessment procedures that accommodate students who develop at different rates, identify potential as well as demonstrated talent, and assess motivation.

Current Status of Education for the Gifted and Talented

The accomplishments of the last 30 years in the education of gifted students since the launching of Sputnik should not be underestimated; the field of education of the gifted, although still historically in its infancy, has emerged as strong, visible, and viable. The most recent *State of the States Gifted and Talented Education Report* (Council of State Directors of Programs for the Gifted, 1994) shows that 47 states, plus Puerto Rico and Guam, have recognized education of the gifted and talented through specific legislation, and the same number of states has assigned state department of education staff to leadership positions in this area. Twenty-nine states have either a policy or position statement from the state board of education supporting the education of the gifted and talented. The report also shows that since 1963, when Pennsylvania first required services for the gifted and talented, 24 other states and Guam have implemented a mandate for services. Twenty-two other states that do not have a mandate support permissive (discretionary) programs for the gifted and talented. This growth has not been constant; however, researchers and scholars in the field have pointed to various high and low points of national interest and commitment to educating the gifted and talented (Gallagher, 1979; Renzulli, 1980; Tannenbaum, 1983). Gallagher described the struggle between support and apathy for special programs for this population as having roots in historical tradition—the battle against an aristocratic elite and our concomitant belief in egalitarianism. Tannenbaum portrays two peak periods of interest in the gifted as the five years following Sputnik in 1957 and the last half of the decade of the 1970s. Tannenbaum described a valley of neglect between the peaks in which the public focused its attention on the disadvantaged and the handicapped. "The cyclical nature of interest in the gifted is probably unique in American education. No other special group of children has been alternately embraced and repelled with so much vigor by educators and laypersons alike" (Tannenbaum, 1983, p.16). Renzulli (1980) raised similar concerns when comparing the gifted child movement with the progressive education movement of the 1930s and 1940s, stating that the field has been alternately embraced and rejected by general educators, parents, and laypeople, and he offers suggestions for dealing with some of the criticisms leveled at proponents of a differentiated education for gifted and talented students. "Simply stated, the field of education for the gifted and talented must develop as strong and defensible a rationale for the practices it advocates as has been developed for those things that it is against" (p. 3).

Excellent educational research continues to be conducted by scholars in the field and at research-based university programs. In the mid-1970s, only one programming model had been developed for gifted programs; by 1986, a textbook on systems and models for gifted programs included 15 models for elementary and secondary programs (Renzulli, 1986b). The Jacob Javits Legislation that was passed in 1990 by the federal government resulted in the creation of a National Research Center for the Gifted and Talented, which involves four universities (the University of Connecticut, the University of Virginia, the University of Georgia, and Yale University), state departments of education in every state, and a consortium of over 300 school districts from across the country.

However, even in the best-funded and most active states, gifted programs are seldom comprehensive. In Connecticut in 1989, more than 150 of 169 cities and towns reported some type of gifted program, yet only a handful of school districts are currently offering continuous services in grades 1 through 12. Recent cutbacks have drastically curtailed these numbers. In addition to this problem, most gifted programs in the United States focus only on the academically able and provide little or no service or attention to artistically talented students or those with potential or demonstrated talent in leadership, specific academic areas, or creative and productive thinking. Most programs for academically able students are concerned with bright students who are also high achievers; few programs are designed for youngsters with high potential who may be underachieving. Many programs take place after school or on Saturdays, ignoring the critical hours that bright students spend in a classroom.

Recently, a reduction or elimination of gifted programs has occurred because of financial restraints or the current emphasis on equity in the nation's schools. During the decades of the 1970s and 1980s, gifted programs flourished and were implemented with increasing frequency. Budget pressures exacerbated by the lingering recession in the 1990s have forced a number of local districts to reconsider the need for programs for gifted students, which are only mandated by less than half of the states in our country. Purcell (1994) found that programs are being affected differentially by the economy and the existence or nonexistence of a state mandate to provide services to high ability students. Programs in states with poor economies and without a mandate (e.g., Delaware, North Dakota, Wyoming) experienced the most dramatic reductions in programs for these students. Twenty nine percent, or one in three programs, were reduced or threatened by reduction in the 1991-1992 academic year. Even programs in states with relatively good economies and a mandate to provide program services (e.g., Alaska, Florida, South Dakota, Utah, Virginia) experienced cutbacks. Fifteen percent of programs in these states experienced reductions or are in jeopardy (Purcell, 1994).

Furthermore, state mandates to provide services for gifted students have been reexamined. Several states, including Oklahoma and Alaska, are reviewing the need for mandates to provide programs or services to high ability students. Services may include diverse offerings such as resource room programs; consultations with classroom teachers by specialists in gifted education; and the provision of programs such as math competitions, problem solving, and training in the invention process. Mississippi and Maine have postponed the implementation of mandates to provide programs. Other states are scrutinizing the need for state-level positions that were designed to ensure coordination of appropriate educational programs for high ability students. For example, in many states the position of gifted education coordinator has been scaled back or eliminated. Other states, including Massachusetts, have eliminated the positions entirely. What is the effect of these program reductions and eliminations on programs and services? The reductions affect programs in several ways. Some districts eliminate grade levels from gifted program services; some eliminate special program components, such as an arts program; and others eliminate personnel (Purcell, 1994). In one district in Connecticut, for example, financial constraints have reduced the gifted program staff from seven teachers to two. Resource room time for independent study, critical and creative thinking skills, and research skills is no longer provided to middle and high school students, and elementary school students spend only one hour a week in a resource room for high ability students. One parent summarized her frustrations about the elimination of her son's program:

> I remember my son coming home and telling me he was upset and angry because they were doing a chapter on telling time in his fourth grade class. He learned to tell time before he entered kindergarten and he said, "I know all of this stuff. I've known all of the math work all year." And I tried to explain that other students needed to learn about time. And he was very angry and said to me, "But, what about me?" And, I didn't know what to say to him (Purcell, 1993).

To compound this problem, some state department of education consultants who have no knowledge of the field are assigned to the position of director or consultant for gifted programs. Many university professors with no background or knowledge about education of the gifted are assigned to teach classes in this area because of local enrollment needs. Ironically, current graduates of doctoral programs in education of the gifted often have difficulty finding university positions because of a lack of substantial commitment to such programs on the part of institutions of higher education.

Problems Facing Talented and Gifted Students

Boredom and Repetition in Our Schools

In the last 20 years, textbook publishers have been accused of "dumbing down" textbooks. In 1982, for example, Kirst indicated that a sample of U.S. publishers agreed that their textbooks had dropped two grade levels in difficulty during the last 10 to 15 years. And, according to Kirst, when California educators tried to find textbooks that would challenge the top one-third of their students, no publisher had a book to present for adoption, and suggested instead the reissuing of textbooks from the late 1960s. The term, "dumbing down of textbooks" was used in 1984 by Terrell Bell, the Secretary of Education during Reagan's first term, as a reaction to *A Nation at Risk*, the well-known national report that was extremely critical of our American schools and provided the impetus for most current educational reform initiatives. However, the problem with the use of easier textbooks did not begin in 1984. A thorough examination (Reis & Westberg, 1994) of the dumbing down problems with textbooks clearly reveals that the trend towards easier materials began in the 1920s when the vocabulary began to decrease because more children of immigrants and the poor began entering schools. Fewer and fewer new words were introduced in textbooks and the words that were introduced were repeated more and more often to meet the needs of a changing population. This trend continued in the 1930s through the 1950s in both reading and social studies. Jeanne S. Chall, director of Harvard University's reading laboratory, has repeatedly questioned the practice of using reading textbooks that were too easy because they were matched to students' *grade placement* and not their reading levels as the problems of less challenging textbooks continued (Chall & Conard, 1991). The problem of dumbing down has been discussed in all content areas, including reading, social studies, mathematics, and science. Some educators and researchers believe that publishers have been forced to make changes to meet state or local readability formulas and to conform to meet the needs of all students in a single text and to reflect the interests of all special interest groups. Such pressure and resulting policies result in textbooks that "mention" much but cover little in depth. Since most schools currently use heterogeneous grouping for instructional purposes, it is clear that textbooks must be written so that the less able students at each grade level can understand them. A recent study (Westberg, Archambault, Dobyns, & Salvin, 1993) of 46 classrooms across the country that were each observed for two days revealed that heterogeneous grouping was used for instructional purposes 79% of the time. What, then, happens to our most advanced students? Imagine, for example, the frustration faced by a precocious reader who enters kindergarten reading at a relatively advanced level and spends the next year learning the letters of the alphabet and beginning letter blends.

Despite parental frustrations, federal reports, and a host of criticisms, textbooks do not seem to have improved. In all content areas, textbooks continue to be accused of superficiality, lack of context, repetition, and limitations in concept learning and connections with the real world. International comparisons provide a sobering reality: American students are not learning at the same pace and depth as their European and Asian counterparts. In Taiwan and Japan, fifth graders study motion problems and elementary algebra (negative numbers, first-degree equations). In Holland, practice in multiplication and division computation are considered completed in third grade. Yet in our country, instruction in addition and subtraction are repeated during every elementary school year, and in many school districts, students are still learning to add and subtract in secondary school. And unfortunately, many studies indicate that textbooks dominate classroom instruction, constituting 75-90% of instructional time (Goodlad, 1984).

Repetition of content and instruction is present in most textbooks, and as a result, little challenge exists for high ability students. In elementary math textbooks, for example, content changes little as students progress through the grades. Content overlaps across grades because topics begun at the end of one grade are continued into the beginning of the next grade. Flanders (1987) investigated three separate mathematics textbook series to examine just how much new content is presented each year. He found that a relatively steady decrease occurs in the amount of new content over the years up through eighth grade, where less than one-third of the material taught is new to students. Overall, students in grades two to five encounter approximately 40% to 65% new content, an equivalent of new material just two or three days per week. By eighth grade, this amount has dropped to 30%, just one and one-half days per week. Flanders found that most of the new content in any text is found in the second half of the book. In grades seven to eight, where the total new content is lowest, new material occurs in less than 28% of the first half of the books. Flanders's study shows that the mathematical content of some textbooks is mostly review of previous topics: "The result is that earlier in the year, when students are likely to be more eager to study, they repeat what they have seen before. Later on, when they are sufficiently bored, they see new material—if they get to the end of the book." He elaborates, "There should be little wonder why good students get bored: they do the same thing year after year. Average or slower-than-average students get the same message, and who could blame them for becoming complacent about their mathematics studies? They know that if they don't learn it now, it will be retaught next year." The problem, however, does not exist solely in mathematics textbooks. In a recent study (Taylor & Frye, 1988) above average readers scored 93% on pretests given to them on comprehension skills *before* these skills were covered in the basal readers.

What Happens to High Ability Students in Classrooms Across the Country

Recently, three studies conducted by the University of Connecticut, site of the National Research Center on the Gifted and Talented, have analyzed what occurs in American classrooms for high ability students and the results portray a disturbing pattern. The Classroom Practices Survey (Archambault et al., 1993) was conducted to determine the extent to which gifted and talented students receive differentiated education in regular classrooms. Approximately 7,300 third- and fourth-grade teachers in public and private schools were randomly selected to participate in this research, and over 51% of this national sample of classroom teachers responded to the survey. Sixty-one percent of public school teachers and 54% of private school teachers reported that they had **never** had any training in teaching gifted students. The major finding of this study is that classroom teachers make only **minor** modifications in the regular curriculum to meet the needs of gifted students. This result was consistent for all types of schools sampled and for classrooms in various parts of the country and for various types of communities.

The Classroom Practices Observational Study (Westberg et al., 1993) examined the instructional and curricular practices used with gifted and talented students in regular elementary classrooms throughout the United States. Systematic observations were conducted in 46 third- or fourth-grade classrooms. The observations were designed to determine if and how classroom teachers meet the needs of gifted students in the regular classroom. Two students, one high ability student and one average ability student, were selected as target students for each observation day and the types and frequencies of instruction that both students received through modifications in curricular activities, materials, and teacher-student verbal interactions were documented by trained observers. The results indicated little differentiation in the instructional and curricular practices, including grouping arrangements and verbal interactions, for gifted students in the regular classroom. In all content areas in 92 observation days, gifted students rarely received instruction in homogeneous groups (only 21% of the time), and more alarmingly, the target gifted students experienced **no instructional or curricular differentiation in 84% of the instructional activities in which they participated**. The daily summaries of these observations completed by the trained observers were also examined. The most dominant theme in this content analysis involved the use of identical practices for all targeted students. For example, phrases such as "no purposeful differentiation" appeared on 51 of the 92 daily summaries. Anecdotal summaries such as the following provided poignant glimpses into the daily experiences of high ability students: "It should be noted that the targeted gifted student was inattentive during all of her classes. She appeared to be sleepy, never volunteered, and was visibly unenthusiastic about all activities. No attempt was made to direct higher order thinking skills questions to her or to engage her in more challenging work. She never acted out in any way."

A third study, the Curriculum Compacting Study (Reis et al., 1993) examined the effects of using a plan called curriculum compacting to modify the curriculum and eliminate previously mastered work for high ability students. The work that is eliminated is content that is repeated from previous textbooks or content that may be new in the curriculum but that some students already know. Four hundred and thirty-six teachers participated in this study as did 783 students. Students took the next chronological grade level Iowa Test of Basic Skills in both October and May. When classroom teachers in the group eliminated between 40 and 50% of the previously mastered regular curriculum for high ability students, no differences were found between students whose work was compacted and students who did *all* the work in reading, math computation, social studies, and spelling. In science and math concepts, students whose curriculum was compacted scored significantly higher than their counterparts in the control group. Accordingly teachers could eliminate as much as 40-50% of material without detrimental effects to achievement scores. And in some content areas, scores were actually higher when this elimination of previously mastered content took place!

Current Educational Trends Having an Impact on Talented Students

In addition to these recent research studies, two other current trends in education may have a negative impact on schooling for high ability students. First, the current movement to eliminate "tracking" students into classes that predetermine what they will learn has resulted in a widespread belief that absolutely all forms of grouping have a negative impact on children. We define tracking as the relatively permanent (at least for the school year) placement of students into a class or group for students of a certain level. Grouping, on the other hand, is viewed as a less permanent arrangement. Students may be grouped by interests, by mastery of a subject as measured by a pretest, or by ability to do advanced work in one or more content areas. While few, if any, educators support the use of tracking, recent research indicates that some forms of ability or instructional grouping are necessary if advanced content is to be provided to high ability students (Kulik, 1992; Rogers, 1991). In fact, most teachers must use some form of flexible instructional grouping if they are to provide appropriate levels of challenge and instruction for the wide range of abilities and interests that exist in their classrooms. Eliminating all math groups and all reading groups results in a "one size fits all" curriculum that is usually tailored to students in the middle of the class, or worse yet, to students who achieve at the lowest level in the classroom. When this occurs, the absence of instructional grouping can have further detrimental effects on high ability students.

The second movement that may be detrimental to the needs of gifted students is cooperative learning, in which groups of students work together on assigned classwork. Cooperative learning is a sound pedagogical strategy that enables youngsters to work in small groups on projects, assignments, and sometimes self-selected study topics. It has been used for decades and has many benefits for all students. Recently, however, some types of cooperative learning may have had a negative impact on high ability students when used for long periods of instruction. For example, in one form of cooperative learning, one bright child, two average children, and one below average student are placed into a group. Within these groups an assumption is made that the smart student will help the others and that all will benefit. Unfortunately, some bright students are not interested in teaching others, and some cannot even explain how they have acquired advanced concepts. And what about the right of an advanced student to be challenged, to work at an appropriate pace in content with appropriate levels of challenge?

The Ramifications of the Lack of Challenge for Gifted Students

Three ramifications clearly exist for gifted students in our nation's schools.

First, they are clearly underchallenged, and therefore, their development is delayed or even halted. If instructional materials are not above the students' level of knowledge or understanding, learning is less efficient and intellectual growth may stop. It is, for example, not surprising to find a very bright first grader in an urban school who reads on a fifth-grade level but who is reading only slightly above grade level when he/she enters fifth grade. Second, too many of our brightest students never learn to work and, consequently, acquire poor work habits. In a current study (Reis, Hébert, Diaz, Maxfield & Rattley, in press) conducted on 35 high ability urban students who underachieve in school, students have provided insight into why they are doing poorly in high school, blaming an elementary school program that is too easy. One student's comment is representative of the attitude of the majority of students involved in the study: "Elementary school was fun. I always got As on my report card. I never studied when we were in class and I never had to study at home."

The problem of not learning to work does not, however, exist only in urban areas. A recent study of America's highest achieving students conducted by Who's Who Among American High School Students found that most high achieving students study an hour or less a day. The current federal report, *National Excellence,* also cites statistics indicating that "... our top-performing students are undistinguished at best and poor at worst when compared with top students in other countries." For example, a recent study, which is cited in *National Excellence,* compares U.S. seniors taking Advanced Placement (AP) courses in science with top students in 13 other countries. U.S. students represented the top 1% of students in the nation in this study, which found that American students were:

- 13 out of 13 in biology;
- 11 out of 13 in chemistry; and
- 9 out of 13 in physics.

When controlled for selectivity (a higher percentage of the total school population in other countries takes advanced classes), American students scored the lowest of the participating nations in all three areas. In mathematics, the top 1% of students in the United States scored very poorly when compared to a similar group of students in 13 countries:

- 13 out of 13 in algebra and
- 12 out of 13 in geometry and calculus.

These sobering statistics may provide one explanation of why graduate school enrollments of American students in mathematics and science have substantially declined in the last two decades and the number of foreign-born students enrolled in graduate school has increased. In 1990, 57% of the doctorates granted in mathematics in the United States went to students from other countries (O'Connell-Ross, 1993).

Identification of High Ability Students

Currently many gifted programs use a modified multiple criteria approach, which usually involves a placement team. The team first decides on a definition, target population and programming model, then a screening procedure is selected and, next, identification instruments (tests, checklists, etc.) are chosen for the final selection process. Some state guidelines require a minimum group or individual IQ score for students to be placed in an academic gifted program; however, most states currently have policies requiring the use of multiple criteria for the identification of talented students. State department personnel should be contacted before extensive work is completed on an identification process that may not comply with state guidelines.

The first step in identification should always be to ask a simple question: identification for what? For what type of program or experience is the youngster being identified? If, for example, an arts program is being developed for talented artists, the resulting identification system must be structured to locate youngsters with either demonstrated or potential talent in art. Therefore, IQ and achievement tests would *not* be appropriate for identifying this population.

With recent expanded conceptions of giftedness, the use of multiple criteria for identification became popular. Using multiple criteria generally means that at least three appropriate criteria will be used in the identification process. For example, a pool of eligible students may be identified by eliminating from consideration all students who score below the 95th percentile on national norms. The next step might be to gather information on this group including: teacher ratings, creativity tests, grades, and evidence of task commitment. The final step might be the administration of an individually administered IQ test. Students scoring over 132 (as determined by some state guidelines) will then be included in the program. One might reasonably ask why the effort was made to gather other information if the *final* decision was based solely on the results of an IQ test. What is *gathered* should be used.

The adoption of the federal definition as state and local policy throughout our country has caused problems in the process of identification. Although the definition clearly states that children capable of high performance may display high achievement or potential in any one of several areas, practice often sets standards that are not implied in the federal definition. For example, many states include high achievement as a criterion for identification. Accordingly, students who are not achieving in the regular classroom and may not be recommended by classroom teachers often will not qualify for admission into the gifted program. Some districts develop a matrix that includes all of the criteria listed in the 1972 federal definition (except psychomotor ability, which was eliminated in 1978). Students are assessed by weighted scores on all the criteria, thus students with outstanding performance in only one academic area or students who may be underachieving are effectively eliminated. This practice is an ineffective implementation of a definition that was designed to be flexible, but perhaps the newer definition will result in more flexibility.

Several states and many local districts mistakenly defend their identification process, insisting that it is based on multiple criteria including variables such as group achievement test scores, group IQ scores, teacher nomination, parent nomination, student interviews, grades, classroom performance, and creativity. After gathering all of this data, students are sent to a school psychologist, who administers an individual IQ test. If the student's score is below the predetermined state or local cut-off score, the youngster is refused admission to the gifted program. This practice is logically flawed. If the youngster has extremely high scores or recommendations on all or many of the other criteria, the multiple criteria are being ignored in favor of a single indicator, an individually administered IQ test. This practice often eliminates students who may not perform well on standardized IQ tests. Disadvantaged, culturally different, bilingual, or minority students may be excluded from taking part in a gifted program because of tests that may be culturally biased against these groups. If we have any doubt that there is more involved in defining giftedness than IQ alone, this practice must be characterized as arbitrary. Recently, Renzulli (1994) has advocated an alternative to traditional identification that includes the use of an instrument called the Total Talent Portfolio in which educators view special programs as opportunities for talent development and identification of behavioral characteristics as the focus of our efforts. Renzulli urges us to discontinue the use of statements such as "Sara is a gifted child," and instead, concentrate on behavioral characteristics such as "Sara is a third grader who reads at a ninth-grade level and has a fascination for science projects."

Underserved Populations in Gifted Education

Unfortunately, the majority of young people participating in gifted and talented programs across the country continues to represent the majority culture in our society. Few doubts exist regarding the reasons that economically disadvantaged and other minority group students are underrepresented in gifted programs. For example, Frasier and Passow (1994) indicate that identification and selection procedures may be ineffective and inappropriate for the identification of these young people. They also indicate that limited referrals and nominations of students who are minorities or from other disadvantaged groups affect their eventual placement in programs. Test bias and inappropriateness have been mentioned as a reason for the continued reliance on traditional identification approaches. Groups that have been traditionally underrepresented in gifted programs could be better served, according to Frasier and Passow (1994), if the following elements are considered: new constructs of giftedness, attention to cultural and contextual variability, the use of more varied and authentic assessment, performance identification, identification through learning opportunities, and attention to both absolute attributes of giftedness, the traits, aptitudes, and behaviors universally associated with talent as well as the specific behaviors that represent different manifestations of gifted potential and performance as a consequence of the social and cultural contexts in which they occur (p. xvii).

In addition to students from economically disadvantaged populations, various minority and cultural groups, as well as gifted students with various disabilities such as learning disabilities, visual and hearing impairments, and physical handicaps, another group of students who are traditionally underrepresented in gifted programs is females who have potential in mathematics and science, as well as gifted females who achieve in school but later underachieve in life (Reis, 1987). Special programs, strategies, and identification procedures have been suggested for many of these groups; however, much progress still remains to be made to achieve equity for these underrepresented groups.

Promising Programs and Exemplary Practices

In the last decade many promising practices have been implemented in the education of gifted and talented students. More primary and secondary programs have been developed since the first programming model, the Enrichment Triad Model (Renzulli, 1977) was developed for gifted students. Other programming models such as the Purdue Three-Stage Enrichment Model (Feldhusen & Kolloff, 1986), Talents Unlimited (Schlichter, 1986), and the Autonomous Learner Model (Betts, 1986) are also widely used throughout the country. National programs such as Future Problem Solving, which was conceived by E. Paul Torrance, have taught hundreds of thousands of students to apply problem-solving techniques to the real problems of our society. Although not developed solely for gifted students, Future Problem Solving is widely used in gifted programs because of the curricular freedom associated with these programs.

The Future Problem Solving Program is a year-long program in which teams of four students use a six-step problem-solving process to solve complex scientific and social problems of the future such as the overcrowding of prisons or the greenhouse effect. At regular intervals throughout the year, the teams mail their work to evaluators, who review it and return it with their suggestions for improvement. As the year progresses, the teams become increasingly more proficient at problem solving. The Future Problem Solving Program takes students beyond memorization. The program challenges students to apply information they have learned to some of the most complex issues facing society. They are asked to *think,* to make decisions, and, in some instances, to carry out their solutions.

A national program called Talent Search actively recruits and provides testing and program opportunities for mathematically precocious youth. Talent Search is an annual effort to identify 12–14 year old students who score in the top 5% of the country in mathematics on the SAT math test. These students generally have scored highly in other standardized tests and are recommended by teachers or counselors to take the SAT–Math. If they do well on this test, they are eligible for multiple options including summer programs, grade skipping, completing two or more years of a math course in one year, taking college courses, or other options. Eleven states have created separate schools for talented students in math and science such as the North Carolina School for Math and Science. Some large school districts have established magnet schools to serve the needs of talented students. In New York City, for example, the Bronx High School of Science has helped to nurture and develop mathematical and scientific talent for decades, producing internationally known scientists and Nobel laureates. In other states, Governor's Schools provide advanced, intensive summer programs in a variety of content areas. It is clear, however, that these opportunities touch a small percentage of students who could benefit from them.

Within the schools that have gifted programs, limited options often exist. Resource room programs in which a student leaves his/her regular classroom and spends a limited amount of time doing independent study or becoming involved in advanced research in a resource room for gifted students with a teacher are commonly found. Independent study projects provide talented students with opportunities to engage in pursuing individual interests and advanced content. Many local districts have created innovative mentorship programs that pair a bright student with a high school student or adult who has an interest in the same area as the student. Some schools use cluster grouping which allows students who are gifted in a certain content area to be grouped in one classroom with other students who are talented in the same area. Therefore, one fifth-grade teacher may have six students who are advanced in mathematics in a classroom instead of having these six students distributed among four different fifth-grade classrooms. Some schools acknowledge that they can do little that is different for gifted students within the school day and provide afterschool enrichment programs or send talented students to Saturday programs offered by museums, science centers, or local universities. Unfortunately, many of these promising strategies seem insignificant when compared with the plight of thousands of bright students who still sit in classrooms across the country bored, unmotivated, and unchallenged.

Acceleration, once a standard practice in our country, is often dismissed by teachers and administrators as an inappropriate practice for a variety of reasons, including scheduling problems, concerns about the social effects of grade skipping, and others. Previously used forms of acceleration, including enabling precocious students to enter kindergarten or first grade early, grade skipping, and early entrance to college, are not commonly used or encouraged by most school districts. And in many schools, the pervasive influence of anti-intellectualism that affects our society has a two-pronged effect. First, policymakers do little to encourage excellence in our schools and less and less attention is paid to intellectual growth. Second, peer pressure is exerted on gifted students. The labels such as "smarty-pants" commonly used to describe bright students in the 1950s and 1960s has been replaced by more negative labels such as "nerd," "dweeb," or "dork." Our brightest students often learn not to answer in class, to stop raising their hands and to minimize their abilities to avoid peer pressures.

A number of challenging curriculum options in science and language arts have been developed under the auspices of the federal Javits Education Act mentioned earlier. Several national programs have been developed or implemented for high ability students in many districts, regional service centers, and states. In addition to Future Problem Solving, programs such as Odyssey of the Mind, a national program in which teams of students use creative problem solving to design structures, vehicles, and solutions to problems such as designing a vehicle that uses a mousetrap as its primary power source. Many high ability students have the opportunity to participate in History Day in which students work individually or in small groups on an historical event, person from the past, or invention related to a theme that is determined each year. Using primary source data including diaries or other resources gathered in libraries, museums, and interviews, students prepare research papers, projects, media presentations, or performances for entry. These entries are judged by local historians, educators, and other professionals and state finalists compete with winners from other states each June. Information about History Day can be obtained from state historical societies. These and other model projects such as mentorships, Saturday programs, summer internships, and computer camps that are of extremely high quality continue to be implemented.

Unfortunately, major alternative strategies involving the regrouping or radical reorganization of school structures have been slow in coming, perhaps due to the difficulty of making major educational changes, because of scheduling, finances, and other issues that have caused schools to substantively change so little in the last half century. Because of this delay, gifted students often do not receive classroom instruction based on their unique needs that place them far ahead of their chronological peers in basic skills and verbal abilities and that enables them to learn much more rapidly and tackle much more complex material than their peers. Our most able students need appropriately paced, challenging instruction and, instead, are often held back in school.

Applying "Gifted" Education Pedagogy to Develop Talents in All Students

Much that has been learned and developed in gifted programs can offer exciting, creative alternatives in instruction and curriculum for all students. A rather impressive menu of exciting curricular adaptations, independent study and thinking skill strategies, grouping options, and enrichment strategies have been developed in gifted programs that could be used to improve schools. The Schoolwide Enrichment Model (Renzulli & Reis, 1985) has been field tested and implemented by hundreds of school districts across the country for the last nine years. Our experiences with schoolwide enrichment led us to realize that when an effective approach to enrichment is implemented, all students in the school benefit and the entire school begins to improve. This led to the development of *Schools for Talent Development* (Renzulli, 1994). This approach seeks to apply strategies used in gifted programs to the entire school population, emphasizing talent development in *all* students through a variety of acceleration and enrichment strategies that have been discussed earlier. Not all students can, of course, participate in all advanced opportunities but many can work far beyond what they are currently asked to do. It is clear that our most advanced students need different types of educational experiences than they are currently receiving and that without these services, talents may not be nurtured in many American students, especially those who attend schools in which survival is a major daily goal.

Reform, restructuring, and innovation are not just the educational catchwords of the 1990s. Efforts to change and improve education have been around for decades, if not centuries; and they will undoubtedly be around as long as thoughtful people have the courage, creativity, and vision to look for better ways of solving the endless array of problems that a changing culture and society places on the doorsteps of the school. But amidst all of the restructuring efforts, we cannot forget talent development in our schools.

The result of the dumbing down of the curriculum and the proliferation of basic skills practice material may result in the creation of the largest percentage of high ability underachievers in the history of public schools in America. Many of these bright students will learn at a very early age that if they do their best in school, they will be rewarded by endless pages of more of the same kind of practice materials. These same young people may also learn that if they display their abilities in a heterogeneous classroom the result may be ridicule from peers and the attainment of one of a multitude of nicknames including: brain, nerd, dweeb, and/or others. Consider the following quotation written by a high school student in support of homogeneous grouping and gifted programs for high ability students:

In my 12 years in Torrington Schools, I have been placed in many "average" classes—especially up until the junior high school level—in which I have been spit on, ostracized, and verbally abused for doing my homework on a regular basis, for raising my hand in class, and particularly for receiving outstanding grades (Peters, 1990).

Our field's pedagogy and innovative programs admittedly will not provide quick-fix solutions to the organizational questions raised by the current school reform movement, but it can offer numerous creative alternatives regarding instruction and curriculum. In our relatively short history we have achieved a rather impressive menu of exciting curricular adaptations, thinking skills applications, methods for teaching independent study, and numerous other innovations. For example, specialists in the area of education of the gifted have concentrated on identifying student interests and learning styles and providing relevant and challenging curricular experiences to individual students instead of identical experiences to 30 students in a classroom without consideration of their previous knowledge or background.

Specialists in the area of gifted education have also gained expertise in adjusting the regular curriculum to meet the needs of advanced students in a variety of ways including: accelerating content, incorporating a thematic approach, and substituting more challenging textbooks or assignments. The present range of instructional techniques used in most classrooms observed by Goodlad (1984) and his colleagues is vastly different than what is recommended in many gifted programs today. The flexibility in grouping that is encouraged in many gifted programs might also be helpful in other types of educational settings.

We can, therefore, make every attempt to share with other educators the technology we have gained in teaching students process skills, modifying the regular curriculum, and helping students become producers of knowledge (Renzulli, 1977). We can extend enrichment activities and provide staff development in the many principles that guide our programming models. Yet, without the changes at the local, state, and national policy-making levels that will alter the current emphasis on raising test scores and purchasing unchallenging, flat, and downright sterile textbooks, our efforts may be insignificant.

Maintaining Our Identity

Because of fiscal constraints, more gifted programs are being eliminated and fewer students are being challenged by these programs. Consider the following correspondence received from a classroom teacher with 10 years of teaching experience and a graduate degree in education of the gifted and talented.

My frustration at not being able to adequately challenge the gifted students in my heterogeneous classroom grows each year. With 28 students of varying levels and abilities and special needs, I often find the most neglected are the brightest. Even though I know what to do for these youngsters, I simply do not have the time to provide the differentiated instruction they need and deserve. Instead, my attention shifts, as it has in the past, to the students in my class with special needs.

While sharing our technology is, indeed, one of our goals, we must continue to create and maintain exemplary programs and practices that serve as models of what can be accomplished for high ability students. Through our professional organizations we must continue to advocate the different needs of high ability students. We must argue logically and forcefully to maintain the programs, the equitable grouping practices, and the differentiated learning experiences that the students we represent so desperately need. To simply allow these youngsters to be placed in classrooms in which no provisions will be made for their special needs is an enormous step backwards for our field. To lose our quest for excellence in the current move to guarantee equity will undoubtedly result in a disappointing, if not disastrous, education for our most potentially able children.

A Change in Direction: From Being Gifted to the Development of Gifted Behaviors

While we believe it is imperative to maintain our identity through our programs and professional organizations, we advocate, as we have in the past (Renzulli, 1980), a slight change in our labeling processes. Up to this time, the general approach to the study of gifted persons could easily lead the casual reader to believe that giftedness is an absolute condition that is magically bestowed upon a person in much the same way that nature endows us with blue eyes, red hair, or a dark complexion. This position is not supported by the research. For too many years we have pretended that we can identify gifted children in an absolute and unequivocal fashion. Many people have been lead to believe that certain individuals have been endowed with a golden chromosome that makes him or her "a gifted person." This belief has further led to the mistaken idea that all we need to do is find the right combination of factors that prove the existence of this "gift." The further use of terms such as "the truly gifted," "the highly gifted," the "moderately gifted," and the "borderline gifted" only serves to confound the issue because they invariably hearken back to a conception of giftedness that equates the concept with test scores. The misuse of the concept of giftedness has given rise to a great deal of criticism and confusion about both identification and programming, and the result has been that so many mixed messages have been sent to educators and the public at large that both groups now have a justifiable skepticism about the credibility of the gifted education establishment and our ability to offer services that are qualitatively different from general education.

Most of the confusion and controversy surrounding the definitions of giftedness that have been offered by various writers can be placed into proper perspective if we examine a few key questions. Is giftedness an absolute or relative concept? That is, is a person either gifted or not gifted (the absolute view), or can varying degrees of gifted behaviors be developed in certain people, at certain times, and under certain circumstances (the relative view)? Is gifted a static concept (i.e., you have or you don't have it) or is it a dynamic concept (i.e., it varies within person and learning/performance situations)?

These questions have led us to advocate a fundamental change in the ways the concept of giftedness should be viewed in the future. Except for certain functional purposes related mainly to professional focal points (i.e., research, training, legislation) and to ease-of-expression, we believe that labeling students as "the gifted" is counterproductive to the educational efforts aimed at providing supplementary educational experiences for certain students in the general school population. We believe that our field should shift its emphasis from a traditional concept of "being gifted" (or not being gifted) to a concern about the development of gifted behaviors in those youngsters who have the highest potential for benefiting from special educational services. This slight shift in terminology might appear insignificant, but we believe that it has implications for the entire way that we think about the concept of giftedness and the ways in which we should structure our identification and programming endeavors. This change in terminology may also provide the flexibility in both identification and programming endeavors that will encourage the inclusion of at-risk and underachieving students in our programs. If that occurs, not only will we be giving these high potential youngsters an opportunity to participate, we will also help to eliminate the charges of elitism and bias in grouping that are sometimes legitimately directed at some gifted programs.

We cannot forget that our schools should be places that seek to develop talents in children. We won't produce future Thomas Edisons or Marie Curies by forcing them to spend large amounts of their science and mathematics classes tutoring students who don't understand the material. A student who is tutoring others in a cooperative learning situation in mathematics may refine some of his or her basic skill processes, but this type of situation does not provide the level of challenge necessary for the most advanced types of involvement in the subject, nor for inspiring our young people to strive to develop their talents.

References

Albert, R.S., & Runco, M.A. (1986). The achievement of eminence: A model based on a longitudinal study of exceptionally gifted boys and their families. In R.J. Sternberg & J.E. Davidson (Eds.), *Conceptions of giftedness* (pp. 332–360). New York: Cambridge University Press.

Alvino, J., McDonnel, R.C., & Richert, S. (1981). National survey of identification practices in gifted and talented education. *Exceptional Children, 18*, 124–132.

Archambault, F.X., Westberg, K.L., Brown, S., Hallmark, B.W., Emmons, C., & Zhang, W. (1993). *Regular classroom practices with gifted students: Results of a national survey of classroom teachers* (Research Monograph No. 93102). Storrs, CT: The National Research Center on the Gifted and Talented.

Bamberger, J. (1986). Cognitive issues in the development of musically gifted children. In R.J. Sternberg & J.E. Davidson (Eds.), *Conceptions of giftedness* (pp. 388–413). New York: Cambridge University Press.

Betts, G. (1986). The autonomous learner model for the gifted and talented. In J.S. Renzulli (Ed.), *Systems and models for developing programs for the gifted and talented* (pp. 27–56). Mansfield Center, CT: Creative Learning Press.

Bloom, B.S. (Ed.). (1985). *Developing talent in young people*. New York: Ballantine Books.

Cattell, R.B. (1971). *Abilities: Their structure, growth, and action*. Boston: Houghton Mifflin.

Chall, J.S., & Conard, S.S. (1991). *Should textbooks challenge students? The case for easier or harder textbooks*. New York: Teachers College Press.

Council of State Directors of Programs for the Gifted. (1987). *The 1987 State of the States Gifted and Talented Education Report*.

Council of State Directors of Programs for the Gifted. (1994). *The 1994 State of the States Gifted and Talented Education Report*.

Csikszentmihalyi, M., & Robinson, R.E. (1986). Culture, time and the development of talent. In R.J. Sternberg & J.E. Davidson (Eds.), *Conceptions of giftedness* (pp. 264–284). New York: Cambridge University Press.

Feldhusen, J., & Kolloff, P.B. (1986). The Purdue three-stage enrichment model for gifted education at the elementary level. In J.S. Renzulli (Ed.), *Systems and models for developing programs for the gifted and talented* (pp. 126–152). Mansfield Center, CT: Creative Learning Press.

Feldman, D.H. (1986). Giftedness as a developmentalist sees it. In R.J. Sternberg & J.E. Davidson (Eds.), *Conceptions of giftedness* (pp. 285–305). New York: Cambridge University Press.

Flanders, J.R. (1987). How much of the content in mathematics textbooks is new? *Arithmetic Teacher, 35*, 18–23.

Frasier, M.M., & Passow, A.H. (1994). *Toward a new paradigm for identifying talent potential* (Research Monograph No. 94112). Storrs, CT: The National Research Center on the Gifted and Talented.

Gallagher, J.J. (1979). Issues in education for the gifted. In A.H. Passow (Ed.), *The gifted and the talented: Their education and development* (pp. 28–44). Chicago: University of Chicago Press.

Gallagher, J.J., Weiss, P., Oglesby, K., & Thomas, T. (1983). *The status of gifted/talented education: United States surveys of needs practices and policies*. Ventura, CA: Office of the Superintendent of Ventura County Schools.

Gardner, H. (1983). *Frames of mind*. New York: Basic Books.

Goodlad, J. (1984). *A place called school*. New York: McGraw Hill.

Gruber, H.E. (1986). The self-construction of the extraordinary. In R.J. Sternberg & J.E. Davidson (Eds.), *Conceptions of giftedness* (pp. 247–263). New York: Cambridge University Press.

Guilford, J.P. (1959). Three faces of intellect. *American Psychologist, 14*, 469–479.

Jenkins-Friedman, R. (1982). Myth: Cosmetic use of multiple selection criteria. *Gifted Child Quarterly, 26*(1), 24–26.

Karnes, F.A., & Collins, E.C. (1978). State definitions on the gifted and talented. *Journal for the Education of the Gifted, 1*(1), 44–62.

Kirst, M.W. (1982). How to improve schools without spending more money. *Phi Delta Kappan, 64*(1), 6–8.

Kulik, J.A. (1992). *An analysis of the research on ability grouping: Historical and contemporary perspectives* (RBDM No. 9204). Storrs, CT: The National Research Center on the Gifted and Talented.

Marland, S.P., Jr. (1972). *Education of the gifted and talented: Vol. 1. Report to the Congress of the United States by the U.S. Commissioner of Education.* Washington, DC: U.S. Government Printing Office.

O'Connell-Ross, P. (1993). *National Excellence, a Case for Developing America's Talent.* Washington, DC: U.S. Department of Education, Government Printing Office.

Peters, P. (1990, July). TAG student defends programs against critic [Letter to the editor]. *The Register Citizen,* p. 10.

Purcell, J. (1994). *The status of programs for high ability students* (CRS 94305). Storrs, CT: The National Research Center on the Gifted and Talented.

Purcell, J. (1993). The effects of the elimination of gifted and talented programs on participating students and their parents. *Gifted Child Quarterly, 37*(4), 177–187.

Reis, S.M. (1987). We can't change what we don't recognize. *Gifted Child Quarterly, 31*(2), 83–89.

Reis, S.M., Hébert, T., Diaz, E., Maxfield, L., & Rattley, M. (in press). Creating culture of achievement: Case studies of talented urban students. Storrs, CT: The National Research Center on the Gifted and Talented.

Reis, S.M., Westberg, K.L., Kulikowich, J., Caillard, F., Hébert, T., Plucker, J., Purcell, J.H., Rogers, J.B., & Smist, J.M. (1993). *Why not let high ability students start school in January?* (Research Monograph No. 93106). Storrs, CT: The National Research Center on the Gifted and Talented.

Reis, S.M., & Westberg, K.L. (1994). An examination of current school district policies: Acceleration of secondary students. *Journal of Secondary Gifted Education, 5*(4), 7–17.

Renzulli, J.S. (1977). *The enrichment triad model.* Mansfield Center, CT: Creative Learning Press.

Renzulli, J.S. (1978). What makes giftedness? Reexamining a definition. *Phi Delta Kappan, 60*(5), 180–184.

Renzulli, J.S. (1980). Will the gifted child movement be alive and well in 1990? *Gifted Child Quarterly, 24*(1), 3–9.

Renzulli, J.S. (1986a). The three-ring conception of giftedness: A developmental model for creative productivity. In R.J. Sternberg & J.E. Davidson (Eds.), *Conceptions of giftedness* (pp. 53–92). New York: Cambridge University Press.

Renzulli, J.S. (Ed.). (1986b). *Systems and models for developing programs for the gifted and talented.* Mansfield Center, CT: Creative Learning Press.

Renzulli, J.S. (1994) *Schools for talent development.* Mansfield Center, CT: Creative Learning Press.

Renzulli, J.S., & Delcourt, M.A.B. (1986). The legacy and logic of research on the identification of gifted persons. *Gifted Child Quarterly, 30*(1), 20–23.

Renzulli, J.S., & Reis, S.M. (1985). *The schoolwide enrichment model: A comprehensive plan for educational excellence.* Mansfield Center, CT: Creative Learning Press.

Rogers, K.B. (1991). *The relationship of grouping practices to the education of the gifted and talented learner* (RBDM No. 9102). Storrs, CT: The National Research Center on the Gifted and Talented.

Schlichter, C.L. (1986). Talents unlimited: Applying the multiple talent approach in mainstream and gifted programs. In J.S. Renzulli (Ed.), *Systems and models for developing programs for the gifted and talented* (pp. 352–390). Mansfield Center, CT: Creative Learning Press.

Stanley, J.C., & Benbow, C.P. (1986). Youths who reason exceptionally well mathematically. In R.J. Sternberg & J.E. Davidson (Eds.), *Conceptions of giftedness* (pp. 361–387). New York: Cambridge University Press.

Sternberg, R.J. (1986). A triarchic theory of intellectual giftedness. In R.J. Sternberg & J.E. Davidson (Eds.), *Conceptions of giftedness* (pp. 223–247). New York: Cambridge University Press.

Sternberg, R.J., & Davidson, J.E. (1986). Conceptions of giftedness: A map of the terrain. In R.J. Sternberg & J.E. Davidson (Eds.), *Conceptions of giftedness*. New York: Cambridge University Press.

Tannenbaum, A.J. (1983). *Gifted children: Psychological and educational perspectives.* New York: Macmillan.

Tannenbaum, A.J. (1986). Giftedness: A psychosocial approach. In R.J. Sternberg & J.E. Davidson (Eds.), *Conceptions of giftedness* (pp. 21–52). New York: Cambridge University Press.

Taylor, B.M., & Frye, B.J. (1988). Pretesting: Minimize time spent on skill work for intermediate readers. *The Reading Teacher*, 42(2), 100–103.

Walters, J., & Gardner, H. (1986). The crystallizing experience: Discovering an intellectual gift. In R.J. Sternberg & J.E. Davidson (Eds.), *Conceptions of giftedness* (pp. 306–331). New York: Cambridge University Press.

Westberg, K.L., Archambault, F.X., Dobyns, S.M., & Salvin, T.J. (1993). *Technical report: An observational study of instructional and curricular practices used with gifted and talented students in regular classrooms.* Storrs, CT: The National Research Center on the Gifted and Talented.

Note

Research for this report was supported under the Javits Act Program (Grant No. R206R00001) as administered by the Office of Educational Research and Improvement, US Department of Education. Grantees undertaking such projects are encouraged to express freely their professional judgement. This report, therefore, does not necessarily represent positions or policies of the Government, and no official endorsement should be inferred.

6

Educational Resilience

MARGARET C. WANG and GENEVA D. HAERTEL
Center for Research in Human Development and Education, Temple University, USA

Finding ways to effectively respond to children with special needs is a critical concern of researchers and school practitioners during the current era of school reform. Resilience is an emerging psychological construct that provides a conceptual base for an intervention that calls for responsive classroom, school, and community environments. This chapter overviews a number of studies of populations who were at risk but who managed to "beat the odds" and demonstrate better than expected outcomes, yielding a rich database on risks and protective factors. Resilience-promoting instructional strategies are examined, including student-centered learning, active teaching, adaptive instruction, and metacognition, among others. Also reviewed are resilience-promoting schoolwide strategies, which include provision of a safe and secure school; a positive, academically oriented school culture; smaller educational units; inclusive classrooms and schools; and others. The chapter concludes that resilience holds much promise as a strategy to overcome adversities and facilitate academic and life success for all students.

Introduction

Finding ways to effectively respond to children with special needs is a critical concern of researchers and school practitioners during the current era of school reform. Resilience is an emerging psychological construct that provides a conceptual base for an intervention that calls for responsive classroom, school, and community environments. Developmental psychopathologists have studied a number of populations who were at risk for a wide range of poor developmental outcomes, but who managed to "beat the odds" and demonstrate better than expected outcomes. These research studies provide a rich database on the vulnerabilities, risks, and protective factors that support healthy development among children in challenging circumstances (Rolf, Masten, Cicchetti, Nuechterlein, & Weintraub, 1990). By studying individuals who navigate through stressful circumstances successfully, processes of adaptation can be identified that will guide interventions with others at risk. Ann Masten (1994, p. 21) concludes, "The study of resilience is a hopeful enterprise." It enlightens and inspires efforts to improve challenging life circumstances based upon the experiences of others in similar situations.

Among the special populations studied using the resilience paradigm are children whose families had histories of mental illness (Goldstein, 1990); children of divorced parents (Watt, Moorehead-Slaughter, Japzon, & Keller, 1990); children exposed to high levels of maternal stress (Pianta, Egeland, & Sroufe, 1990); drug-addicted children (Newcomb & Bentler, 1990); children born at medical risk (O'Dougherty & Wright, 1990); children exposed to domestic violence (Straus, 1983); children exposed to early parental death (Brown, Harris, & Bilfulco, 1986); and children reared in poverty (Elder, van Nguyen, & Caspi, 1985). Masten (1994) cites Horowitz (1989), Kopp (1983), and Masten and Garmezy (1985), who identify other studies that focused on children who are born prematurely, those with low birth weight, as well as children suffering from chronic illness and malnutrition. These various populations include children exposed to adverse environmental liabilities (e.g., poverty, domestic violence, high levels of familial stress) and children with particular biological and psychological attributes (e.g., prematurity, chronic illness, drug-addiction).

A developmental model of psychopathology that addresses vulnerability and resistance to disorders has emerged from resilience research in recent years. Studies of resilience identified processes that underlie adaptation and promote successful pathways from early childhood to adulthood. Because some children thrive in difficult circumstances, it suggests that protective mechanisms might be identified and promoted. Constellations of environmental and psychological factors can be identified that contribute to successful adaptation and inform the design of interventions. Family, educational, and community-based practices and policies can be identified that are congruent with findings from studies of resilience and are likely to promote the life success of children with special needs. In particular, educational practices can be identified that are likely to deny versus promote schooling success.

Schools are challenged to serve an increasingly diverse body of students as nations across the world strive to achieve the goal of providing educational opportunities for all citizens, regardless of racial, ethnic, socioeconomic, and cultural differences. Accommodating student diversity has placed complex demands on schools, particularly those in urban communities. In many urban schools, there is a high concentration of students from economically disadvantaged families and a large percentage of them are from diverse ethnic and language backgrounds. Diversity is further increased by the number of children who fail to learn the basic skills in regular school programs, or exhibit unacceptable social behaviors, or low self-esteem which often results in placement in "special" or other categorical programs (Reynolds, 1994). In many schools, in the United States, for example, approximately 50% or more of all students are placed in a special categorical program at some time during their academic career (Reynolds, Zetlin, & Wang, 1993). The use of resilience-promoting practices and policies is one avenue to accommodating the growth of student diversity. In combination with powerful instructional approaches, resilience-promoting strategies can lead to educational success for all students.

This chapter briefly reviews research on resilience, including a definition of the construct, and profiles characteristics of resilient children. The construct of resilience is extended to the educational realm and specific practices and policies that are likely to promote resilience among students at-risk of school failure are identified. The role of resilience-promoting practices in refocusing educational and social policy is discussed.

Definition of Resilience

Psychologists have advanced a number of definitions of resilience. One of the most widely cited definitions, put forth by Masten, Best, and Garmezy (1990), is "[resilience is the] capacity for or outcome of successful adaptation despite challenging or threatening circumstances" (p. 425). Masten et al. describe the adaptation as behavioral and based upon internal states of well-being, effective functioning in the environment, or both. Constructs such as vulnerability, protective factors, adversity, assets, and risk are used to explain the mechanism whereby an individual makes a successful adaptation in the face of challenging circumstances and is regarded as resilient.

In the resilience literature, physical conditions, such as blindness, deafness, and lack of mobility are regarded as vulnerabilities or risk factors; they are not treated as stressors and adversities. A vulnerability is typically an attribute of an individual that makes them more susceptible to a particular threat, such as risk of academic failure. Risk factors are characteristics of a group of individuals that have been associated with undesirable outcomes (e.g., being a member of an impoverished, ethnic-minority family that lives in a neighborhood with frequent criminal activity places children at-risk for a variety of poor developmental outcomes). Stressors and adversities are discrete events that can interfere with an individual's normal performance. Masten (1994) explains that adversities can be natural events or manmade. Examples of adversities are earthquakes, hurricanes, and other acts of God. Manmade adversities include war and other violent acts, including child abuse. Adversities produce stress that frequently overwhelms the resources an individual has available for meeting the challenge.

Children in at-risk circumstances often face multiple challenges and co-occurring risks. Garmezy and Masten (1994) observe that as children are exposed to more risks, the greater the negative influence on their cognitive and socioemotional development. Special and remedial educators can cull from the resilience literature and design interventions that minimize adversity and promote the development of resiliency among children with special needs.

Characteristics of Resilient Children

In the first decade of research on resilience, empirical studies were conducted that identified the attributes, dispositions, and circumstances of individuals who thrived amid adversity. One landmark study conducted by Werner and Smith (Werner, 1989; Werner & Smith, 1992) followed a cohort of children born on the Hawaiian Island of Kauai for 30 years. About 33% of the children were designated as high risk due to exposure to four or more risk factors, including poverty, perinatal stress, family discord, and under-educated parents. Ten percent of these high-risk children were later identified as resilient because of their successful adaptation in childhood and adolescence. The resilient adolescents displayed more responsibility, maturity, achievement motivation, and social connectedness than the non-resilient adolescents. Protective factors were also identified, including good physical health, positive relations with caregivers (more attention and less separation), less family conflict, and exposure to fewer life stressors. Based on this longitudinal study, resilient children were described as "vulnerable but invincible."

Other socioemotional characteristics of resilient children include strong interpersonal skills, responsiveness to others, high activity levels, goal-direction, maintaining healthy expectations, and an internal locus of control (Benard, 1991). Rutter (1990) identified resilient children's high activity level as central to escaping from the adversities and risks that surround them. Children who are proactive and participate in a number of activities and relationships are more likely to find opportunities that promote success. Garmezy (1974) concludes that resilient children possess well-developed self-systems, including an internal locus of control, high self-esteem, high self-efficacy, and autonomy. Peng (1994) used the National Education Longitudinal Study (NELS:88) electronic database to examine differences in resilience rates for children in various environments. He tested eight models combining measures of student attributes, family support, and school environment for each of three community types. Peng concludes there are individual differences in capacity to cope with adverse circumstances. Students with high educational aspirations, high self-esteem, including a more positive and optimistic personality or attitude, and an internal locus of control are more likely to be resilient than students with low scores on these attributes.

Close, supportive family relationships play a key role in fostering resilience (Masten, 1994; Rutter, 1990; Taylor, 1994). Resilient children's temperament and interpersonal skills affect the quality of relationship they experience within their families. Even temperament, high malleability, predictable behavior, mild-to-moderate emotional reactions, approaching rather than withdrawing from novel situations, and a sense of humor are attributes of resilient children that entice positive attention from adults (Rutter, 1990). Rutter has observed that non-resilient children often exhibit behaviors and emotional reactions, such as sloppiness, eating and sleeping irregularities, low malleability, and moodiness that can generate hostility and a lack of support from adults. Some resilient children, however, achieve success by defying family circumstances. Chess (1989), for example, describes "adaptive distancing" as the process that allows resilient children to stand apart from their disordered families and to accomplish their goals. Adaptive distancing may facilitate children's interacting with peers and adults outside the family in constructive ways that contribute to their development and learning.

Resilient children exhibit cognitive skills that mitigate against risk and stress. Rutter's (1990) research on the problem-solving capabilities of resilient children provides evidence of their ability to plan, change their environment, and successfully alter their lives. Other cognitive correlates of resilience include: above-average intelligence, high verbal communication skills, divergent thinking, humor, and an ability to think reflectively about problems (Hauser, Vieyra, Jacobson, & Wertlieb, 1989; Wang, Haertel, & Walberg, 1994a).

Overall, social competence, good problem-solving skills, independence, and goal-direction are among the most commonly cited characteristics of resilient children. This brief review provides groundbreaking findings on what contributes to resilience and makes clear why these children are described as "invincible," "hardy," "invulnerable," and "superkids."

The Importance of Environment

Behavior is the product of an interaction between an individual's genetic make-up, temperament, developmental status, and their environment. All behavior has a context. Masten and Braswell (1991) recognize the contribution of systems theory in identifying the many interrelated systems that influence an individual's behavior. Bronfenbrenner (1979) has advanced an ecological model that identifies the microsystems, mesosystems, and macrosystems in which all individuals exist. From the perspective of systems theory, children's behavior, including the emergence of resilient behaviors, depends upon the interaction among these many interrelated systems. Psychologists, sociologists, and educational researchers have paid increasing attention to the influence of family, school, and community environments when studying children's adaptation to adversity (Rigsby, Reynolds, & Wang, 1995; Wang & Gordon, 1994). The perspectives of individuals from the various environments—parents, siblings, teachers, peers, and social workers—can be brought to bear on children's behaviors, attitudes, and competencies.

Environments differ in terms of the assets and liabilities they present. Assets and liabilities can be tangible commodities, such as the amount and quality of reading material in the home, availability of nutritious food, or the presence of a neighborhood playground. Assets and liabilities can also be services. The availability of medical and psychological assistance or transportation or access to day care are all services whose presence or absence dramatically alters the environment in which a child lives. Still other environmental assets and liabilities have to do with the interpersonal relationships available and their quality. The behaviors of family members, peers, teachers, and other significant adults profoundly influence the environment in which children live. In terms of interpersonal assets, some children live in families with two parents present and little family discord; others are members of extended families where a strong sense of belonging is present. Some children attend schools with well-trained, knowledgeable teachers who believe all children can succeed. Other children participate in social and religious activities with caring adults. Interpersonal liabilities are present in environments characterized by the absence of relationships or the presence of destructive relationships. Examples of children who are reared in environments characterized by interpersonal liabilities include: children who live in families with frequent domestic violence, who attend large high schools where instruction is impersonal, or live in communities where there are few opportunities to participate in after-school programs supervised by trained, caring adults.

Thus, environments differ in the commodities and tangible resources available, in the services available, and in the quality and abundance of interpersonal relationships. These environments provide the context in which children must thrive. Interventions to promote resilience must focus on the entire environmental system and the assets and liabilities that characterize it. Cultural anthropologists have long recognized the importance of kinship in traditional societies. Kinship is the sharing of characteristics or origins (Hawkins & Allen, 1991). It provides a sense of belonging for children and is regarded as essential to acquiring self-identity. Beyond kinship, children participate in a variety of institutions that involve social relationship, including: schools, churches, and community clubs and social organizations. These networks of familial and social relationships are interdependent, although in many communities they have become disconnected and fragmented. To promote resilience the interdependence of these environmental contexts must be acknowledged and re-established.

Educational Resilience: An Emergent Construct

Wang, Haertel, and Walberg (1995a) extended the definition of resilience to educational performance. Educational resilience is defined as the heightened likelihood of success in school and in other aspects of life, despite environmental risks and adversities, brought about by an individual's disposition, conditions, and experiences. Increased understanding of the lives and educational potential of children who are members of a family beset by adverse circumstances and thereby placed at risk of academic failure, dropping out, or graduating from high school ill-prepared for employment or future learning can be gleaned from studies of educationally resilient children, schools, and communities (Wang & Gordon, 1994; Werner & Smith, 1992).

The Appendix presents an ecological framework that can guide research and program development efforts to foster educational resilience. For each of three contexts (the family, the school, and the community) the following information is presented: potential risk factors, adversities, and vulnerabilities; protective factors that mitigate against school failure; and indicators that can be used to verify the effectiveness of resilience-promoting interventions.

The Appendix was prepared to address the need for research on educational resilience among diverse student populations, including racial and ethnic minorities, the poor, non-English speaking, and the physically and mentally challenged. The information presented in the Appendix applies to a variety of educational settings: grade levels K-12; urban, suburban, and rural districts; elementary, middle school, junior and senior high; children from diverse racial and ethnic backgrounds; and regular vs. categorical programs. The principle understandings gathered from prior studies of resilience apply across these settings, however, the resilience-promoting strategies must be tailored to fit the specific educational contexts.

Prior research provides information on the crucial roles of the family, school, and community in promoting educational resilience. This chapter focuses on the role of schools in promoting resilience; a detailed summary of research findings for home and community contexts is presented in Wang et al. (1995a).

Promoting Resilience to Achieve School Success

Results from studies of resilience provide guidelines for educational interventions that are facilitative in achieving learning success of students at risk of academic failure. Currently, narrowly framed "categorical" programs make up the menu of educational opportunities available to children with special needs. Children are labeled according to the type of special need they exhibit and attend programs subsidized by state and federal moneys. Most programs are organized around children's learning ability. The following children at risk of academic failure are among those served by categorical programs: the learning disabled (LD) and mentally retarded (MR), poor children who are low achievers (Chapter 1 program participants), non-English and limited English proficiency speaking children (NEP and LEP program participants), children who are frequently expelled and suspended (oftentimes minority students who are LD, MR, or enrolled in Chapter 1 and LEP programs) and the gifted and talented (Wang, Reynolds, & Walberg, 1995). These students are all learners who challenge teachers' commitments and competencies. Most are poor achievers.

Limitations of Current Categorical Programs

The use of programs designed to provide "special" educational and related services for students with special needs have a number of well-documented limitations. The use of such programs produces social and academic segregation. According to Wang and Kovach (1995):

> Minority students frequently are reassigned to integrated schools but then resegregated by their placement in "special" programs. Here we refer to a somewhat different form of segregation—educational segregation via the so-called second systems programs (e.g., Chapter 1, Title 1, and special education programs) initiated to respond to the diversity of student needs (p. 12).

The content and instruction delivered in categorical programs is radically different from that presented in regular classrooms and is not appropriate to the needs of these students (Allington & Johnston, 1989; Oakes, 1985). Many teachers assigned to categorical programs hold low expectations of students' skills. Thus, oftentimes fundamental content is not covered. Additionally, they rarely incorporate higher order thinking skills or advanced content in their curriculum. Frequently, when more challenging work is presented, its introduction is delayed. These conditions fail to provide a motivating context for special needs students. Baker, Wang, and Walberg (1995) argue that proto-typical remedial or compensatory education programs often contribute to these children's learning problems. Wang, Reynolds, and Walberg (1988) note that students with special needs may actually receive inferior instruction when schools provide them with specially designed pull-out programs to meet their greater-than-usual learning needs. These special programs often ignore funda-mental content and emphasize drill rather than higher order, advanced skills.

Resilience-Promoting Instructional Strategies

As researchers and educators rethink, restructure, and recreate systems of educational and related service delivery to meet the needs of diverse student populations, including and especially the large number of children at risk of school failure, a number of instructional strategies have been identified that are likely to promote resilience. Selected key instructional strategies are described below:

Student-Centered Learning

Advances in cognitive psychology have contributed much to an under-standing of learning as an active process of integrating new information into existing prior knowledge. Learners are not passive recipients of information, but rather they construct their own interpretation of the information that is presented (American Psychological Association, 1993). Individuals differ in the amount of knowledge they have and in the types of strategies and approaches they use in problem solving (Means & Knapp, 1991).

Actively involving students in learning requires student-centered teaching. "Student-centered learning," as used by Anderson and Pellicer (1994), refers to a number of instructional strategies. These strategies include: teachers interact and challenge students frequently; students and teachers share responsibility for the selection of materials and classroom activities; teachers hold students accountable for learning the material presented; students are provided with strategies to stay on task; teachers rarely use drill and practice activities; students work collaboratively with their peers and teachers; teachers are familiar with each student's strengths and weaknesses and use this information to tailor instruction.

These student-centered strategies focus on students setting their own goals, exploring ideas and knowledge bases, and regulating their own learning. Students work on challenging tasks and make use of external supports (e.g., the teacher, peers, text, and technology) to gather information and solve complex problems. Students interact frequently, with many information sources, as they complete their tasks. The tasks are authentic, real-life problems that are topical, multidisciplinary, and require advanced skills. The emphasis is on developing students' literacy, composition, mathematical reasoning, and scientific inquiry skills (O'Day & Smith, 1993). Teachers facilitate students' learning; the traditional role of teacher as transmitter of knowledge does not apply in student-centered classrooms. In student-centered learning, classrooms become communities of learners. These environments foster meaningful, collaborative learning that is based on an understanding of a child's developmental level within a domain of content (Brown, 1994).

Student-centered learning strategies increase student engagement and thereby reduce the risk of students disengaging from their academic work and failing or dropping out. High engagement is likely to produce learning and increased motivation for continued learning (Finn, 1989; Wehlage, Rutter, Smith, Lesko, & Fernandez, 1989). Student-centered learning provides many opportunities for students to participate and succeed in their educational pursuits. High engagement in educational pursuits is likely to promote resilience, especially among children who are at risk of school failure.

Active Teaching

Students with special needs, like all students, achieve more in classes where most of their time is engaged in instructional interactions with teachers—as opposed to working passively on seat work or not working at all. Productive classrooms include a wide range of activities by teachers and students. In effective classrooms, teachers present well-prepared lessons during which new principles, concepts, and skills are introduced, students are actively involved in learning through discussion, demonstration, exploration, cooperative learning, and individualized practice. Progress is monitored, and the teacher provides frequent and constructive feedback and engages in reteaching when needed.

The construct of "active teaching" evolved out of empirical research findings, psychological theories, and educational policy. Active teaching has its origins in a cognitive psychological model of learning and teaching (Brown, 1994); the research literature on students at risk of school failure (Wang & Gordon, 1994; Wong & Wang, 1994); the systemic educational reform movement (O'Day & Smith, 1993); Vygotsky's (1978) developmental theory; the research literature on collaborative learning (Johnson, Johnson, & Holubec, 1986); and emerging educational technologies (Means et al., 1993).

Active teaching involves a large investment of teacher time interacting with students. Teachers act as facilitators to students' learning activities. They help students set goals, gather information, clarify purposes, question, provide feedback, and analyze and generate problem solutions. Teachers design curriculum and inquiry units, but do not direct all the learning that occurs. Learning activities dominate the classroom as opposed to managerial and procedural activities. Teachers spend much time organizing collaborative activities that meet the needs of diverse students. Instructional activities are designed to provide students with developmentally appropriate opportunities for learning (i.e., within the zone of proximal development, defined by Vygotsky [1978] as the upper and lower limits of a student's potential for learning depending on the degree of external support available).

In these classrooms, there are many external supports and aids that students can use in performing complex, rigorous learning activities. The teacher provides support, as do peers, adult mentors, and instructional aids including texts and new technologies. These external supports provide scaffolding for tasks that require expertise beyond what a student possesses. By providing these supports, and an accompanying philosophy about how they are to be applied, students can solve complex problems, expand their knowledge base, and develop intellectual and collaborative skills.

All students, including low achievers and/or those otherwise considered to be academically at risk, benefit from active teaching and the opportunity to engage in inquiry-based activities that facilitate students' construction of new knowledge. For students with learning difficulties, a rigorous, structured, academically oriented program is more likely to promote academic success and resilience.

Adaptive Instruction and Instructional Mediation

Adaptive instruction approaches alter classroom instruction to respond to the diverse needs of students (Glaser, 1976; Wang & Walberg, 1985). Teachers who are effective in providing for student diversity tailor learning environments to maximize each student's opportunities for learning success (Wang, 1992).

Individual differences among students are one basis for adapting instruction. Differences among learners' intellectual ability and prior knowledge, cognitive and learning styles, cultural backgrounds, and academic motivation as well as other related personality traits, such as self-concept, school-related anxiety, and locus of control are reasons for tailoring instruction.

Instruction can be adapted by varying: classroom organizational structures (short-term instructional groups, learning centers); methods of presenting new material (amount of time spent on review, number of examples, use of summaries, points of emphasis); the supports used (aides, peer tutoring, media); level, form, and number of questions asked; nature and amount of reinforcement for correct answers; amount of time devoted to instruction; use of "inductive teaching strategies" to foster inquiry skills; amount of information presented to prompt students to generate their own examples; and use of self-regulating techniques (Corno & Snow, 1986).

The Adaptive Learning Environments Model (ALEM) is an example of an educational program designed to ensure the learning success of each child using principles of adaptive education to respond to students' diverse social and academic needs. ALEM is based on the premise that students learn in a variety of ways and require different amounts and rates of instruction. In the ALEM classroom, student differences are accommodated by promoting a variety of instructional methods, alternative learning sequences, and options that teachers feel free to employ with diverse students. ALEM promotes student self-responsibility for learning by engaging students in the planning and management of their own educational tasks. It is designed to be implemented in regular classrooms and can serve the needs of both students who are low achievers as well as the academically talented without the deleterious effects of negative labeling and school segregation (Wang, Nojan, Strom, & Walberg, 1984; Wang & Zollers, 1990).

ALEM involves the use of individualized progress plans, a diagnostic-prescriptive monitoring system, and a self-scheduling system that helps students take responsibility for their own behavior and learning. It also features a program delivery system that includes: staff development, school and classroom organizational support, and family and community involvement. In ALEM programs, specialized staff (Chapter 1, special education teachers) collaborate frequently with regular classroom teachers to help in the implementation of ALEM, including diagnosis of student needs, instruction, and as peer coaches and consultants. Evidence collected over two decades indicates that when the program is fully implemented, there are positive increases in student-teacher interactions for instructional purposes, decreases in interactions for purposes of classroom management, and increases in time spent in cooperative learning and peer tutoring (Wang & Zollers, 1990).

Instructional mediation also adapts instruction to different student attributes. Instructional mediation can greatly enhance student cognition and behavior by adapting instructional treatments in terms of the information-processing demands and degree of self-control required of students. Higher level thinking places many demands on a student's cognitive and meta-cognitive processing. When the instruction carries more of the information processing load, then the student's intellectual abilities and prior knowledge becomes less important. Likewise, when instruction becomes more structured and the intensity of behavioral control increases, a student's level of self-regulation matters less. According to Corno and Snow (1986) instruction can be mediated to assist low-ability children, those with little prior knowledge, low motivation or poor self-regulation, as well as for children who are very knowledgeable, talented, and gifted. Thus, a student's special needs can be accommodated by designing instruction that varies the information-processing burden and the degree of behavioral self-control required.

Instructional approaches that emphasize discovery learning and inquiry approaches and some of the advanced subject matter areas place much of the information-processing burden on the student. They require significant amounts of prior knowledge and use of metacognitive processes. Scaffolding of instructional activities provides students with more information and structure, supplying them with the vocabulary, principles, and concepts that are needed to actively construct meaning and generate higher order, intellectual understandings. Classrooms become the settings for multiple zones of proximal development (Brown, 1994).

Adaptive instruction and instructional mediation provide an approach to responding to diversity in the classroom. Teachers can vary the amount of time, pace, use of instructional supports, cultural content, and other features of instruction to create learning environments tailored to individual or groups of students. When students are successful at school, they develop competence and confidence that empower them to overcome adversities and develop resilience.

Metacognition and Self-Regulated Learning

Metacognitive skills, self-regulated learning, and executive thinking skills are integral to successful problem solving. Metacognition facilitates "attention, motivation, learning, memory, comprehension, as well as remediation of some learning disabilities" (Wittrock, 1986, p. 310). Growth in metacognition is developmental and is a significant facet of an individual's advancing cognitive capabilities (Brown, Bransford, Ferrara, & Campione, 1983). Children who are low achievers or have learning difficulties are likely to benefit from instruction that teaches metacognitive skills and learning strategies (Keefe & Walberg, 1992; Wang & Palincsar, 1989).

Glaser and Bassok (1989) conclude that individuals differ in the availability and use of metacognitive and self-regulatory skills. They argue that individuals with performance difficulties have less well-developed skills. Studies of expert performance document the importance that executive control and cognitive strategies play in competent performance. Experts have well-developed executive skills; they can set goals, plan strategically, check for plan execution, monitor progress, and evaluate and revise strategies.

Learning and retention of content taught in educational settings has been improved through the use of learning strategies (Levin, Shriberg, & Berry, 1983). Instructional approaches that teach metacognition and self-regulated learning have shown to be facilitative to low-achieving children whose repertoire of learning strategies may be ill-developed (Brown & Campione, 1990, 1994).

Several intervention programs have demonstrated the power of metacognitive processes and self-regulated learning in improving student achievement. Reciprocal teaching and jigsaw classrooms are such examples. Reciprocal teaching is a collaborative approach that engages each participant in the group to lead part of the teaching activity—to ask questions, summarize, clarify problems, and, if appropriate, to make predictions. To participate and successfully complete the activity students must be engaged in mindful learning that generates understanding (Palincsar & Brown, 1984). Jigsaw classrooms rely upon collaborative teams that conduct research on a subtopic and then teach what they learned to other class members. Each subtopic supports a central theme that is developmentally appropriate (Vygotsky, 1978) and sensitive to student interests. The class needs information from each team to complete the class project (Brown, 1994). Brown (1992) describes these and other types of sophisticated interventions to develop metacognitive and self-regulated learning processes for use in classroom settings.

Metacognitive processes and self-regulated learning provide learners with strategies that enable them to be more in control over their learning tasks and increase their likelihood of acquiring new knowledge, mastering skills, and becoming independent learners—all of which promote resilience.

Challenging Curriculum Content and Higher Order Thinking Skills for All Children

Educational reform efforts have drawn attention to the superficiality of much of what children learn in school (Perkins, 1991). Using principles of cognitive psychology, educational researchers have been identifying approaches to schooling that should enhance learner understandings and insights among students (Gardner, 1991). The curriculum students are exposed to influences how thoughtfully they approach problems. If the curriculum is composed of generative topics, students' engagement increases and they are encouraged to form connections among knowledge bases from different disciplines. In addition to the curriculum students are exposed to, there are ways of learning that develop students' higher order thinking skills. For example, teachers can construct problems that require information from several subject-matter areas and expect student performances that involve higher order thinking skills, including interpreting, generalizing, using logic, making analogies, and transforming the content. These skills can be generated and facilitated through the use of meaningful classroom activities such as debates, developing mathematical theorems, and designing science experiments.

Newman and Wehlage (1991) identified five standards for authentic instruction. Authentic instruction engages students in higher order thinking that produces significant and meaningful achievement. Based on a constructivist model of learning, Newman and Wehlage propose that engaging students in disciplined inquiry results in thoughtful discourse, performances, and products. Their standards for authentic instruction include: the use of higher order thinking skills, such as manipulating data and ideas in order to synthesize, generalize, explain, hypothesize, and draw conclusions; the systematic use of central topics that are discipline-based and connected to other critical ideas; connectedness to the world, such that students solve real-world problems and can use their prior experience when applying knowledge; substantive dialogues between students and teachers; and social support for student achievement.

Although the Newman and Wehlage framework and standards have yet to be rigorously tested for their impact on student outcomes, other evidence demonstrating positive effects of teaching for higher order thinking skills has been collected (Brown & Palincsar, 1989; Carpenter & Fennema, 1992; Knapp, Shields, & Turnbull, 1992). A rigorous curriculum accompanied by student pursuits that demand high-level thinking is desirable for all students, but especially students at risk of school failure. If poor achievers need more basic skills instruction, it should be done in the context of a challenging curriculum, not as isolated drills and practice worksheets.

Technologies to Enhance Learning

Technological advances have significant implications for improving student learning, particularly for those with special needs. Types of technology that have been found to be particularly facilitative include tutorial assistance, opportunities for exploratory learning, applications and tools for use in completing projects, and communication among individuals and groups.

Tutorials for learning are among the most common technology available to students (Becker, 1992). Computer-based tutorials include computer-assisted instruction (CAI), integrated learning systems (ILS), and intelligent computer-assisted instruction (ICAI). Other types of tutorials include distance learning with one-way transmission and videodiscs. Extant tutorial learning programs tend to be based on direct instructional models that focus on transmission of information from a source to the learner; tutorial programs based on a constructivist model of learning are lacking (Means et al., 1993).

The use of technology to promote exploratory learning, which is based upon a constructivist model of learning, has received increasing attention. Exploratory learning promotes discovery learning processes to acquire new facts, principles, concepts, and skills and elaborate students' existing knowledge structures. Examples of exploratory technologies include: information retrieval systems (e.g., electronic databases); microworlds, microcomputer-based laboratories (MBLs), hypermedia stacks, and simulations; and interactive videos. According to Means et al. (1993), each of these exploratory computer technologies presents a context which learners access, discover, and use to construct knowledge through self-directed learning. Software, such as "Where in the World is Carmen Sandiego?," "Sim City," and "Sim Earth" are examples of microworlds. Each provides a detailed context which students explore en route to solving a complex problem that requires reasoning skills. The Cognition and Technology Group at Vanderbilt University (1991) has designed a series of interactive video applications, known as the "Adventures of Jaspar Woodbury." These videos support higher order thinking in a cross-disciplinary context. Stanley Pogrow's (1990) "Higher Order Thinking Skills" (HOTS) program is also an example of an exploratory technology that focuses on solving complex problems within a novel context, such as a balloon flight simulation.

Technologies can also be used as applications or tools that facilitate learning. Word processing, spelling and grammar checking, graphing and charting, calculating, desktop publishing, image processing, and multimedia productions are such examples. All of these applications can be incorporated as routine classroom learning tools that support learning in subject areas and advance technology skills.

Technologies for communication also facilitate learning. These include computer networks, such as the Internet and World Wide Web. They provide users with access to electronic mail services, bulletin boards, and learning circles. Not all communication technologies are computer based. For example, there are also smaller scale methods to conduct interactive learning at a distance, such as videotapes, two-way video/two-way audio tapes, and one-way video/two-way audiotapes, as well as telephone and voice mail services (Means et al., 1993). Telephone and voice mail services are being used not only to support student learning, but also to link homes and schools. These communication services keep parents informed about homework assignments and school activities.

In the early 1980s, teachers used computers primarily for purposes of computer literacy, to teach about programming, and for drill-and-practice

(Becker, 1985). As the 1980s progressed, the emphasis was altered and computers began to be promoted as tools that could facilitate student work. This trend continues and is accompanied by the continued use of computers for basic skills instruction among students at risk of school failure (Becker, 1990). Mageau (1990) reported that in 1990 there were approximately 10,000 ILS systems in use in U.S. schools, typically in Chapter 1 programs. While ILS is an advance beyond CAI applications, its focus is primarily basic skills instruction.

Computer availability in public schools has been rapidly growing. According to Mageau (1991), by the mid-1990s nearly every school would have at least one computer. Although the number of computers per school and per student dramatically increased in the 1990s, students still have limited access to computers. Research on access to computers indicates that high-SES, white, male, and high-ability students use computers more than low-SES, minority, female, and low-ability students (Means et al., 1993). Other technologies, including videos, CD-ROMs, and satellite technologies are also on the increase. However, there is a concern that students who are enrolled in schools that serve economically disadvantaged neighborhoods have unequal access to these technologies as well (Sutton, 1991). Furthermore, in the case of technologies to develop complex, higher order skills, the differences among groups of students may be even greater. In general, children in at-risk circumstances have less opportunity to work with emerging technologies. Therefore, school personnel must acquire technologies that allow students in at-risk circumstances to engage in collaborative, interactive, inquiry, and problem-solving tasks, and not just basic skills drill and practice.

High Expectations of All Students

Teacher expectations are one of several school and classroom climate variables that are related to student outcomes. In a secondary analysis of the 1988 National Educational Longitudinal Study (NELS:88), urban and suburban school administrators estimated that more students in poverty (as opposed to those not in poverty), were instructed by teachers who held negative attitudes about their students. These teachers were not as likely to encourage poor children to do their best or complete their homework (Peng & Lee, 1994). Teachers' expectations of student performance are positively correlated with student attendance and academic performance (Rutter, Maughan, Mortimore, & Ouston, 1979). High expectations for students' academic performance also impact school effectiveness. Purkey and Smith (1983) identified the use of clear goals and high expectations for student achievement in their portrait of an effective school. Edmonds (1979) identified teacher expectations that all students can attain a minimum mastery of content as a correlate of effective schools. Teddlie and Stringfield (1993) intensively studied a small number (N=6) of low- and middle-SES schools that were classified as effective, typical, and less effective based on student achievement scores. Three characteristics of teachers in effective low-SES schools were: holding high present educational expectations for students' academic work; ensuring that their students believed they could perform at current grade levels; and allowing future educational goals to develop at a later date. Effective low-SES school teachers hold firm present academic expectations for their students, but had more modest future expectations. Their students, however, extrapolated their teachers' beliefs in their current academic performance to their future educational pursuits and believed they would succeed in further schooling.

Teddlie and Stringfield (1993) also conducted an eight-year longitudinal study of eight matched pairs of effective and ineffective elementary schools. Teachers in effective schools demonstrated higher expectations for students than did teachers in ineffective schools.

Holding high expectations for all students promotes resilience because it assures that learning opportunities will be provided to all students, not just those with high ability or extensive prior knowledge or cultural advantages. Advanced material, complex topics, rigorous problem-solving activities, and well-developed communication skills must be available to all students, especially those with special needs. Expressing high expectations for student accomplishment also affects the students' self-system. Students develop information about their self-efficacy in part through persuasion and exhortation by others to engage in the new experiences (Winne, 1991). Mastering these new experiences increases students' sense of self-esteem and contributes to their willingness to engage in new experiences. High expectations reinforce students' beliefs that they are regarded as capable of academic performance, good behavior, and creative contributions. High expectations contribute to a climate in which resilience can be nurtured. Children in categorical programs often suffer low self-esteem because of the deleterious labels attached to them. The families of these children often feel stigmatized by the negative labels as well (Reynolds, 1994). For these children, the setting of expectations for academic and life success is essential. Setting high expectations provides much needed assurance to students at risk of academic failure that they too can contribute to society in a meaningful way.

Role Modeling

Another form of instructional mediation is role modeling (Corno & Snow, 1986). When a student is presented with a learning opportunity without the benefit of instructional mediation, he or she must bring all the intellectual power, cognitive strategies, and motivation to the task. The students' self-direction determines the likelihood of achievement. When a teacher models cognitive strategies, attitudes, and dispositions, which if imitated would increase the likelihood of successful completion of the task, the teacher is providing instructional mediation. A teacher modeling a successful problem-solving strategy provides external support that encourages students to take the next step and apply the strategy to their own work. In some instructional settings, the modeling is accompanied by guided practice which provides additional instructional mediation (Means & Knapp, 1991).

Adult and race-sex role models can not only model complex problem solving, but also provide students with assistance from a caring adult of the same race and sex. The use of race-sex role models increases the likelihood that minority students will have a contact with supportive adults with cultural backgrounds and attributes similar to their own. Similarity between the adult role model and the student can help establish the close bond that facilitates the development of positive behaviors and prosocial attitudes.

Some adult role models are teachers, others are community volunteers who function in mentor and advocate roles (Flaxman, Ascher, & Harrington, 1988; Freedman, 1991). Results from mentoring and advocacy programs reveal that students at risk of school failure can develop more positive attitudes toward school and better attendance. These effects, however, require a combination of weekly face-to-face sessions and discussion of crucial issues and must be sustained for a period of time. Gordon and Song (1994, p. 36) conclude:

A meaningful relationship with a "significant other" is an almost universal factor in the career development of defiers of negative predictions. ... It would appear that autonomous individual effort, although important, may not be sufficient to overcome the odds without the support of a significant other. The experience of accountability to, or identification with, another person is viewed as a universal factor in human development. For persons at high risk of failure, it may well be necessary.

Role modeling is an instructional strategy that is likely to increase resilience. It provides children who may have had limited opportunities to observe complex problem solving with more occasions to witness higher order problem solving. Oftentimes, the instructional setting provides students with further opportunities to apply the observed behaviors and strategies, and practice their newly acquired skills. Adult role models also foster student resilience by supporting the development of prosocial attitudes and behaviors.

Positive Peer Relationships

The value of friendly, supportive peer relationships has been demonstrated by research on cooperative learning strategies and peer tutoring (Natriello, McDill, & Pallas, 1990). These strategies are designed to be integrated into elementary and secondary classrooms. Cooperative learning uses peer groups, typically in small teams, to attain academic and prosocial group goals. Class-room teams compete so that each student's efforts contribute to the group goal, rather than individual students competing for grades. Thus, the classroom norms support each student doing well. The five basic features of the Johnson and Johnson Cooperative Learning Model are: positive interdependence among group members; face-to-face interactions; individual accountability; interpersonal and small-group skills; and group processing. In the Johnson and Johnson Model, the teacher manages the classroom. Small groups are formed either through random selection or teacher selection and are composed of students of mixed ability and cultural and ethnic backgrounds. Teachers instruct students in the necessary interpersonal and small-group skills so that genuinely collaborative interactions may occur. Both academic and collabora-tive skills goals are specified for each lesson (Johnson & Johnson, 1987). Regardless of the learning activity, cooperative learning experiences promote higher achievement than do competitive groups or individual learning activities. In addition, in cooperative groups the students have more positive regard for one another and share information readily with children from different cultural, social class, and ethnic backgrounds and ability levels. Peer-teaching techniques also promote higher order thinking skills. Students experience less evaluation anxiety and more motivation to learn (Johnson et al., 1986).

The beneficial effects of cooperative learning strategies on student achieve-ment and prosocial attitudes have been documented. The strategy's success is mediated by factors such as the incentive structure used, composition of the group, and the subject matter (Stallings & Stipek, 1986). Beyond enhanc-ing achievement in academic subject areas, cooperative learning is the single most effective school-based intervention for reducing alcohol and drug use (Bangert-Downs, 1988).

One-on-one peer tutoring is a strategy that has been used to assist students in at-risk circumstances. Peers are able to provide underachieving students with intensive instruction that is typically only available to children from advantag-ed backgrounds. Natriello et al. (1990) argue that peer tutoring contributes to the achievement of both tutors and tutees. Cohen (1986) demonstrated that peer tutoring also has a moderate, positive effect on achievement and attitudes to-ward subject matter. Results of peer tutoring programs have been compared with other interventions—such as computer-assisted instruction—and, while the res-ults are mixed, there is some evidence that peer tutoring has a positive effect on students' reading and mathematics achievement (Montgomery & Rossi, 1994).

Positive peer relations increase students' sense that they belong to a caring, supportive school community. Positive peer influences provide a strong advantage for students as they reinforce parent, school, and community norms and values. When family, classroom, school, and community values are in concert, it is more likely that students will achieve and behave in desirable ways and resilience will be fostered.

Resilience-Promoting Schoolwide Strategies

The instructional strategies listed above can promote resilience within classrooms, but there are also schoolwide strategies that can be used to establish a school culture that is likely to promote student achievement, positive behavior, and resilience. Based upon a review of the school effects literature (Wang, Haertel, & Walberg, 1995b) and recent research on inclusive schools (Baker et al., 1995), eight schoolwide strategies were selected that are likely to increase resilience. (See the school environment context section of the Appendix.)

Safe and Secure School

Safe and orderly schools are necessary for serious learning to take place. The deleterious effect of school violence is being described with increased frequency in the national media and in educational journals (Sautter, 1995). Results from a recent survey entitled *First Things First: What Americans Expect from the Public Schools* (Public Agenda, 1995) reveal that U.S. residents, in all parts of the nation and across every demographic category, believe their local public schools are not providing the safe conditions that are required for teaching and learning.

Early in the school effectiveness movement, Edmonds (1979) identified an orderly, safe school climate as a correlate of school success. Data from many school effects studies have since confirmed that an orderly environment is a characteristic of effective schools, regardless of socioeconomic status (Teddlie & Stringfield, 1993). Resilience is promoted by an orderly school setting where children and youth can devote their attention and energies to accomplishing school work without fear of violence or disorder. A safe and secure school environment promotes resilience among students by protecting their physical and mental well-being and confirming their belief that their school, teachers, and community care about them.

Positive, Academically Oriented School Culture

Over the past 15 years, researchers have documented that some schools in economically disadvantaged neighborhoods, which serve large numbers of children at-risk of academic failure, have achieved beyond what is expected given their resources and student population (Brookover, Beady, Flood, Schweitzer, & Wisenbaker, 1979; Coleman & Hoffer, 1987; Edmonds, 1979; Mortimore, Sammons, Stoll, Lewis, & Ecob, 1988; Rutter et al., 1979; Teddlie & Stringfield, 1993). All of these studies have highlighted the importance of school culture. Three features of school culture that were viewed as essential in promoting student achievement and good behavior were: an emphasis on academic achievement; a safe and orderly school environment; and a shared vision of the school's purpose held by school personnel, students, and parents.

The role of the principal as instructional leader was also a key ingredient of a positive, academically oriented school. John Gardner (1990) put forth the analogy of "teacher as leader." He believes that all great teachers are leaders, and the responsibility for creating a positive school culture extends to all school personnel. Students need to know where their school is headed and the school's vision must be aligned with their personal goals and aspirations.

School climate is not linked to tangible features of the school, such as age of the building or use of portable classrooms, rather it is a product of several co-occurring processes that include: classroom practices (e.g., high standards for all students; proportion of time devoted to instruction) and aggressively maintained school policies on, for example, discipline, attendance, and recognition of student achievement.

Joyce and Calhoun (1995) persuasively argue that principals in self-renewing schools motivate teachers to examine their teaching and focus on best practices. They recommend that teachers be connected to the increasing knowledge base on teaching and learning in order to generate solutions to instructional problems. Establishing a democratic governing body (Glickman, 1993), an atmosphere of collective inquiry, time for reflection on practice, and a focus on the relationship between the learning environment and student outcomes will contribute to a school culture that renews itself and promotes student learning.

A positive school culture that is academically oriented will reward children and youth for learning. The maintenance of a structured and orderly school climate will eliminate distractions that interfere with children's opportunity to learn. A school culture with a well-developed vision, a principal and teachers who provide leadership, and an emphasis on improved practice and student learning creates an environment that promotes high achievement for all children.

Smaller Educational Units Where Students Remain with the Same Teacher and Peers for Several Years

Recent research demonstrates that smaller schools have achievement advantages (Fowler & Walberg, 1991). Smaller schools also are more amenable to school improvement and systemic reform efforts (Klonsky & Ford, 1994). Klonsky and Ford base their conclusions on research reporting that schools with less than 350 students moved toward systemic reforms more readily, generated less contentious school politics, and supported more democratic action in their governance. In addition, evidence on small schools has revealed that they are more cost effective than larger schools. According to Lee, Bryk, and Smith (1993), results from case studies and surveys also point to negative effects related to large school size. Large schools foster socially stratified learning opportunities—larger schools can provide more specialized services for children with exceptional needs; these services, however, marginalize special-needs students and deny them access to the academic and social resources available to other students. In addition, large schools create a sense of alienation among students and teachers and less involvement with the school's vision and aims. Oxley (1994) has also argued that smaller schools are an alternative to homogeneous grouping. She presents the advantages of smaller educational units for low achievers who are enriched by exposure to more rigorous curricula, intellectually demanding instructional activities, and contacts with diverse groups of students who share their knowledge, skills, and social acceptance.

Increasingly, educators are focusing on the importance of communitarian values in the organization, practices, and policies of schools. Minischools, charters, or houses are ways to reorganize schools into smaller units so that educators, students, peers, and families can develop closer relationships that can be sustained over several years. Reducing the impersonal and bureaucratic relations that often characterize larger schools also reduces students' and teachers' senses of alienation and disengagement. Smaller educational units enhance the strength of relationships that can be established. Additionally, when students and teachers work together for two or three years, students feel more supported and their educational needs can be better met. When teachers spend several years teaching the same students, they are more likely to correctly diagnose student learning difficulties and tailor instruction to accommodate problems. Thus, resilience is promoted in smaller educational units through the increased quality of student-teacher relationships, improved learning opportunities, increased teacher morale, and increased teacher time devoted to individual student needs.

Inclusive Classrooms and Schools

The inclusion of children with special needs into regular classrooms and schools has received increasing support as a systemic educational improvement strategy (Commission on Chapter 1,1992; National Association of State Boards of Education Study Group on Special Education [NASBESGSE], 1992; U.S. Department of Education, 1994). For decades, human rights advocates fought to eliminate the segregation of students from ethnic- and language-minority backgrounds in our nation's schools. Thus, many schools now include large numbers of children from diverse sociocultural and economic backgrounds. However, these efforts to desegregate have done little to increase social and academic integration (Yancey & Saparito, 1995). In schools where racial integration has occurred, students from minority groups are frequently reassigned to "special" pull-out remedial or compensatory education programs (Heller, Holtzman, & Messick, 1982). This exemplifies educational segregation via the so-called second systems programs (e.g., special education programs, Chapter 1). These categorical programs receive special funding streams and have their own eligibility criteria. Their implementation results in added layers of bureaucracy and disjointed implementation within the school's curriculum and instructional program. One of the most disconcerting outcomes of pull-out programs and separate classrooms is the segregation of large numbers of children from ethnic and language minorities (NASBESGSE, 1992; Wang & Kovach, 1995).

Increasingly, research results are being brought to bear on the effectiveness of inclusive programs. The integration of students enrolled in special education classes into regular classes is positively related to academic learning and social relations (Allington, 1987; Baker et al., 1995). In light of these research findings, inclusion is a resilience-promoting strategy that enhances the school performance and life success of students' at risk of school failure.

Active Parent and Community Involvement Programs

The key role that parents play in their children's educational accomplishments has been the subject of much psychological research and received much attention from policymakers. Parental participation is one of the National Educational Goals adopted by Congress in the *Goals 2000: Educate America Act* (National Education Goals Panel, 1994).

Research has documented the variety of positive effects that are produced by parent involvement programs. Parent involvement programs enhance student performance; increase student achievement and school attendance; and decrease student drop out, delinquency, and pregnancy rates (Peterson, 1989). Liontos (1991) concludes that children in adverse family circumstances are especially likely candidates to benefit academically and socially from their family's involvement in education-related programs. Parents who participate in these programs feel better about themselves and are more likely to enroll in courses to advance their own education (Flaxman & Inger, 1991). Educational intervention programs that target all family members are more effective than programs focused only on students (Walberg, 1984). Many educators and researchers believe these programs hold much promise for students from adverse family circumstances. Lee et al. (1993) present the following five conclusions from empirical studies of parent involvement: the degree of parent involvement that occurs is positively related to parents' socioeconomic status and, in particular, to their level of education; the degree of involvement is related to their sense of being informed and how much they believe they can contribute to their children's learning; parents are less involved as children advance through the grades, but they continue to express interest in children's schoolwork through high school; the degree to which parents help children with their homework and are involved in curricular decisions is positively correlated with achievement and enrollment in academic coursework, even when controlling for parents' social class; and parents express an interest in being more engaged in their children's education. Chubb (1988) concludes that when parents are highly involved, cooperative, and informed, their schools become more effective organizationally, as compared to schools in which parents do not exhibit these characteristics.

Parent involvement programs vary in design. Three basic approaches can be identified. One type of program assists parents in becoming improved home educators. These programs teach parenting skills, such as monitoring homework, providing academic assistance to students if needed, and reducing television time. Parents develop better communication skills and help students develop good study habits and high expectations. Empirical results from evaluations indicate that programs that teach specific learning strategies to parents impact students' academics in a strong and positive manner (Lee et al., 1993). A second type of parent involvement involves families directly in school management and decision making, and encourages their presence in the school and at school activities (Bast & Walberg, 1994). The third type of program provides direct services to families and children, including home visits, job training, career counseling, health care, mental health, and social support.

Family literacy programs are another popular activity. Models of these programs are diverse and often intergenerational. Some are home-based; others are in schools, libraries, and other community organizations, including prisons (Nuckolls, 1991). They are based on a premise that students are more likely to improve their literacy skills if their parents are interested. Among families beset by significant adversities, parenting programs provide practical help in improving family members' lives. Parent involvement programs provide strategies on how parents can best assist children who face significant academic, mental health, and physical challenges. For a review of collaborative programs that includes parent involvement programs see Wang, Haertel, and Walberg (1995c).

Community involvement programs are also resilience-promoting. Such programs typically involve local businesses and community groups, such as the chamber of commerce. Local corporations may contribute human or material resources to, for example, a science program, or co-sponsor a school or district event, such as a science and technology fair. Community involvement programs, in addition to making more material resources available to students, also bring adult role models with expertise into contact with students. Community involvement programs can increase a school's material and social resources and provide new experiences to children at risk of school failure.

Facilitate Normative and Unexpected Transitions Between Schools

Family relocations and migration result in school transitions. School transition poses a serious and pervasive risk factor for student learning among poor and minority children (Long, 1975; Straits, 1987). At the elementary level, frequent family relocations are related to declining school achievement, higher probabilities of being retained a grade, and dropping out of school (Lash & Kirkpatrick, 1994). Normative transitions among schools, such as the transition from elementary to junior high school, can produce a decline in grades (Finger & Silverman, 1966; Simmons & Blyth, 1987). (This decline in grades is not paralleled by a comparable decline in achievement scores. Thus, the decline may reflect changes in grading practices between K-8 schools vs. junior high schools). Regardless, the change between schools challenges students' academic and social skills and may alter their motivational orientation toward academics (Eccles et al., 1993).

A change of schools has important consequences for students. It disrupts students' social networks with peers and teachers; contributes to students feeling as if they have little control over important dimensions of their lives; requires students to learn new rules and procedures for participation in their new classrooms and schools; and demands that students accommodate to changes in curriculum, instructional materials, classroom activities, and grouping practices.

Given the deleterious effects of frequent school transitions, educational resilience can be promoted among children by having schoolwide programs to facilitate normative and unexpected transitions. For example, normative transitions can be facilitated through orientation programs that familiarize students with the new school before the school year begins. The orientation can provide students with information on procedures, schedules, and facilities. Handbooks for students and parents can further reduce stress and promote educational resilience.

In terms of unexpected school transitions, such as family moves, the classroom teacher who receives the new student can implement a variety of strategies to make the student's transition less stressful. For example, the teacher can determine the curriculum, instructional materials, and classroom practices and procedures that the student had been exposed to in the prior classroom. The teacher also needs to determine the student's level of prior knowledge and skills. Having this information, the teacher can then assist the new student in accommodating to the new classroom and school environment. This accommodation may be as simple as providing a list of classroom rules and procedures, assigning a peer to answer questions and provide hands-on help, or providing homework that will allow the student to learn the new skills needed to perform in the current classroom (Lash, Haertel, & Kirkpatrick, 1995).

Classroom teachers could receive professional development to raise their awareness about the effects of family relocations and strategies to limit the negative effects. A schoolwide policy on the value of facilitating transition programs would assure that all teachers assist transition students and help them feel at home in their new classroom and school. The use of strategies to facilitate transitions promotes resilience. These strategies ease stress, enhance a student's sense of belonging, and ultimately promote learning.

Provide School-Linked Child and Family-oriented Services

Currently, government and private agencies undertake separate, unco-ordinated service programs aimed to support the healthy development and learning of children and families, especially those in adverse circumstances. Medical, mental health, legal, child care, and job training assistance are among the types of services typically provided. These services, however, are often fragmented and regarded by clients as a maze of resources in different locations, with different eligibility criteria that require extensive and complicated documentation and paperwork (Boyd & Crowson, 1993; Kirst, Koppich, & Kelley, 1994; Wang, Haertel, & Walberg, 1995). If a client does not have transportation readily available, the hardship of traveling to several different service locations causes some, perhaps those most in need, to give up. Illiterate or non-English speaking clients may get overwhelmed by the array of complicated forms that must be completed to receive the services, increasing the probability that clients will drop out of the intake process. In addition, each separate site that must be visited to acquire a particular service is a new, unfamiliar location with service providers who are strangers. For some clients there is a stigma attached to seeking services. Thus, enduring each separate intake process becomes demoralizing.

A number of barriers are eliminated when social services are integrated and linked to local schools. The school becomes the organizational hub around which social services are provided. Schools do not deliver the services, but make them "available, accessible, meaningful, and appropriate for children and their families" (Kirst et al., 1994, p. 199). Kirst and Kelley (1992) describe an interagency system that would link schools and social service organiza-tions—both public and private—with community businesses and resources. This network of institutions and agencies could provide educational, social, medical, and psychological services to children and their families.

The Urban Institute (1992) identified some elements of school-linked service programs that are likely to promote success, including: consensus that caring relationships are essential; identification of coping strategies that families can apply; and the use of non-monetary incentives to encourage children and youth to make positive choices. Soler and Shauffer (1993) identified the following attributes of programs that have potential for success:

a clear value statement; a family orientation; involvement of a variety of community organizations; consistent and meaningful involvement of the local educational system; an emphasis on high-quality services; and careful intake and evaluation of families.

In the fall of 1994, educators, social service practitioners, and policymakers attended a working conference entitled "School-linked Comprehensive Services for Children and Families: What We Know and What We Need to Know" (U.S. Department of Education and the American Educational Research Association [AERA], 1995). Conference participants concluded that although school-linked service programs are not a new development, the number of such programs is rapidly increasing and their design and features are very diverse. Conference participants acknowledged that the evaluative evidence on the effectiveness of school-linked services is scanty. What evidence has been collected tends to be positive—children and families do enter the system and become empowered, the programs are culturally sensitive and well connected to the community. Current practice indicates that most programs are established with the clear intent to increase the flexibility of the services provided and encourage interprofessional collaboration among service providers. Leaders who are committed to collaborative services are essential, but few have the skills needed to guide such complex endeavors. Current experience with existing programs suggests that the funding base for these programs is unstable and fragile. Based upon limited research evidence and practical experience, the consensus is, however, that school-linked services can provide many benefits to those families and students who are at risk of school failure.

There are many practical advantages of schoolwide services. For example, the services are readily accessible even to those families with no transportation. Having the services in a single setting eliminates the need for multiple, repetitive intake procedures. When the services are provided in a neighborhood school, there are school personnel present who know the student and family and can provide background information and documenta-tion for purposes of eligibility. Translators are often available in schools that serve large numbers of non-English speaking students. Finally, visiting the neighborhood school for assistance is not as threatening or stigmatizing an experience as multiple visits to unfamiliar sites. Increased availability of services promotes educational resilience by raising the likelihood that a families' medical, psychological, employment, legal, and economic problems will be addressed and healthy development can occur.

Districtwide Efforts to Improve Quality of Schooling

School districts can enhance or constrain the effectiveness of their schools. School districts allocate funds for different purposes. Some districts use their financial resources to enhance classroom instruction. They fund items such as the development of a teacher evaluation system, innovative and effective programs, and professional development activities. Other districts allocate more funds toward financial management. In particular, districts can negatively influence school effectiveness by not allocating resources toward schools in poor neighborhoods (Teddlie & Stringfield, 1993). There is empirical evidence that district support can significantly increase the likelihood of school improvement (Hill, Wise, & Shapiro, 1989; Lezotte, 1990).

When a district commits itself to systemwide improvement, it increases the likelihood that students will receive enhanced instruction. However, as districts decentralize, schools gain more control of resources. For example, some schools have control of whether principals are retained, which budget items are funded, and which teachers are hired. Thus, district influence can be diluted or enhanced by district governance policies. Pollack, Chrispeels, Watson, Brice, and McCormick (1988) reported district effects when superintendents were actively involved in the selection and hiring of building principals. Thus, the role of districts in improving schooling depends, in part, on the degree and nature of decentralization. However, the allocation of district funds for instructional and schoolwide improvement and district involvement in personal decisions may be key functions that can improve the quality of schooling and thereby promote resilience.

Resilience-Promoting Family and Community Involvement Strategies

Schools are only one of the subsystems that influence the education of children and youth. Two other crucial subsystems, the family and the community, also impact students' learning success. To achieve significant school improvement, it is necessary to forge working connections with all the systems that influence the development of children. The capability of school systems can be greatly enhanced when expertise from multiple disciplines, practitioners, and family and community resources are combined to forge a coordinated approach.

Garmezy (1991) stated that a clear sign of a caring and supportive community is the presence of social organizations that provide for healthy human development. During the past five years, policymakers have become aware that although many communities often have numerous services available (including counseling, financial assistance, legal aid, medical and mental health treatment, religious institutions, job training, recreational opportunities, and libraries), they are often delivered in a fragmented and disconnected manner (Boyd & Crowson, 1993; Council of Chief State School Officers, 1989; Wang et al., 1995).

The learning problems of students cannot be tackled by schools alone. Family resources, schools, and available community resources must be linked together via interagency, collaborative programs. The construction of social policy that integrates interagency collaboration, school-linked services, and family involvement is a central reform issue of the 1990s (Kirst & Kelley, 1995; National Center on Education in the Inner Cities, 1990; Wang & Reynolds, 1995). Policy documents (Committee for Economic Development, 1994), enacted legislation (U.S. Department of Education, 1994), and research reports (Rigsby, Reynolds, & Wang, 1995) document the importance of this agenda. Wang and Kovach (1995) and Wang, Haertel, and Walberg (1994b) identify a number of strategies that are likely to enhance resilience among students and their families. These family-, school- and community-based strategies are briefly described below.

Consistent, Frequently Expressed Cultural Norms and Values

Communities with high expectations for good citizens and educated members provide protective mechanisms against adversities by sending a highly visible message of strongly valued social and cultural norms. This strategy has been demonstrated in studies of the effect of cultural norms on student alcohol and drug abuse (Bell, 1987; Long & Vaillant, 1989). Nettles (1991) analyzed the effectiveness of community-based programs with African-American youth. She found school-based clinics are only partially effective in reducing substance abuse; the most effective programs fostered resilience through the use of more support and adult assistance, concrete help on tasks, and opportunities to develop new interests and skills. In other words, when community values and school values reinforce one another, the potential for change increases. Epstein (1987) established a theory of family-peer group-school-community connections. The more consensus there is among the values, norms, and goals of each subsystem, the greater the potential impact on the developing person. When the home, school, peer culture, and community consistently express support for the same values and behaviors, it is more likely that children and families in at-risk circumstances will engage in healthy, resilience-promoting behaviors.

Prevention-Oriented, School-Linked, Integrated Services Addressing Multiple Needs

School-linked, integrated services should be designed so that they are preventive and address the multiple needs of the entire family (Dryfoos, 1994). Case management is the method of choice to serve the needs of the client and reduce the fragmentation of service delivery (Wang, Haertel, & Walberg, 1994b). School-linked services facilitate an efficient response to the multiple needs of students and their families who are at risk for a variety of poor life outcomes. Whether these services are located at a school, or in a nearby area that is readily accessible to students, the community should provide ready availability of medical, mental health, financial, legal, and other services to the entire family. In some communities, especially those that are economically disadvantaged, basic needs such as food, shelter, and transportation, as well as emergency services, may also need to be provided.

The use of prevention-oriented, comprehensive services reduces the fragmentation of services and makes them more available to students at risk of school failure and their families. This approach to service delivery promotes wellness, and provides crisis intervention if needed. The availability and accessibility of these comprehensive services will promote resilience for children and families beset by adverse circumstances who may not have the personal resources to find the numerous sources needed to ameliorate their difficulties (Kirst et al., 1994). This approach not only meets the needs of economically disadvantaged families and families of children with learning difficulties, but is well suited to the needs of displaced children, including immigrant, migrant, and homeless youth (Buckner, Bassuk, & Brooks, 1994).

Feedback to Teachers from Integrated-Service Providers

Teachers of children at risk of academic failure can benefit from the feedback provided by integrated service providers—including social workers, mental health specialists, and medical practitioners (Wang, Haertel, & Walberg, 1994b). Their feedback can be used by teachers to better tailor instruction and classroom management techniques to students with special needs (King, 1994). The sharing of information among teachers and service providers raises questions, however, of confidentiality. Policies and procedures need to be established among the collaborative agencies that protect the privacy of students and their families and still allow for flexibility among the agencies that deliver services (U.S. Department of Education & AERA, 1995).

Use of Evaluation Data to Select Collaborative Approaches That Match Community Contexts

As educators and other stakeholders select approaches to link family, school, and community resources, they should examine formative and summative evaluations of other collaborative programs. Stakeholders should seek out evaluations that were conducted in communities where the culture is similar to that in their own communities as context can impact the processes and outcomes of collaborative programs (U.S. Department of Education & AERA, 1995).

Evaluations should make use of multiple and diverse process and outcomes measures. Stakeholders should review evaluations that take into account the goals of the school system, participating agencies, and the community. Family-based outcomes should also be included (Wang, Haertel, & Walberg, 1994b, 1995).

Because collaborative programs are interdisciplinary, individuals from different professions may be involved. Thus, an evaluative approach that emphasizes participatory evaluation involving stakeholders and clients, cultural sensitivity, longitudinal evaluative designs, and developmental evaluations that focus on formative feedback may be especially crucial. Evaluators might include multiple case studies, cost benefit and cost effectiveness studies, exemplary practices, reports, etc., and employ both qualitative and quantitative approaches (Knapp, 1994; Lopez & Weiss, 1994; U.S. Department of Education & AERA, 1995).

The use of evaluation data to help communities design collaborative integrated services is a wise investment. Reviewing evaluation data can ensure the design and implementation of a program that is well suited for a local community and, thereby, more likely to promote resilience among children and their families.

Opportunities for Children and Youth to Participate in Meaningful Community Activities

Communities that have well-developed and integrated social organizations have fewer social problems (Miller & Ohlin, 1985). These organizations provide opportunities for children and youth to develop new knowledge, skills, and interests. When a community provides these opportunities, children and youth experience care and support. In communities with an abundance of such activities, children and youth can participate in the cultural, civic, religious, and, in the case of older youth, the economic life of the community.

Youth organizations, religious institutions, and even neighborhoods are "mediating structures" that can mitigate against the dangerous effects of living in economically disadvantaged neighborhoods (Woodson, 1981). Although participation in youth programs is perceived as building the talents of middle class children, in poor neighborhoods participation can be seen as an effort to discourage drug use and delinquency (Littel & Wynn, 1989). Thus, poor children and youth may feel stigmatized when participating in these programs—if you are poor or a minority, you are not expected to do well. Thus, programs must be designed and promoted in ways that all children will feel welcome and inclined to participate.

Community empowerment programs and business–community collaborations are a response to the deterioration of urban communities. Programs such as Rebuild L.A. (1993) and the Atlanta Project (1992) are identified by Montgomery and Rossi (1994) as efforts to address underlying community economic and social problems. These types of programs stress volunteerism and talent development. Churches and religious groups also provide many activities in which young children and youth can be involved. These organizations teach values and skills that promote healthy development and constructive involvement in the community.

Coleman (1961), in *The Adolescent Society*, warned communities about the dangers of failing to provide youth with real opportunities for meaningful involvement in their community. Adolescent society needs to be aligned with the values and norms of the larger community. When there are too few opportunities for genuine participation, children and youth are placed at greater risk for alienation and disengagement from community life. Thus, abundant opportunities for participation in the community are likely to promote resilience among children and youth who are in at-risk circumstances.

Improved Public Safety

Improved public safety promotes resilience. Improved safety in a community benefits all residents, but especially those individuals who are in economically disadvantaged circumstances. When a neighborhood is beleaguered by delinquent and criminal activity it becomes a risk for children and youth to travel to and from school, to cultural and recreational activities, or even play in neighborhood parks. Improved public safety confirms the value of all community residents and increases the likelihood that children and youth will not be victims of violence. Children and youth need a safe and secure neighborhood that is free of drugs and crime. Community-based teams composed of local business representatives, school personnel, local law enforcement, and community residents can provide the necessary support.

Role of Resilience in Refocusing Educational Policy for Students At Risk of School Failure

Resilience is a social science construct. It cannot be measured precisely and the conditions that promote it are context-specific. The promotion of resilience cannot be a substitute for social policy, because there is no singular course of action that unquestionably leads to resilience. Rather, for educational purposes, the promotion of resilience provides a goal toward which educators, researchers, practitioners, and policymakers can focus their efforts.

Research projects examining resilience-promoting strategies are being conducted in a variety of geographic sites, socioeconomic settings, and educational contexts. These research projects, which focus on the role of the family, school, and community in fostering resiliency, have been designed to identify differential effects of resilience-promoting strategies on children and youth from different risk groups and those experiencing different degrees of risk. This chapter identifies strategies that promote resilience in classrooms, schools, homes, and communities.

Classroom and school environments are likely to foster resilience when students are highly engaged in educational pursuits that are challenging, sensitive to students' abilities, prior achievement, cultural background, and interests. In resilience-promoting classrooms, the teacher's role is that of a facilitator, as well as a transmitter of knowledge. Teachers foster resilience by encouraging students to take responsibility for their own learning. Resilience-promoting classrooms provide many external supports, including assistance from teachers and peers, texts, and technology, to scaffold challenging content to students' cognitive levels.

Resilience is also promoted by the nurturing attitude that teachers and school personnel demonstrate. High expectations are held for all students, not for only the gifted or high-SES students. School safety, the provision of school-linked services, inclusive classrooms, and adequate resources create a school and classroom environment that is supportive of students. Family-school-community partnerships promote resilience when each environmental system works in concert to reinforce a set of positive values that promote healthy development.

As we move into the 21st century, resilience holds much promise as a strategy to overcome adversities. It conveys hope and holds out the promise of academic and life success for all students. Collaborative efforts among families, schools, and local communities bring together resources that can mitigate against the vulnerabilities and adversities many students face. Thus, this construct is well suited to refocus educational policy in the 1990s, when diversity, limited resources, and increasing demands for intellectual and social competencies characterize our nation's schools and workplaces.

References

Allington, R.L. (1987). Shattered hopes. *Learning, 87*, 67-68.

Allington, R.L., & Johnston, P. (1989). Coordination, collaboration, and consistency: The redesign of compensatory and special education intervention. In R. Slavin, N. Madden, & N. Karweit (Eds.), *Preventing school failure: Effective programs for students at risk* (pp. 320-354). Boston: Allyn & Bacon.

American Psychological Association. (1993) *Learner-centered psychological principles: Guidelines for school redesign and reform* [report produced by the Presidential Task Force on Psychology in Education of the American Psychological Association]. Washington, DC: American Psychological Association and the Mid-continent Regional Education Laboratory.

Anderson, L.W., & Pellicer, L.O. (1994). Compensatory education in South Carolina: Lessons from the past visions for the future. In K.K. Wong & M.C. Wang (Eds.), *Rethinking policy for at-risk students* (pp. 91–121). Berkeley, CA: McCutchan Publishing Corporation.

Atlanta Project. (1992). *The Atlanta Project*. Atlanta, GA: Author.

Baker, E.T., Wang, M.C., & Walberg, H.J. (1995). The effects of inclusion on learning. *Educational Leadership, 52*(4), 33–35.

Bangert-Downs, R. (1988). The effects of school-based substance abuse education. *Journal of Drug Education, 18*(3), 1–9.

Bast, J.L., & Walberg, H.J. (1994). Free market choice: Can education be privatized? In C.E. Finn, Jr., & H.J. Walberg (Eds.), *Radical Education Reforms* (pp. 149–171). Berkeley, CA: McCutchan Publishing Corporation.

Becker, H.J. (1985). How schools use microcomputers: Results from a national survey. In M. Chen & W. Paisley (Eds.), *Children and microcomputers: Research on the newest medium* (pp. 87–107). Beverly Hills, CA: Sage.

Becker, H.J. (1990, April). *Computer use in United States schools: 1989. An initial report of U.S. participation in the I.E.A. Computers in Education Survey.* Paper presented at the annual meeting of the American Educational Research Association, Boston.

Becker, H.J. (1992). Computer-based integrated learning systems in the elementary and middle grades. *Journal of Educational Computing Research, 8*(1), 1–41.

Bell, P. (1987). Community-based prevention. *Proceedings of the National Conference on Alcohol and Drug Abuse Prevention: Sharing knowledge for action.* Washington, DC: NICA.

Benard, B. (1991, August). *Fostering resiliency in kids: Protective factors in the family, school and community.* Portland, OR: Northwest Regional Education Laboratory.

Bereiter, C., & Scardamalia, M. (1989). Intentional learning as a goal of instruction. In L.B. Resnick (Ed.), *Knowing, learning, and instruction: Essays in honor of Robert Glaser* (pp. 361–392). Hillsdale, NJ: Lawrence Erlbaum Associates.

Boyd, W., & Crowson, R. (1993). Coordinated services for children: Designing arks for storms and seas unknown. *American Journal of Education, 101,* 140–179.

Bronfenbrenner, U. (1979). *The ecology of human development: Experiments by nature and design.* Cambridge, MA: Harvard University Press.

Brookover, W., Beady, C., Flood, P., Schweitzer, J., & Wisenbaker, J. (1979). *School social systems and student achievement: Schools make up a difference.* New York: Praeger Press.

Brown, A.L. (1990). Domain-specific principles affect learning and transfer in children. *Cognitive Science, 14,* 107–133.

Brown, A.L. (1992). Design experiments: Theoretical and methodological challenges in creating complex interventions in classroom settings. *The Journal of the Learning Sciences, 2*(2), 141–178.

Brown, A.L. (1994). The advancement of learning. *Educational Researcher, 23*(8), 4–12.

Brown, A.L., Bransford, J., Ferrara, R., & Campione, J.(1983). Learning, remembering, and understanding. In J. Flavell & E. Markman (Eds.), *Carmichael's manual of child psychology (Vol. 6)* (pp. 77–166). New York: Wiley.

Brown, A.L., & Campione, J.C. (1990). Communities of learning and thinking, or a context by any other name. *Contributions to Human Development, 21,* 108–125.

Brown, A.L., & Campione, J.C. (1994). Guided discovery in a community of learners. In K. McGilly (Ed.), *Classroom lessons: Integrating cognitive theory and classroom practice* (pp. 229–270). Cambridge, MA: MIT Press/Bradford Books.

Brown, A.L., & Palincsar, A. (1989, March). *Coherence and causality in science readings.* Paper presented at the annual meeting of the American Educational Research Association, San Francisco.

Brown, G.W., Harris, T.O., & Bifulco, A. (1986). The long-term effects of early loss of parent. In M. Rutter, C.E. Izard, & P.B. Read (Eds.), *Depression in young people* (pp. 251-296). New York: Guilford Press.

Buckner, J., Bassuk, E., & Brooks, M. (1994). *Displaced children: Meeting the educational and service needs of immigrant, migrant, and homeless youth.* Paper prepared for the Invitational Conference on School-Linked Comprehensive Services for Children and Families, Leesburg, VA.

Carpenter, T.P., & Fennoma, E. (1992). Cognitively guided instruction: Building on the knowledge of students and teachers. *International Journal of Educational Research* [Special issue]. 457–470.

Chess, S. (1989). Defying the voice of doom. In T. Dugan & R. Coles (Eds.), *The child in our times* (pp. 179-199). New York: Brunner/Mazel.

Chubb, J.E. (1988). Why the current wave of school reform will fail. *The Public Interest, 90,* 28–49.

Cognition and Technology Group at Vanderbilt. (1991). Technology and the design of generative learning environments. *Educational Technology Journal, 31*(5), 34–40.

Cohen, E.G. (1986). *Designing group work: Strategies for the heterogeneous classroom.* New York: Teachers College Press.

Coleman, J.S., with the assistance of J.W.C. Johnston & K. Jonassohn. (1961). *The adolescent society: The social life of the teenager and its impact on education.* Glencoe, New York: Free Press.

Coleman, J.S., & Hoffer, T. (1987). *Public and private high schools: The impact of communities.* New York: Basic Books.

Commission on Chapter 1. (1992). *Making schools work for children in poverty.* Washington, DC, Council of Chief State School Officers.

Committee for Economic Development. (1994). *Putting learning first: Governing schools for high achievement.* New York: Author.

Corno, L., & Snow, R.E. (1986). Adapting teaching to individual differences among learners. In M.C. Wittrock (Ed.), *Handbook of research on teaching* (3rd ed., pp. 605–629). New York: Macmillan.

Council of Chief State School Officers. (1989). *Family support, education, and involvement: A guide for state action.* Washington, DC: Author.

Dryfoos, J. (1994). *School-linked comprehensive services for adolescents.* Paper presented at the Invitational Conference on School-Linked Comprehensive Services for Children and Families, Leesburg, VA.

Eccles, J.S., Midgley, C., Wigfield, A., Buchanan, C.M., Reuman, D., Flanagan, C., & Mac Iver, D. (1993). Development during adolescence: The impact of stage environment fit on young adolescents, experiences in schools and in families. *American Psychologist, 48*(2), 90–101.

Edmonds, R.R. (1979). Effective schools for the urban poor. *Educational Leadership, 37*(1), 15–27.

Elder, G.H., van Nguyen, T., & Caspi, A. (1985). Linking family hardship to children's lives. *Child Development, 56,* 361–375.

Epstein, J.L. (1987). Toward a theory of family-school connections: Teacher practices and parent involvement. In K. Hurrelamn, F. Kaufmann & F. Losel (Eds.), *Social intervention: Potential and constraints* (pp. 121–136). New York: W. De Gruyler.

Finger, J.A., & Silvermen, M. (1966). Changes in academic performance in the junior high school. *Personnel and Guidance Journal, 45,* 157–164.

Finn, J. D. (1989). Withdrawing from school. *Review of Educational Research, 59,* 117–142.

Flaxman, E., Ascher, C., & Harrington, C. (1988). *Youth mentoring: Programs and practices.* New York: ERIC Clearinghouse on Urban Education, Columbia University, Teachers College.

Flaxman, E., & Inger, M. (1991). Parents and schooling in the 1990's. *ERIC Review, 1*(3), 2–6.

Fowler, W.J., & Walberg, H.J. (1991). School size, characteristics, and outcomes. *Educational Evaluation and Policy Analysis, 13*(2), 189–202.

Freedman, M. (1991). *The kindness of strangers: Reflections on the mentoring movement.* Philadelphia, PA: Public/Private Ventures.

Gardner, H. (1991). *The unschooled mind.* New York: Basic Books.

Gardner, J. (1990). *On leadership.* New York: The Free Press.

Garmezy, N. (1974). *The study of children at risk: New perspectives for developmental psychopathology.* Paper presented at the 82nd annual meeting of the American Psychological Association, New Orleans, LA.

Garmezy, N. (1991). Resiliency and vulnerability to adverse developmental outcomes associated with poverty. *American Behavioral Scientist, 34*(4), 416–430.

Garmezy, N., & Masten, A.S. (1994). Chronic adversities. In M. Rutter, E. Taylor, & L. Hersov (Eds.), *Child and adolescent psychiatry: Modern approaches* (pp. 191–208). Oxford: Blackwell Scientific Publishers.

Glaser, R., & Bassok, M. (1989). Learning theory and the study of instruction. *Annual Review of Psychology, 40,* 631–666.

Glickman, C.D. (1993). *Renewing America's schools: A guide for school-based action.* San Francisco: Jossey-Bass.

Goldstein, M.J. (1990). Factors in the development of schizophrenia and other severe psychopathology in late adolescence and adulthood. In J. Rolf, A.S. Masten, D. Cicchetti, K.H. Nuerchterlein, & S. Weintraub (Eds.), *Risk and protective factors in the development of psychopathology* (pp. 408–423). New York: Cambridge University.

Gordon, E.W., & Song, L.D. (1994). Variations in the experience of resilience. In M.C. Wang & E.W. Gordon (Eds.), *Educational resilience in inner-city America: Challenges and prospects* (pp. 27–44). Hillsdale, NJ: Lawrence Erlbaum Associates.

Hauser, S.T., Vieyra, M.A., Jacobson, A.M., & Wertlieb, D. (1989). Family aspects of vulnerability and resilience in adolescence: A theoretical perspective. In T. Dugan & R. Coles (Ed.), *The child in our times* (pp. 109-133). New York: Brunner/Mazel.

Hawkins, J.M., & Allen, R. (Eds.). (1991). *The Oxford encyclopedic English dictionary*. Oxford: Clarendon Press.

Heller, K., Holtzman, W., & Messick, S. (1982). *Placing children in special education: A strategy for equity.* Washington, DC: National Academy of Science Press.

Hill, P.T., Wise, A.E., & Shapiro, L. (1989). *Educational progress: Cities mobilize to improve their schools.* Santa Monica, CA: Rand Center for the Study of the Teaching Profession.

Horowitz, F.D. (1989, April). *The concept of risk: A re-evaluation.* Invited address before the Society for Research on Child Development, Kansas City, MO.

Johnson, D.W., & Johnson, R. (1987). *Cooperation and competition.* Hillsdale, NJ: Lawrence Erlbaum Associates.

Johnson, D.W., Johnson, R., & Holubec, E.J. (1986). *Circles of learning: Cooperation in the classroom.* Edina, MN: Interaction Book Co.

Joyce, B., & Calhoun, E. (1995, April). School renewal: An inquiry, not a formula. *Educational Leadership, 52*(7), 51–55.

King, A. (1994). *Challenges facing the successful implementation of a full-service elementary school.* Paper prepared for the Invitational Conference on School-Linked Comprehensive Services for Children and Families, Leesburg, VA.

Kirst, M., & Kelley, C. (1992). *Changing the system for children's services.* Washington, DC: Council of Chief State School Officers.

Kirst, M., & Kelley, C. (1995). Collaborating to improve children's services: Politics and policymaking. In L.C. Rigsby, M.C. Reynolds, & M.C. Wang (Eds.), *School-community connections: Exploring issues for research and practice* (pp. 21–43). San Francisco: Jossey-Bass.

Kirst, M., Koppich, J.E., & Kelley, C. (1994). School-linked services and Chapter 1: A new approach to improving outcomes for children. In K. Wong & M.C. Wang (Eds.), *Rethinking policy for at-risk students* (pp. 197–220). Berkeley, CA: McCutchan Publishing Corporation.

Klonsky, M., & Ford, P. (1994, May). One urban solution: Small schools. *Educational Leadership, 51*(8), 64–66.

Knapp, M.S. (1994). *How shall we study comprehensive, collaborative services for children and families?* Paper prepared for the Invitational Conference on School-Linked Comprehensive Services for Children and Families, Leesburg, VA.

Knapp, M.S., Shields, P.M., & Turnbull, B.J. (1992). *Academic challenge for the children of poverty: Summary report.* Washington, DC: U.S. Department of Education, Office of Policy and Planning.

Kopp, C.B. (1983). Risk factors in development. In M.M. Haith & J.J. Campos (Eds.), *Mussen's handbook of child psychology: Vol. 2. Infancy and developmental psychology* (pp. 1081–1188). New York: Wiley.

Lash, A.A., Haertel, G.D., & Kirkpatrick, S.L. (1995). *Learning in schools: A framework accounting for effects of educational transitions.* Presentation to the annual conference of the American Educational Research Association, San Francisco, CA.

Lash, A.A., & Kirkpatrick, S.L. (1994). Interrupted lessons: Teacher views of transfer student education. *American Educational Research Journal, 31*(4), 813–843.

Lee, V.E., Bryk, A.S., & Smith, J.B. (1993). The organization of effective secondary schools. In L. Darling-Hammond (Ed.), *Review of Research, Vol. 19,* (pp. 171–267). Washington, DC: American Educational Research Association.

Levin, J.R., Shriberg, L.K., & Berry, J.K. (1983). A concrete strategy for remembering abstract prose. *American Educational Research Journal, 20,* 277–290.

Lezotte, L.W. (1990). Lessons learned. In B.O. Taylor (Ed.), *Case studies in effective schools research* (pp. 195-199). Madison, WI: National Center for Effective Schools Research and Development.

Liontos, B. (1991). *Involving the families of at-risk youth in the educational process.* Eugene, OR: ERIC Clearinghouse on Educational Management (ERIC Digest Series Number EA58).

Littel, J., & Wynn, J. (1989). *The availability and use of community resources for young adolescents in an inner city and a suburban community.* Chicago: Chopin Hall Center for Children, University of Chicago.

Long, J.V.F., & Vaillant, G. (1989). Escape from the underclass. In T. Dugan & R. Coles (Eds.), *The child in our times* (pp. 200–213). New York: Brunner/Mazel.

Long, L.H. (1975). Does migration interfere with children's progress in school? *Sociology of Education, 45*(Summer), 369–381.

Lopez, M., & Weiss, H. (1994). *Can we get here from there? Examining and expanding the research base for comprehensive, school-linked early childhood services.* Paper prepared for the Invitational Conference on School-Linked Comprehensive Services for Children and Families, Leesburg, VA.

Masten, A.S. (1994). Resilience in individual development: Successful adaptation despite risk and adversity. In M.C. Wang & E.W. Gordon (Eds.), *Educational resilience in inner-city America: Challenges and prospects* (pp. 3–26). Hillsdale, NJ: Lawrence Erlbaum Associates.

Masten, A.S., Best, K.M., & Garmezy, N. (1990). Resilience and development: Contributions from the study of children who overcome adversity. *Development and Psychopathology, 2,* 425–444.

Masten, A.S., & Braswell, L. (1991). Developmental psychopathology: An integrative framework. In P.R. Martin (Ed.), *Handbook of behavior therapy and psychological science: An integrative approach* (pp. 35–56). New York: Pergamon Press.

Masten, A.S., & Garmezy, N. (1985). Risk, vulnerability, and protective factors in developmental psychopathology. In B.B. Lahey & A.E.Kazdin (Eds.), *Advances in clinical child psychology* (Vol. 8, pp. 1–51). New York: Plenum.

Means, B., Blando, J., Olson, K., Middleton, T., Morocco, C.C., Remz, A. R., & Zorfass, J. (1993). *Using technology to support education reform* (Contract no. RR91172010). Washington, DC: Office of Research, U.S. Department of Education.

Means, B., & Knapp, M.S. (Eds.). (1991). *Teaching advanced skills to educationally disadvantaged students* (Final Report, Contract No. LC89089001). Washington, DC: U.S. Department of Education.

Miller, A., & Ohlen, L. (1985). *Delinquency and community: Creating opportunities and controls.* Beverly Hills, CA: Sage.

Mageau, T. (1990). ILS: Its new role in schools. *Electronic Learning, 10*(1), 22–32.

Mageau, T. (1991, Spring). Computers using teachers. *Agenda, 1,* 51.

Montgomery, A., & Rossi, R. (1994, August). *Educational reforms and students at risk: A review of the current state of the art* (contract no. RR91-1172011). Washington, DC: U.S. Department of Education, Office of Research.

Mortimore, P., Sammons, P., Stoll, L., Lewis, D., & Ecob, R. (1988). *School matters: The junior years.* Somerset, England: Open Books.

National Association of State Boards of Education Study Group on Special Education. (1992). *Winners all: A call for inclusive schools.* Alexandria, VA: Author.

National Center on Education in the Inner Cities. (1990). *Center for education in the inner cities: A technical proposal.* Philadelphia, PA: Temple University Center for Research in Human Development and Education.

National Education Goals Panel. (1994). *The national education goals report: Building a nation of learners.* Washington, DC: Author.

Natriello, G., McDill, E., & Pallas, A. (1990). *Schooling disadvantaged children: Racing against catastrophe.* New York: Teachers College Press.

Nettles, S.M. (1991). Community contributions to school outcomes of African-American students. *Education and Urban Society, 24*(1), 132–147.

Newcomb, M., & Bentler, P. (1990). Drug use, educational aspirations, and involvement: The transition from adolescence to young adulthood. *American Journal of Community Psychology, 14*(3), 303–321.

Newman, F.M., & Wehlage, G.G. (1991, October). Five standards of authentic instruction. *Educational Leadership, 49*(2), 8–12.

Nuckolls, M.E. (1991, September). Expanding students' potential through family literacy. *Educational Leadership, 49*(1), 45–46.

Oakes, J. (1985). *Keeping track: How schools structure inequality.* New Haven, CT: Yale University Press.

O'Day, J., & Smith, M.S. (1993). Systemic reform and educational opportunity. In S. Fuhrman (Ed.), *Designing coherent educational policy* (pp. 250–312). San Francisco: Jossey-Bass Inc.

O'Dougherty, M., & Wright, F.S. (1990). Children born at medical risk: Factors affecting vulnerability and resilience. In J. Rolf, A. S. Masten, D. Cicchetti, K. H. Nuechterlein, & S. Weintraub (Eds.), *Risk and protective factors in the development of psychopathology* (pp. 120–140). New York: Cambridge University Press.

Oxley, D. (1994). Organizing schools into small units: Alternatives to homogeneous grouping. *Phi Delta Kappan, 75*(7), 521–526.

Palincsar, A.S., & Brown, A.L. (1984). Reciprocal teaching of comprehension-fostering and monitoring activities. *Cognition and Instruction, 1*(2), 117–175.

Peng, S.S. (1994). Understanding resilient students: The use of national longitudinal databases. In M.C. Wang & E.W. Gordon (Eds.), *Educational resilience in inner-city America: Challenges and prospects* (pp. 73–84). Hillsdale, NJ: Lawrence Erlbaum Associates.

Peng, S.S., & Lee, R.M. (1994). Educational experiences and needs of middle school students in poverty. In K.K. Wong and M.C. Wang (Eds.), *Rethinking policy for at-risk students* (pp. 49–64). Berkeley, CA: McCutchan Publishing Company.

Perkins, D.N. (1991, October). Educating for insight. *Educational Leadership, 49*(2), 4–8.

Peterson, D. (1989). *Parent involvement in the educational process.* Urbana, IL: ERIC Clearinghouse on Educational Management, University of Illinois. (ED 312 776)

Pianta, R.C., Egeland, B., & Sroufe, L.A. (1990). Maternal stress and children's development: Prediction of school outcomes and identification of protective factors. In J. Rolf, A.S. Masten, D. Cicchetti, K.H. Nuechterlein, & S. Weintraub (Eds.), *Risk and protective factors in the development of psychopathology* (pp. 141–163). New York: Cambridge University Press.

Pogrow, S. (1990, January). Challenging at-risk students: Findings from the HOTS program. *Phi Delta Kappan,* 389-397.

Pollack, S., Chrispeels, J., Watson, D., Brice, R., & McCormick, S. (1988, April). *A description of district factors that assist in the development of equity schools.* Paper presented at the annual meeting of the American Educational Research Association, New Orleans, LA.

Public Agenda. (1995). *First things first: What Americans expect from the public schools.* Washington, DC: Author.

Purkey, S.C., & Smith, M.S. (1983). Effective schools: A review. *Elementary School Journal, 83,* 427–452.

Rebuild L.A. (1993, May 17). The renaissance of Los Angeles. *Fortune* [special advertising section].

Reynolds, M.C. (1994). Special education as a resilience-related venture. In M.C. Wang & E.W. Gordon (Eds.), *Educational resilience: Resilience in inner-city America* (pp. 131–137). Hillsdale, NJ: Lawrence Erlbaum Associates.

Reynolds, M.C., Zetlin, A.G., & Wang, M.C. (1993). 20/20 analysis: Taking a close look at the margins. *Exceptional Children, 59*(4), 294–300.

Rigsby, L.C., Reynolds, M.C., & Wang, M.C. (1995). *School-community connections: Exploring issues for research and practice.* San Francisco: Jossey-Bass, Inc.

Rutter, M.B. (1990). Psychosocial resilience and protective mechanisms. In J. Rolf, A.S. Masten, D. Cicchetti, K.H. Nuechterlein, & S. Weintraub (Eds.), *Risk and protective factors in the development of psychopathology* (pp. 181–214). New York: Cambridge University Press.

Rutter, M.B., Maughan, P., Mortimore, P., & Ouston, J. (1979). *Fifteen thousand hours: Secondary schools and their effects on children.* London, England: Open Books.

Sautter, R.C. (1995, January). Standing up to violence. (Special insert). *Phi Delta Kappan*, K1-K12.

Simmons, R.G., & Blyth, D.A. (1987). *Moving into adolescence: The impact of pubertal change and school contexts.* Hawthorne, NY: Aldine de Gruyter.

Soler, M., & Shauffer, C. (1993). Fighting fragmentation: Coordination of services for children and families. *Education and Urban Society*, 25(2), 129–140.

Stallings, J.A., & Stipek, D. (1986). Research on early childhood and elementary school teaching programs. In M.C. Wittrock (Ed.). *Handbook of research on teaching* (3rd ed.) (pp. 727–753). New York: Macmillan.

Straits, B.C. (1987). Residence, migration and school progress. *Sociology of Education*, 60(January), 34–43.

Straus, M. (1983). Ordinary violence, child abuse and wife beating: What do they have in common? In D. Finkelhor, R. Gelles, G. Hotaling, & M. Straus (Eds.), *The dark side of families: Current family violence research.* Beverly Hills, CA: Sage.

Sutton, R.E. (1991). Equity and computers in the schools: A decade of research. *Review of Educational Research*, 61(4), 475–503.

Taylor, R. (1994). Risk and resilience: Contextual influences on the development of African-American adolescents. In M.C. Wang & E.W. Gordon (Eds.), *Educational resilience in inner-city America: Challenges and prospects* (pp. 119–130). Hillsdale, NJ: Lawrence Erlbaum Associates.

Teddlie, C., & Stringfield, S. (1993). *Schools make a difference: Lessons learned from a 10-year study of school effects.* New York, NY: Teachers College Press.

Urban Institute. (1992). *Confronting the urban crisis.* Washington, DC: Author.

U.S. Department of Education. (1994). *The Goals 2000: Educate America Act: A strategy for reinventing our schools.* Washington, DC: Author.

U.S. Department of Education and American Educational Research Association. (1995, April). *School-linked comprehensive services for children and families: What we know and what we need to know.* Washington, DC: U.S. Department of Education.

Vaughan, E.D., Dytman, J.A., & Wang, M.C. (1987). Implementing an innovative program: Staff development and teacher classroom performance. *Journal of Teacher Education*, 36, 40–47.

Vygotsky, L.S. (1978). *Mind in society: The development of higher psychological processes.* (M. Cole, V. John-Steiner, S. Scribner, & E. Souberman, Eds. and Trans.). Cambridge, MA: Harvard University Press.

Walberg, H.J. (1984). Families as partners in educational productivity. *Phi Delta Kappan*, 65, 397–400.

Wang, M.C., & Gordon, E.W. (Eds.). (1994). *Educational resilience in inner-city America: Challenges and prospects.* Hillsdale, NJ: Lawrence Erlbaum Associates

Wang, M.C., Haertel, G.D., & Walberg, H.J. (1995). The effectiveness of collaborative school-linked services. In E. Flaxman & A.H. Passow (Eds.). *Changing populations/changing schools: The 94th yearbook of the National Society for the Study of Education.* Chicago: NSSE.

Wang, M.C., Haertel, G.D., & Walberg, H.J. (1994a). Educational resilience in inner cities. In M.C. Wang & E.W. Gordon (Eds.), *Educational resilience in inner-city America: Challenges and prospects* (pp. 45–72). Hillsdale, NJ: Lawrence Erlbaum Associates.

Wang, M.C., Haertel, G.D., & Walberg, H.J. (1994b). *Effective features of collaborative school-linked services for children in elementary schools: What do we know from research and practice?* Paper prepared for the Invitational Conference of School-Linked Comprehensive Services for Children and Families, Leesburg, VA.

Wang, M.C., Haertel, G.D., & Walberg, H.J. (1995a). *Educational resilience: An emergent construct*. Paper presented at the annual meeting of the American Educational Research Association, San Francisco, CA.

Wang, M.C., Haertel, G.D., & Walberg, H.J. (1995b). *Research on school effects in urban schools*. Unpublished manuscript, Center for Research in Human Development, Temple University, Philadelphia.

Wang, M.C., Haertel, G.D., & Walberg, H.J. (1995c). The effectiveness of collaborative school-linked services. In L.C. Rigsby, M.C. Reynolds, and M.C. Wang (Eds.), *School-community connections: Exploring issues for research and practice* (pp. 283–310). San Francisco: Jossey-Bass.

Wang, M.C., & Kovach, J.A. (1995). *Bridging the achievement gap in urban schools: Reducing educational segregation, separation and advancing resilience-promoting strategies* (CEIC Publication Series No. 95–9). Philadelphia: National Center on Education in the Inner Cities.

Wang, M.C., Nojan, M., Strom, C.D., & Walberg, H.J. (1984). The utility of degree of implementation measures in program implementation and evaluation research. *Curriculum Inquiry, 14*(3), 249–286.

Wang, M.C., & Palincsar, A.S. (1989). Teaching students to assume an active role in their learning. In M.C. Reynolds (Ed.), *Knowledge base for the beginning teacher* (pp. 71–84). Oxford, England: Pergamon Press.

Wang, M.C., & Reynolds, M.C. (1995). *Making a difference for students at risk: Trends and alternatives*. Thousand Oaks, CA: Corwin Press.

Wang, M.C., Reynolds, M.C., & Walberg, H.J. (1988). Integrating the children of the second system. *Phi Delta Kappan, 70*(3), 248–251.

Wang, M.C., Reynolds, M.C., & Walberg, H.J. (December 1994/January 1995). Serving students at the margins. *Educational Leadership, 52*(4),12–17.

Wang, M.C., & Walberg, H.J. (Eds.). (1985). *Adapting instruction to individual differences*. Berkeley, CA: McCutchan.

Wang, M.C., & Zollers, N.J. (1990). Adaptive instruction: An alternative service delivery approach. *Remedial and Special Education, 11*(1), 7–21.

Watt, N.F., Moorehead-Slaughter, O., Japzon, D.M., & Keller, G.G. (1990). Children's adjustment to parental divorce: Self-image, social relations, and school performance. In J. Rolf, A.S. Masten, D. Cicchetti, K.H. Nuechterlein, & S. Weintraub (Eds.), *Risk and protective factors in the development of psychopathology* (pp. 281–304). New York: Cambridge University Press.

Wehlage, G.G., Rutter, R.A., Smith, G.A., Lesko, N., & Fernandez, R.R. (1989). *Reducing the risks: Schools as communities of support*. New York: Falmer Press.

Werner, E.E. (1989, April). Children of the garden island. *Scientific American*, 107–111.

Werner, E.E., & Smith, R.S. (1992). *Overcoming the odds: High risk children from birth to adulthood*. Ithaca, NY: Cornell University

Winne, P.H. (1991). Motivation and training. In H.C. Waxman and H.J. Walberg (Eds.), *Effective training: Current research* (pp. 295–314). Berkeley, CA: McCutchan Publishing Company.

Wittrock, M.C. (1986). Students' thought processes. In M.C. Wittrock (Ed.), *Handbook of research on teaching* (3rd ed.) (pp. 297–314).

Wong, K.K., & Wang, M.C. (Eds.) (1994). *Rethinking policy for at-risk students*. Berkeley, CA: McCutchan Publishing Corporation.

Woodson, R.L. (1981). *A summons to life: Mediating structures and the prevention of youth crime*. Washington, DC: American Enterprise Institute.

Yancey, W. & Saporito, S. (1995). Ecological embeddedness of educational processes and outcomes. In L.C. Rigsby, M.C. Reynolds, & M.C. Wang (Eds.), *School-community connections: Exploring issues for research and practice* (pp. 193–227). San Francisco: Jossey-Bass.

Appendix—An Ecological Framework to Guide Research on Educational Resilience

Contexts	Risk Factors/Adversities and Vulnerabilities	Protective Factors That Mitigate Against School Failure	Resilience-Promoting Indicators
Home Environment	Malnutrition Poverty Toxic Environment Unemployment Chronic physical and mental illness Divorce/family dissolution Limited parental education Frequent family relocations Perinatal stress Unsafe and unhealthy neighborhoods Child maltreatment (severe neglect, abuse) Limited transportation Little or no health care Poor parenting skills Poor communication skills	**Home Environment** • Stable and organized family environment • At least one strong relationship with adult (not always parent) • Absence of discord • Family warmth • Family cohesion • Children perform chores to help family • Family nurtures physical growth • Family provides information • Family provides learning opportunities • Family provides behavioral models • Family provides connections to other resources • Family nurtures self-esteem, self-efficacy • Family nurtures mastery motivation • Family holds high academic expectations for children's behavior • Family involvement in programs and courses that advance their skills	**Child:** achievement; school satisfaction; self-efficacy; academic self-concept **Family Background:** income; maternal and parental occupations; presence of physical/mental illness; presence of father in home; degree of parental education **Family Behaviors:** participation in school and community programs; quality of family relationships; opportunities at home for children to learn; family members support for children's education; organized home environment
School Environment negative labeling of children	Academic underachievement Low expectations for student achievement Few resources Large numbers of low-SES/minority students Inadequate teaching staff Poor leadership Unsafe school Poor instructional quality Use of pull-out programs and with special needs Basic skills curricula with little higher level content presented Too little time devoted to instruction Large class size	**School Environment** *Instructional* • Student-centered learning • Active teaching • Adaptive instruction and instructional mediation • Metacognitive and self-regulated learning • Advanced curriculum content and higher order thinking skills for all children • Technologies to enhance learning • High expectations of all students • Role-modeling • Positive peer relationships *Schoolwide* • Safe and secure school • Positive academically oriented school culture • Smaller educational units where students remain with the same teacher and peers for several years • Inclusive classrooms and schools • Active parent and community involvement programs	**Demographic:** school size; % AFDC families; attendance rate; racial/ethnic mix; % free breakfast and lunch; dropout rate **Student:** achievement; school satisfaction; self-efficacy; academic self-concept; number of health and mental health services provided; **Classroom:** quality and quantity of teacher-student interactions; number of higher-order questions initiated by teacher vs. lower-order questions; participation in extracurricular activities; amount of interaction with teachers, tutors and other school-related adults; classroom climate (i.e., cohesiveness, competitiveness, cooperativeness); quantity of time devoted to instruction; % of time on higher learning; % of time on basic and remedial skills; use of student background in selecting materials and activities; use of goal-setting and other strategies for self-regulated learning; use of techniques and practices to build self-esteem; use of adaptive learning techniques; use of direct instruction; use of cooperative learning techniques

Appendix (*continued*)

Contexts	Risk Factors/Adversities and Vulnerabilities	Protective Factors That Mitigate Against School Failure	Resilience-Promoting Indicators
		• Facilitate normative and unexpected transitions between schools • Provide school-linked child- and family-oriented services • Districtwide efforts to improve quality of schooling	**School:** safe school environment; strong instructional leadership by principal; schoolwide culture emphasizing achievements; variety of student clubs and extra-curricular activities; active parent involvement program; active community involvement program; provision of health, mental health, and other services to students; programs to facilitate transitions between school and grades **Teacher:** attitudes and beliefs about all students' abilities to learn; teacher knowledge about subject matter; teacher pedagogical knowledge; teacher knowledge about students' background and culture; teacher years of experience; opportunities for teacher inservice training
Community Environment	High crime rate Unemployment Substance abuse Teenage pregnancy Few community services Fragmented community services Barriers to services (language, eligibility, cost, transportation) Unsafe neighborhood	**Community Environment** • Consistent, frequently expressed social norms and values • Prevention-oriented, school-linked integrated services addressing multiple needs • Feedback to teachers from integrated-service providers • Use of evaluation data to select collaborative approaches that match community contexts • Opportunities for children and youth to participate in meaningful community activities • Improved public safety	**Demographic:** community crime rate; unemployment rate; number of substance abusers; number of teenage pregnancies; delinquency rates **Available services:** number and types of services provided; number of children, youth, and families served; degree of service integration (i.e., school-linked services); accessibility (relaxed eligibility criteria); availability of translators; use of vans to transport clients to services **Opportunities for support and involvement:** availability of programs that mentor children and youth; opportunities for apprenticeships and job training; number of church-based activities available **Expressed community norms:** evidence of rules and regulations expressing norms; degree of consensus among community members about expressed norms; multiple settings and opportunities for student exposure to norms; evidence of community programs that provide incentives and rewards for school achievement, good behavior, and accomplishment

Section 2

Distinct Disabilities

The emphases in Section 2 are on several program areas in which the disabilities of students are distinct and in which the educational program must be correspondingly discrete. In these special education areas, disordinal interactions exist between student characteristics and treatments. For each of the subtopics/chapters in this section, coverage is intended for the full range of life experience from early childhood to adult education. For each area, the authors give attention to general trends, points of consensus and conflict, organizational arrangements, and areas where well-confirmed knowledge and literature are lacking. Also, they make observations about the extent to which prevailing practices tend to reflect the state of the art or the professional knowledge base in the field.

The first chapter of Section 2 is concerned with students with visual impairments. Coverage includes incidence and prevalence, diagnosis and planning, curriculum, instructional materials, organizational arrangements, working with parents, and more. Particular emphasis is given to early education, since that aspect of education seems so critical to the long-term development of the students considered in this chapter. A second chapter treats a similar set of topics concerning learners who are deaf or hard-of-hearing. In both of these chapters there are difficult issues about the place of residential schools and of general adherence to the "least restrictive environment" principle. The limited extent of research activities in these low-incidence areas is a major problem.

The third chapter treats programs and issues relating to students who show severe emotional or behavioral disabilities. This is an area of growing importance, possibly because of increasing incidence, high expense, high teacher "burn-out" rates, and general discouragement by educators and others. Assuredly, this is an area that requires increased attention by researchers and practitioners. Problems in this domain are not "going away" in any basic way even though the students showing these problems are often those expelled or suspended from the schools at a high rate.

A fourth chapter covers a range of topics concerned with severe/profound attenuation in mental abilities. In recent years programs for students showing severe/profound retardation have accelerated within the schools. A corresponding set of developments is occurring in the community as well; more and more often students and adults who show very significant limitations in intellectual abilities are accommodated in ordinary institutions of the community. Fortunately, an impressive cadre of researchers is dedicated to work in this field and it is advancing at a relatively good pace.

The fifth chapter in this section treats programs, research, and issues relating to speech problems and language differences among students. This is a large and complex field of enormous size and complexity. It is not always included in treatments of special and remedial education; or only parts of it may be included. So, its inclusion here is arguable. It is, nevertheless, the case that much of what is accomplished in work with students who are exceptional in speech and/or language will depend upon the same general trends as observed in other areas of education as represented in this volume.

7

Learners with Visual Impairment and Blindness

JAMIE DOTE-KWAN
Division of Special Education, California State University, Los Angeles, USA

DEBORAH CHEN
Department of Special Education, California State University, Northridge, USA

The lack of vision creates numerous challenges that could limit an individual's ability to access and learn from his or her environment. These challenges create unique educational needs that should be addressed in a disability-specific curriculum. Whether all these instructional needs can be met in integrated or inclusive settings is an area of much concern and debate in recent years. Other major issues of concern in the past decade have been in response to changes in the population of visually impaired learners. Both the increases in the numbers of infants and toddlers who are visually impaired and learners with multiple disabilities have prompted a number of studies to identify the impact of visual impairment on early development and to identify effective instructional strategies and procedures. Concerns about the decline in braille readers have caused: (a) the creation of ad hoc committees and position papers, (b) studies to improve the braille code and its use, and (c) a proliferation of legislative mandates across the nation. These trends and issues have caused teachers and personnel preparation programs in the area of visual impairment to re-examine and expand their roles to meet these challenges.

Introduction

Of all the senses, vision provides the greatest quantity and quality of information about the world. It is the primary sensory channel for obtaining information that is beyond one's own body (Barraga, 1986). It is a distance sense. Vision not only allows one to see objects in their wholeness, but provides the sensations for the interpretation of color, dimensional quality of objects, impressions of distance, and the ability to experience movement while the body remains stationary (Barraga, 1986). Hearing can inform the listener about the distance and direction of an object, but gives little information about the physical characteristics or appearance of that object. The blind child has no control over the presence or absence of sound in his or her environment; "voices come out of nothingness and return to nothingness when they cease" (Cutsforth, 1951, p. 5). Although hearing provides information about things that are beyond one's reach, for a blind child, sounds are unconfirmed by visual impressions.

More incidental learning occurs through vision than through any other sense (Barraga, 1986). Without incidental learning many social behaviors such as facial expressions, body language, and gestures essential to effective communication and interaction may be absent. The inability to see and imitate age-appropriate dress, manners, and hairstyle may impact an individual's ability to develop ongoing and mutually satisfying peer relationships (Huebner, 1986).

Although the sense of touch can provide the most complete and accurate information about a blind child's environment, it has limitations too (Barraga, 1986). Touch requires active placement of one's hands and is therefore limited by the size of the perceptual window (that is, the fingertips or hands) and the length of one's reach. This means that stars that are too far away, buildings that are too large to touch as a whole, or a rainbow that can be experienced only visually are more difficult or impossible for the learner with severe visual impairment to experience and understand through touch.

According to Piaget (1952) vision is the primary sense used in the construction of sensorimotor intelligence. Thus, a child with any significant impairment of the visual system would be expected to experience delays in the development of sensorimotor understanding of physical causality, spatial relations, and object permanence (Cutsforth, 1951). Research has supported this notion of delay. Sighted infants accomplish object permanence tasks between the ages of 6 to 12 months, while infants with visual impairments do not complete these items until they are 16 and 21 months of age (Rogers & Puchalski, 1988)

From the above discussion, it is apparent that learners with visual impairment and blindness are faced with numerous challenges that could limit or prohibit their ability to access and learn from their environment. These challenges also present several unique educational needs that must be addressed. Prior to examining these unique needs and educational programming and placement options, it is important to define this population as well as the number of children and youth who are visually impaired.

Population Description

Children with visual impairment are extremely heterogeneous. Approximately half of all children who are visually impaired have sufficient functional vision to see large print, while approximately 25% are totally blind, and the other 25% have some light perception (Buncic, 1987). The age of onset has significant implications for the type of intervention required. The learner who is visually impaired from birth or before the age of five has different learning needs from the learner who is adventitiously impaired at a later time. The latter, depending on the age of onset, may have a better conceptual understanding of the world through visual assimilation, while the congenitally visually impaired learner needs much more deliberate input. To learn concepts, the information needs to be brought to them and tactile observances need to be guided hand-over-hand with verbal mediation to be meaningful.

Besides the amount of functional vision, the conditions resulting in the loss of vision vary considerably across age groups. For example, once a rare ocular anomaly, optic nerve hypoplasia is a major cause of visual loss in infancy (Hoyt, 1986). It is associated with growth deficiency, mental retardation, seizures, hypoglycemia, motor deficiency, and behavioral problems. The incidence of blindness has increased as birthweight has decreased resulting in retinopathy of prematurity (ROP) as another leading cause of visual impairment. ROP primarily affects infants with a birthweight of less than 1,000 g (about 2.2 lb). ROP is caused by high levels of oxygen administered in incubators leading to bleeding, scarring, and eventually retinal detachment. In the early 1980s the rate of ROP was similar to the 1943–1953 epidemic years with about 2,100 infants reported as having some degree of ROP annually (Trief, Duckman, Morse, & Silberman, 1989). Approximately 23% of these infants became severely visually impaired. Finally, cortical visual impairment (CVI), which is the bilateral loss of vision with normal pupillary response and no other ocular abnormalities present upon examination, is another major cause of vision loss (Good et al., 1994). CVI is most commonly caused by perinatal hypoxia-ischemia, but it may occur as a result of trauma, epilepsy, infections, drugs, or poisons.

Prevalence

Identifying the total number of individuals with visual impairment and blindness is an extremely complicated task. This is due in part to the various definitions and criteria used to collect this type of data and the heterogeneity of the population. There is a tendency for the field of visual impairment and blindness to rely on the National Society to Prevent Blindness (NSPB) for prevalence and incidence rate data (Kirchner, 1989). One reason for this is NSPB is one of the few sources that use an official definition of legal blindness. This definition is in accordance with the federal regulations for Social Security Administration programs, that is, "visual acuity of 20/200 or worse in the better eye with best correction as measured on the Snellen Chart, or a visual field of 20 degrees or less" (Kirchner, 1989, p. 138). The NSPB prevalence rates for 1980 of legally blind children (in the United States) under 20 years of age were 41,500 with 6,900 under the age of five. The American Printing House for the Blind (APH; 1993) reported there were 52,791 legally blind students registered with APH to receive accessible educational materials. This number included 26,120 students in postsecondary or nongraded academic settings resulting in 26,671 children and youth who are legally blind birth through 12th grade.

In addition, professionals have sought different sources of prevalence data in order to identify the larger population of children and youth who are visually impaired, but not necessarily legally blind. Figures based on the Health Interview Survey conducted by the National Center for Health Statistics in the United States in 1990 estimated an age-specific rate for individuals birth through 17 years of age as 1.5 per 1,000 persons, or 95,410 individuals with severe visual impairment (Nelson & Dimitrova, 1993).

The annual count of children and youth with disabilities receiving special education and related services under special education mandates as reported by the U.S. Department of Education is another source of data related to the number of children with visual impairments. In 1991–92, there were 24,169 children ages 6 through 21 who were enrolled in programs for the visually impaired and 1,423 who were deaf-blind (U.S. Department of Education, 1993). This total of 25,592 represented 0.6% of the total number of children and youth with disabilities. Interestingly, an examination of these data over time has shown significant decreases from 1979–80 to 1991–92 in the number and percentages of students with visual impairments (12%, from 20,821 to 18,296) and deaf-blindness (42%, from 1,341 to 773) served in the nation's schools. The number of students with multiple disabilities has dramatically increased (nearly 80% from approximately 45,000 to 80,396) in the same period. It is important to note that children who have multiple disabilities including visual impairment are categorized as having "multiple disabilities" not "visual impairments." The nature of unduplicated counting may result in the lower prevalence rate than reported by other sources (Kirchner, 1989). Furthermore, national estimates of children whose multiple disabilities include visual impairment vary according to differences in samples and differences in definitions of disabilities (Chen, 1995). Research with visually impaired young children (birth to five years) indicates that 40 to 60% have additional disabilities (Bishop, 1991).

The U.S. Department of Education data provide information on children six years or older. The lack of disability-specific data under the age of six may potentially represent an additional 5,000 children with visual impairments. A survey of preschool-age children who were visually impaired from 30 states indicated more than 4,100 children were divided across two age groups—birth to two years and three to five years (Bishop, 1991).

Unique Educational Needs

Learners with visual impairment and blindness have several unique educational needs (Hazekamp & Huebner, 1989). These include:

- concept development and academic needs such as understanding concepts, developing listening skills, and developing aural comprehension;
- communication needs such as reading skills in an alternative medium (i.e., braille, large print, or aural) or writing skills (i.e., braille, print, typewriting, script writing, or slate and stylus);
- low vision training for those learners with some remaining vision, so they are able to utilize their low vision to the maximum extent possible;
- social emotional needs such as acceptable social behavior (i.e., facial expressions, body language, social distances) for communicating with others or psychological implications (i.e., knowledge about one's own eye condition, accepting limitations of one's visual impairment);
- sensory-motor needs such as the development of alternative sensory awareness, discrimination, and sensitivity and appropriate posture, balance, strength, and movement;
- orientation and mobility needs such as independent travel, spatial concepts, body imagery, sighted guide skills, use of the long cane;
- daily living skills needs including personal hygiene, dressing skills, housekeeping skills, organizational skills, and so forth that are essential to create independence and acceptance; and
- career and vocational needs beginning with career awareness, career exploration, and eventually vocational preparation as well as adaptive skills including advocacy and assertiveness training.

In addition to the traditional general education curriculum, these instructional needs should be addressed in a disability-specific curriculum, including orientation and mobility training, instruction in adaptive living skills, use of assistive devices to maximize visual efficiency, braille instruction, accessing information in alternative formats, and so forth. It also encompasses skills and knowledge needed to complement and support the general education for learners with visual impairments. This curriculum when combined with the traditional general education curriculum demonstrates the concept of the "dual curriculum" (Curry & Hatlen, 1988).

In the dual curriculum, the general education component is determined by state or local policy and includes as a minimum: those courses and competencies required for advancement to the next grade and eventually graduation (Curry & Hatlen, 1988). The disability-specific component is determined based on individual learner needs. The type of instruction (i.e., script writing, typing, braille writing) and the intensity of instruction will vary based on a myriad of variables including but not limited to the extent of vision loss, age of onset, grade level, learning rate, and age of the learner. The concept of "most appropriate placement" (MAP) requires that placement considerations be based on where the learner's needs in each of the unique skill areas as well as the general academic needs of all learners can be best provided (Curry & Hatlen, 1988). Learners with visual impairment might be placed at any level of a continuum of organizational arrangements for special education, depending upon what is judged to be most appropriate for each individual.

Least Restrictive Environment

The trend to educate all children with disabilities in the "least restrictive environment" often overshadows the need to educate students, especially learners with visual impairment, in environments where all their educational needs will be met. For the last decade a movement from segregated to inclusive settings has been increasing in special education. The full inclusion position advocates that all children regardless of their disabilities should be educated in general education classes (Stainback & Stainback, 1990). Others advocate for maintaining and using the full continuum of placement options as required by federal law.

Although residential schools have been criticized for being too restrictive, they are essential in providing comprehensive services to certain learners, particularly those living in isolated or rural areas. The low incidence nature of visual impairment often forces local school programs to provide only itinerant services. This means that learners with visual impairment are often placed in general education classrooms and a certified teacher of the visually impaired provides consultation, materials and adaptive equipment, and/or supplemental instruction on a regular basis. For the majority of students, this service delivery model has proven to be effective, yet this model cannot begin to support a beginning braille reader, who is provided only two hours of braille instruction per week. This model cannot continue to be effective if itinerant teachers in urban and suburban areas have high caseloads and those in rural and isolated areas have large service regions so that neither can provide enough direct instructional time to meet the unique needs of their students.

According to the U.S. Department of Education (1993), children and youth (age 6 to 21 years) who are visually impaired were more likely than disabled students in other categories to be placed in regular education classes (42.1%). Students with visual impairment had the second highest placement rate in regular education classes, second only to students with speech or language impairments. The remaining educational placements for students who are visually impaired were resource room (23.2%), separate class (19.9%), separate school (5.0%), residential facility (8.8%), and homebound or hospital (1.0%). Although these statistics clearly demonstrate the desire to place students in local and neighborhood schools and programs when appropriate, the

the placement in residential facilities was the third highest for students with visual impairment as compared with other disabilities categories. Deaf-blindness (25.2%) and hearing impairments (11.0%) were the two categories with the highest rates for placements in residential facilities.

Residential schools historically were responsible for providing comprehensive educational programs. They provided instruction in the general education curriculum but also stressed the dual curriculum needs of learners with visual impairment. Areas such as dressing, eating, grooming, and orientation and mobility could be taught in the natural environment and within natural routines, rather than teaching these skills at unrealistic times or settings, such as making a bed in the middle of the day in the classroom. In addition, visually impaired learners could participate actively in extra-curricular sports and recreational activities with the availability of adaptive equipment and accessible facilities as part of the extended-day instruction.

Today, residential schools for the blind and visually impaired are no longer just one of several options along the continuum of possible educational placement options (Miller, 1989). They have expanded their role to offer a variety of service delivery models. This includes serving as a primary resource to local school programs by providing technical assistance, diagnostic and evaluation services, consultation, and direct services to children and families as well as a central depository for instructional materials and books.

Head, Maddock, Healey, and Griffing (1993) conducted a survey of the residential schools for the visually impaired. Of the 43 schools, 34 chief administrative officers responded (79% response rate). Head et al. found that the total number of students for all schools was 4,704, with enrollments ranging from 31 to 410 students. Only half of these students (50%) were actually residential, 18% were on-campus day students, 12% were in-home teaching, and 20% were served in off-campus classes. Approximately one-third of the population was comprised of infants and toddlers or preschoolers, 11 and 18% respectively. Twenty-eight percent were elementary-aged students, and 43% were secondary school students.

Research Trends

The low incidence of visual impairment and the heterogeneity of the population severely limits research in this field. Traditional research paradigms are not usually possible and qualitative studies with small samples are not funding priorities. As a result, there are few researchers in the area of visual impairment. Furthermore, in personnel preparation, there are fewer than 50 full-time university faculty in the area of visual impairment nationally. The heavy teaching loads, committee work, and practicum supervision required of these single faculty members limit their motivation and opportunities to conduct research (Silberman, Corn, & Sowell, 1989). Those rare individuals who are able to conduct research have focused on intervention or instructional issues across various areas and age groups.

Researchers from related fields such as psychology and linguistics have primarily focused on examining the effects of visual impairment on development through a comparative approach as will be discussed later. This differential approach, however, does not provide an empirical base for developing instructional practices.

Research questions have been drawn from different theoretical perspectives and available research has examined a variety of narrowly defined areas with varying age groups. These practices add to the difficulty of comparing or generalizing findings across studies. As a result the field lacks a cohesive or substantive body of empirical evidence.

The following sections will discuss the major issues related to research and practice in the field of visual impairment and blindness that has emerged in the past decade in the field. This discussion is not intended to be a comprehensive review of the literature but will identify current research themes and cite representative research findings.

Infancy and Early Childhood

Recent research has identified key considerations in developing early intervention and early childhood special education programs for infants, toddlers, and preschoolers with visual impairments. Ferrell et al. (1990) conducted a retrospective study through parent interview to identify the age at which infants with visual impairments (N=39) and infants with multiple disabilities including visual impairments (N=43) attained selected developmental milestones. Findings indicate that most milestones were achieved later by the group of children with multiple disabilities and that the sequence of achievement is different for infants with visual impairments than for sighted infants. The acquisition of several skills (walks up and down stairs with alternating feet, follows two step directions, removes T-shirt independently, sings a song from memory, and toilet trained without diapers) was reported to occur earlier for infants with visual impairments than the median age for sighted children. In addition, gross motor skills were acquired at a lower median age than observed in Fraiberg's earlier study (1977). Given methodological considerations, these findings should be interpreted cautiously.

Visual impairment in infancy affects the quality and quantity of early interactions with caregivers. Infants with visual impairments demonstrate a limited repertoire of behaviors for initiating and maintaining social interactions (Urwin, 1984; Rogers & Puchalski, 1984). In addition, mothers of infants with visual impairments spent less time looking at their babies than mothers of sighted infants (Rogers & Puchalski, 1984). These findings emphasize the powerful effect of gaze and eye contact. Kekelis and Anderson (1984) found that an infant's vision loss can create difficulties in establishing joint attention, interpreting communicative intent, and expanding on a mutual topic. Similarly, Rowland (1984) reported although three mothers spoke to their blind infants (11 to 16 months) frequently, it was not in response to infant vocalization. When infants vocalized, these mothers tended to respond by touching or making sounds to their infant. After their mothers vocalized, these infants were less likely to vocalize and were more likely to smile.

With 18 older legally blind toddlers (20 to 36 months) with no other disabilities, Dote-Kwan (1995) found that maternal responsiveness to child initiations was positively related to the child's development. Maternal behaviors included responding to the child's request for help or for an object, paraphrasing or repeating child's communicative behaviors, adding new information to child's sounds or words, and pacing the rate of speech and length of pauses to promote turntaking in vocal interactions. These mothers repeated or rephrased children's communicative behaviors 75% of time. This finding differs from Rowland's observation (1984) that mothers responded to infant vocalization with touch and nonverbal sound. However, Rowland's three infants were totally blind, younger, and two were developmentally delayed with other disabilities. Dote-Kwan (1995) reported that mothers of children with less developed language used more attentional cueing (saying the children's names more than eight times during the hour observation). Research indicates that the absence of visual referents makes it difficult for parents to understand the toddlers early communications (Kekelis & Anderson, 1984). These data emphasize the need for early intervention programs to assist parents of infants with visual impairments in identifying, interpreting, and responding to the child's communicative behavior.

One of the primary tasks in early intervention is identifying environmental considerations that can promote an infant's development (Head, Bradley, & Rock, 1990). For this purpose, the Home Observation for Measurement of the Environment (HOME; Caldwell & Bradley, 1984) is a tool that is used commonly in early intervention and early childhood programs. Recently, two studies have examined the usefulness of the HOME in assessing the home environments of infants, toddlers, and preschoolers who are visually impaired. Rock et al. (1994) used modified version of the HOME with 31 visually impaired young children (six months to six years). Nineteen of these children had other disabilities in addition to visual impairments. Although more capable preschoolers received higher levels and variety of language and academic stimulation, parents of more capable infants and preschoolers seemed more likely to restrict their activities. Levels of stimulation and support in families of children with visual impairments did not seem different from the mean levels of families in general.

Another study using the HOME with 18 toddlers (20–36 months) who were visually impaired with no other disabilities (Dote-Kwan & Hughes, 1994) found a positive relationship between emotional and verbal responsiveness of mothers and the expressive language abilities of their children. In general, the home environments were favorable regardless of the socio-economic status levels of the families or the visual acuities of the toddlers. The researchers speculate that the child's visual impairment may require family and environmental adaptations and that participation in an early intervention program may contribute positively to the home environment. Although further research is needed to establish the validity of the HOME for families of young visually impaired children, preliminary findings indicate that the HOME identifies aspects of the social and caregiving environment that are likely to support the development of very young children with visual impairments.

Other research has focused on creating a responsive physical environment to promote learning in blind infants. Neilsen (1991) specifically designed an environment, the "Little Room," to enable congenitally blind infants to develop an awareness of spatial relations, object permanence, and cause effect in making sounds from objects. The "Little Room" is designed to fit over the child and is composed of a plexiglass structure with objects attached to the top and the sides within the child's reach. Twenty blind infants (5–19 months) including nine with significant developmental delays were observed in a number of sessions in the "Little Room" and under a frame with available objects. In the "Little Room" infants attended to sounds emerging from their own activities and were not distracted by environmental sounds. They were very interested in touching objects with different surfaces and sharp points and not interested in exploring objects with smooth surfaces. Their tactile search was encouraged by a high degree of tactile discrepancy within an object, for example, rattles and rubber pad. They compared details when two objects were moderately different in tactile qualities, for example, spoons and keys. In the "Little Room" infants began to produce sounds, to compare self-produced sounds, and to play sequence games with objects. The physical components of the "Little Room" encouraged infants to repeat activities with objects because the objects could be found in the same position. Many teachers have implemented the "Little Room" approach especially with infants who are visually impaired with other disabilities. However, this practice is controversial for following reasons: (a) the use of the "Little Room" is an artificially created environment that does not promote learning that can be transferred easily into everyday situations, (b) the use of this physical apparatus in a setting with nondisabled peers may stigmatize a young visually impaired child, (c) this equipment may be used incorrectly, that is, for prolonged periods without supervision, and (d) most importantly, the use of the "Little Room" may preclude a focus on social interactions and the development of a responsive caregiving environment.

A major goal of preschool programs is encouraging the development of play. Thus a primary task of teachers serving young children with visual impairment is identifying appropriate toys with interesting tactile and auditory consequences, developing games that motivate interaction between visually impaired and sighted children, and using strategies that facilitate play interactions. Several studies have identified specific considerations in serving preschoolers who are visually impaired.

Priesler and Palmer (1989) found that blind toddlers in a nursery school setting remained relatively uninvolved with toys and others. Schneekloth (1989) found that young blind children spent almost twice as much time in solitary play as low vision children and four times as much as sighted children. Sighted preschoolers spent most of their time with other children whereas visually impaired children spent one-third time interacting with adults. When visually impaired toddlers and preschoolers do interact with toys, Parsons (1986b) found that they used toys more for unintended and stereotypic actions, such as mouthing, waving, and banging. Frequently, these children wandered away from toys and tried to engage adults in interaction. Parsons (1986a) suggests that delays in imaginative and exploratory play of low vision children may be related to their limitations in accessing incidental learning.

On the other hand, Olson (1983) found similar exploratory behaviors when visually impaired and sighted children (ages two to six years) were exposed to a variety of toys. However, children with more school experience engaged in more exploratory behaviors than those who had less school experience. Troster and Brambring (1994) used a parent questionnaire to examine the play behaviors of 91 children with visual impairments (ages 4 to 72 months). Findings indicated that visually impaired children interacted less with peers than sighted children, preferred tactile-auditory games and toys, and rarely engaged in symbolic play. Sighted children engaged in more complex play at younger ages than blind children. Blind children preferred noisemaking objects, household objects, natural objects (i.e., stones), and musical toys. Sighted children preferred symbolic and construction toys, picture-touch books, paints, crayons, and play dough. When playing with others, blind children listened to cassettes or used picture-touch books while sighted engaged in complex construction play and painting. When playing with parents, blind children and their parents engaged in preschool songs, cuddling, noisemaking while sighted children and parents looked at books and read stories, and participated in complex construction games, puzzles, and sorting games.

Observations have revealed that strategies used by visually impaired preschoolers to initiate peer interactions were not necessarily successful in maintaining them (Erwin, 1993). Visually impaired preschoolers may have difficulty interpreting nonverbal communication which results in a communication breakdown with sighted peer. As a result, preschoolers with visual impairments tended to spend most of their time in solitary play (Erwin, 1993). Similarly, observations of sighted and visually impaired kindergartners and first graders found that these children did not repair breakdowns in communication between them (Kekelis & Sacks, 1992). Furthermore, childen with visual impairments tended to use physical contact rather than eye contact with peers. This strategy was not always seen as appropriate by peers.

Workman (1986) found that teachers' use of descriptions of the social environment and direct and indirect prompts encouraged social interactions between preschoolers who were visually impaired and their sighted peers. Skellenger and Hill (1994) found that following the child's lead to identify topics for play, modeling, and participating in play were effective teacher strategies in encouraging play skills in young blind children who were at the beginning level of play. Specific strategies included choosing play materials and activities similar to those used by the child, making general comments about toys, for example, "I wonder if I can get all these people on the bus," offering specific play suggestions, for example, "Will you help me put this necklace together?" (popbeads), and using hand-over-hand modeling, for example, "Let's feed the baby together." These children (ages five to seven years) demonstrated increases in more age-appropriate play skills in both shared play and spontaneous play sessions.

In summary, most research on young children with visual impairments has used a comparative approach to identify the impact of visual impairment on early development. These studies (Ferrell et al., 1990; Kekelis & Anderson, 1984; Olson, 1983; Rogers & Puchalski, 1984; Troster & Brambring, 1994; Urwin, 1984) have examined the differences in a variety of developmental areas in young blind and sighted children. However, this comparative approach does not provide an empirical base for the specific types of intervention that may be needed for individual children. Warren (1994) suggests using a within-group analysis to identify contributing factors that lead to positive developmental outcomes for visually impaired children. This position requires a method which enables identification of hereditary and environmental factors that influence the development of young children with visual impairments. The identification of these potential influences can then be used to guide the development of appropriate interventions.

Learners with Visual Impairments and Multiple Disabilities

When learners with severe and multiple disabilities have a visual impairment (Cress et al., 1981) it can be argued that they require the services of a teacher certified in the area of visual impairment. However, the issue of serving learners with severe and profound disabilities is difficult and the subject of much debate. Some question whether these learners will benefit from what the specialized teacher has to offer while proponents take the position that all children with visual impairments have a right to specialized services (Erin, 1986). In serving learners with severe and multiple disabilities, the typical responsibility of the teacher certified in the area of visual impairments is to provide consultation to other professionals, and/or direct instruction that encourages the learner's use of functional vision and/or compensatory strategies and skills.

Currently, it appears that effective instruction for learners with severe and multiple disabilities operates best within naturally occurring activities and through a team approach (Bailey & Head, 1993; Chen & Smith, 1992; Erhardt, 1987; Kelley, Davidson, & Sanspree, 1993). Such an approach requires familiarity with the learner's daily routine and real-life situations in order to develop effective programming in specific areas. Research has shown that such a functional approach promotes communication skills (Downing & Seigel-Causey, 1988), functional vision skills (Downing & Bailey, 1990; Goetz & Gee, 1987a), and orientation and mobility skills (Chen & Smith, 1992) of learners with severe disabilities and visual impairments. These learners will benefit from the specialized skills of the teacher certified in the area of visual impairments if the teacher is willing to participate in a team approach using a functional curricular model.

Research in the area of mental retardation has found that highlighting sensory input to create salient features increases the probability that a learner with severe cognitive disabilities will attend to distinctive features (Patton, Payne, & Bierne-Smith, 1986). This simple but powerful strategy has significant implications for teaching learners whose severe and multiple disabilities include visual impairment. Teachers certified in the area of visual impairments are trained to adapt materials to compensate for vision loss. Adding within stimulus instructional cues to natural environmental cues would attract visual attention, for example, providing a color outline to restroom signs. Similarly, distinctive tactile stimuli can be used to facilitate learning. Murray-Branck, Udvari-Solner, and Bailey (1991) found that a learner with vision and hearing loss unable to use braille, manual signs, or tangible objects to make choices was able to acquire a vocabulary of over 20 different texture symbols paired with meaningful referents. Each symbol had a different tactual dimension, for example, round/hard–sandpaper, smooth/bumpy–dried glue dots, and soft/dense–carpet. Distinctive tactile features facilitated learning the relationship with specific referent. Other communication programs have enhanced features of visual saliency through manipulation of size, color, contrast, shape, and graphic patterns based on the individual learner's type of visual impairment and personal preferences (Bailey & Downing, 1994).

A primary task of teachers certified in the area of visual impairment is the implementation of a visual stimulation or vision training model with low vision learners. However, there is little research on the efficacy of visual stimulation and training programs for young children with visual impairments (Lundervold, Lewin, & Irvin, 1987; Tavernier, 1993).

Research has found that contingent stimulation increases overall visual abilities in infants with visual impairments including some with additional disabilities (Sonksen, Petrie, & Drew, 1991) and increases visual fixation on an object in preschoolers who multiple disabilities includes severe visual impairments (Goetz & Gee, 1987a; Utley, Duncan, Strain, & Scanlon, 1983). Research has also demonstrated that learners with severe or profound mental retardation can be taught visual-motor tasks through specific prompting and shaping procedures (Goetz & Gee, 1987a; Mosk & Bucher, 1984). These data indicate that a systematic program of vision training is needed for learners with severe and multiple disabilities. This program should be implemented within the instructional context of other age-appropriate activities particularly when there is a natural and motivating consequence to the learner's use of vision (Goetz & Gee, 1987b).

Hall and Bailey (1989) have differentiated between a vision stimulation program that is designed to promote visual attention and a vision training program that is designed to increase specific skills including visual examining behaviors and visually guided motor behaviors. They have developed a model for training infants, toddlers, and preschoolers with visual impairments and older children with multiple disabilities to use their functional vision. This model identifies the importance of using visual cue control across visual environment management, visual skills training, and visually dependent task training. The components of visual cue control include visual coding, that is, the addition of visual cues to assist task completion, and visual conspicuity, that is, the capacity of a stimulus to engage visual attention. Clearly, teachers certified in the area of visual impairments have specialized skills that will benefit learners whose multiple disabilities include visual impairment.

Social Skills

Recent research has suggested that competent social skills play a significant role in successful outcomes for learners with visual impairments in integrated school and later vocational settings (Bina, 1986; Bishop, 1986). As discussed earlier, learners with visual impairments have demonstrated a lack of social skills in verbal and other social interactions (Erwin, 1993; Kekelis & Sacks, 1992; Van Hasselt, Hersen, & Kazdin, 1985). These findings have underscored the need for social skills training in programs serving learners with visual impairments (Erin, Dignan, & Brown, 1991). As a result, inservice training programs have been established for teachers and rehabilitation counselors as well as for learners with visual impairments (Sacks & Pruett, 1992).

A training program for four adolescent students resulted in an increase in social skills related to job interview situations (Howze, 1987). This program involved the students role-playing responses to various questions asked during simulated interviews. This study also found that long-term practice was needed to maintain these skills.

Peer-mediated training with visually impaired elementary-aged children was found to be more effective than teacher-directed training. Increases in social behaviors such as initiations, group participation, and sharing with others (Sacks & Gaylord-Ross, 1989) were more likely to be maintained and generalized across environments with the peer-mediated approach.

These studies indicate that social skills training can be effective. However, these programs require specific and ongoing instruction through modeling, peer-mediation, role-playing, and opportunities for feedback and discussion. It is unlikely that an itinerant or teacher-consultant service delivery model can provide sufficient time and resources to implement an effective social skills training program. Teachers certified in the area of visual impairment should have competencies in developing and implementing an individualized social skills training program in collaboration with general education personnel. This unique educational need of learners with visual impairments further expands the instructional responsibilities of their teachers and challenges the resources of personnel preparation programs and service delivery options.

Braille Literacy

One area of much debate and controversy over the last decade is the decrease in braille literacy among the visually impaired population. American Printing House for the Blind (APH) has reported a consistent decline in braille readers and a more disturbing increase in nonreaders. In 1968, of 19,902 individuals registered by APH, 40% were braille readers, 45% were reading large print or regular print, and 4% were reading both braille and print. According to the latest APH reported quota figures, there are 52,791 legally blind students registered with APH, but only 5,451 or approximately 10% read Braille (APH, 1993). The percentage of nonreaders among students who are legally blind and registered with APH has steadily increased from 20% in 1985 when the category of nonreader was first established (Mullen, 1990) to 31% (16,348) in 1993.

Although, professionals, consumers, and producers of braille agree there is a problem, varying opinions regarding the cause have been given. The seven most commonly cited reasons are: (a) the increase in children with multiple disabilities who are nonreaders; (b) the emphasis on increasing visual functioning; (c) the increased dependency on technology to provide an alternative communication mode to braille; (d) the complexity and intricacy of the braille code; (e) the attitude that braille is an inferior reading medium; (f) the lack of teacher competency in personnel preparation programs or the lack of qualified teachers serving learners with visual impairments; and (g) the increase in the teacher-consultant model over the self-contained model to serve learners with visual impairments in public schools (Mullen, 1990; Rex, 1989; Schroeder, 1989; Spungin, 1990).

Braille is a tactile reading and writing system originating from the work of Louis Braille in 1829. This system is based on a cell of six raised dots that provide for 63 different combinations to form the alphabetical, numerical, and grammatical characters. English Braille is comprised of Grade 1 and Grade 2 braille and its rules and usages are similar in English-speaking countries (Krebs, 1993). Grade 1 or uncontracted braille consists of the alphabet, punctuation, numbers, and specific braille composition signs. Grade 2 or contracted braille consists of Grade 1 and 189 additional contractions and short-form words.

In the early 1990s, to battle against this growing functional illiteracy rate among blind individuals, several position papers were written and ad hoc committees were established to state the importance of literacy skills, particularly braille literacy (Council of Executives of American Residential Schools for the Visually Handicapped, 1990; Committee to Develop Guidelines for Literacy, 1991). Although many speculated as to the cause of braille illiteracy, few have sought empirical evidence to support or reject any of these commonly cited reasons. Of those individuals seeking empirical truth, the areas most often dealt with in the literature are the complexity and intricacy of the Braille code, attitudes about braille, and teacher competency.

Improving the Braille Code and Its Use

One the major disadvantages of braille is the slow reading speed. Over the years researchers have attempted to improve braille reading speed through a variety of methods (Olson, 1976; Wormsley, 1978) and devices (Kederis, Morris, & Nolan, 1964; Kusajima, 1974; Nolan & Kederis, 1969). In recent years, this quest to improve reading speed still continues. Cates and Sowell (1990) used a computer-generated tachistoscope-like program and although they increased their subjects reading speeds during treatment those gains were not maintained and comprehension did not improve.

Some researchers have sought to reduce the number of contractions as a means of enhancing the acquisition of braille. In Norway, Bruteig (1987) investigated the readability of contracted versus uncontracted braille with 35 adventitiously blinded adults. She examined the reading rate in continuous, normal contracted and less contracted text and compared these reading rates with those obtained using uncontracted continuous text. The results indicated that contracted braille was significantly faster (18.5%) than uncontracted (38.7 versus 32.7 words per minute), although there were considerable individual variations in rate.

Teacher Attitudes and Training Programs.

Wittenstein's (1994) nationwide survey of 1,663 teachers of students who are blind and visually impaired found that, as a group, these teachers were confident in their braille abilities, recognize braille's importance, and strongly support its use for their students. There was a relationship between the types of braille training the teachers received in their preservice training programs and their attitudes towards braille. Preservice programs that emphasized methodology of teaching braille in addition to the teaching of the code and rules of braille produced teachers who felt more competent to teach braille and had more positive attitudes towards braille than did programs in which only braille transcription was taught. Unfortunately, only 353 (21.2%) of the respondents received the more promising type of training. Nearly 80% of the respondents reported dissatisfaction with their preparation to teach braille reading and writing. Wittenstein concluded that these results showed a strong need for teacher preparation programs to emphasize the methods of braille reading and the concepts of tactual perception in addition to the teaching of the code.

Reactions by the Field

The early 1990s has seen a proliferation of legislation related to braille literacy. There are currently 16 states that have passed "braille bills" and 11 additional states with bills under consideration (Schroeder & Hatlen, 1993). Some states merely require the individualized education program (IEP) team to consider the need for braille instruction (e.g., Arizona and Virginia) while other states require the IEP team to provide reasons if such instruction is not included (e.g., California, Maine, Minnesota, and Wisconsin).

In eight states (Illinois, Kansas, Kentucky, Louisiana, Maryland, Missouri, New Mexico, and South Carolina), in addition to the legal definition of blindness, the braille bills stipulate specific conditions under which a student who is functionally blind should receive braille instruction. These conditions may include the inability to read and write at a level commensurate with sighted peers or having a progressive vision loss that may result in functional blindness. Two states (South Dakota and Texas) have also specified procedures for conducting braille literacy assessments to determine if braille is the appropriate reading medium.

Furthermore, the majority of the states have mandated that teachers of students who are visually impaired demonstrate competency in braille reading and writing. The competency can be demonstrated by identified certification standards or examinations to be determined by either the state boards of education (e.g., in Illinois, Kansas, Maine, New Mexico, South Carolina, South Dakota, Texas, and Wisconsin,) or consistent with the National Library

Services for the Blind and Physically Handicapped, Library of Congress (e.g., in Kentucky, Louisiana, and Missouri).

The common theme across all of these laws and proposed bills is mandated braille instruction to those students who may require this learning medium. What these laws and proposed bills attempt to do is to remove as many of the so-called causes of braille illiteracy as possible. If teachers are inadequately trained or have let their braille skills diminish, then requiring them to demonstrate competency in braille reading and writing should correct this problem. Mandating braille instruction for qualified students should also override any negative teacher attitudes that might inhibit the student's access to instruction. In addition, if teachers' caseloads or the teacher-consultant model for service delivery prevent the teaching of braille reading and writing, then braille legislation should result in a different service delivery model.

In summary, the population of learners with visual impairment has significantly changed. The increase in learners with multiple disabilities, the increase in infants and toddlers, and the decline in braille readers has taxed the educational system and caused redefinition and expansion of the roles of teachers. Currently, most children are served in their home or neighborhood schools on a teacher-consultant or itinerant service delivery model. There are few children in self-contained classes and at the residential schools for the blind. Teachers certified in the area of visual impairments are required to serve the heterogeneous school-age population of learners with visual impairment as well as infants and toddlers and learners with multiple disabilities, while not neglecting their traditional role of teaching braille reading and writing. As the role of teachers expand and change, so must the role of teacher training programs.

Issues in Teacher Training

The low incidence of visual impairment, inadequate funding of preservice and inservice training programs, a trend towards a generic or noncategorical approach to education of students with disabilities, and lack of incentives for obtaining specialized certification has resulted in a severe shortage of teachers certified in the area of visual impairments (Head, 1989; Jones, 1991; Parsons, 1990; Stolarski & Erwin, 1991). Available teachers are faced with classrooms or caseloads of students representing a range of abilities and with varied and challenging roles. Some states do not required specific certification in visual impairment (e.g., Maine, Montana, South Dakota, Vermont, and Washington) while another state, New Mexico has eliminated its certification requirement (Huebner & Paige Strumwasser, 1987). Only 24 states offer teacher preparation programs in visual impairments. To meet current needs, existing personnel preparation programs need to double their capacity (Head, 1989). However, the majority of these personnel preparation programs have low student enrollment and as a result are staffed by only one full-time faculty member (Silberman, Corn, & Sowell, 1989). The majority of teachers certified in the area of visual impairments have received little if any preservice training on working with learners with severe multiple disabilities or very young children (Erin, Daugherty, Dignan, & Pearson, 1990; Huebner & Paige Strumwasser, 1987; Seitz, 1994). Data from a national survey indicated that only one state offered special certification for early childhood teachers of visually impaired students (Huebner & Paige Strumwasser, 1987).

The new and multiple responsibilities of teachers certified in the area of visual impairments demand a re-examination of personnel preparation programs and the service delivery practices. Given the lack of specific skills with infants and preschoolers who are visually impaired or with learners who are severely and multiply disabled and the nature of consultation and itinerant services, how can the teacher certified in the area of visual impairments best serve these populations?

Summary

Currently, the field of visual impairment is faced with several critical challenges. These include: (a) the shortage of teachers to serve a more diverse population (i.e., increases in the numbers of infants, toddlers, and preschoolers and increases in the numbers of learners with multiple disabilities); (b) a service delivery model that may not be conducive to meeting the learners' unique educational needs (i.e., braille instruction, functional visual training, and social skills training); (c) limited resources for research and teacher preparation programs; and (d) controversy in the field related to the least restrictive environment, braille legislation, the roles and responsibilities of teachers certified in the area of visual impairments, and specific intervention practices (i.e., the use of the "Little Room" and passive vision stimulation programs).

These challenges can be met only through collaborative efforts that involve pooled resources and a shared vision. A regional approach to personnel preparation could involve a consortium of university training programs to develop comprehensive curricula that can be delivered in nontraditional ways. These may include a variety of inservice and preservice opportunities involving distance education, intensive summer programs, and ongoing mentoring relationships for teachers of learners with visual impairments. A similar approach could be developed to promote collaborative research efforts that will result in larger samples and a more cohesive empirical base. These proposals require external funding to support the time, resources, and leadership necessary to develop innovations. This may be the most effective strategy for addressing current and future challenges in the field of visual impairments.

References

References

American Printing House for the Blind. (1993). *Distribution of Federal Quota Report.* Louisville, KY: Author.

Bailey, B.R., & Downing, J. (1994). Using visual accents to enhance attending to communication symbols for students with severe multiple disabilities. *RE:view, 26,* 101–118.

Bailey, B.R., & Head, D.N. (1993). Providing O&M services to children and youth with severe multiple disabilities. *RE:view, 25,* 57–66.

Barraga, N. (1986). Sensory perceptual development. In G.T. Scholl (Ed.), *Foundations of education for blind and visually handicapped children and youth: Theory and practice* (pp. 83–94). New York: American Foundation for the Blind.

Bina, M. (1986). Social skills development through cooperative group learning strategies. *Education of the Visually Handicapped, 13,* 27–40.

Bishop, V.E. (1986). Identifying the components of success in mainstreaming. *Journal of Visual Impairment & Blindness, 80,* 939–946.

Bishop, V.E. (1991). Preschool visually impaired children: A demographic study. *Journal of Visual Impairment & Blindness, 85,* 69–74.

Bruteig, J.M. (1987). The reading rates for contracted and uncontracted braille of blind Norwegian adults. *Journal of Visual Impairment & Blindness, 81,* 19–23.

Buncic, J.R. (1987). The blind child. *Pediatric Clinics of North America, 34,* 1403–1414.

Caldwell, B.M., & Bradley, R.H. (1984). *Home observation for measurement of the environment.* Little Rock: University of Arkansas, Center for Child Development and Education.

Cates, D.L., & Sowell, V.M. (1990). Using a braille tachistoscope to improve braille reading speed. *Journal of Visual Impairment & Blindness, 84,* 556–558.

Chen, D. (1995). Who are young children whose multiple disabilities include visual impairment? In D. Chen & J. Dote-Kwan, *Starting points: Instructional practices for young children whose multiple disabilities include visual impairment.* (pp. 1–14). Los Angeles, CA: Blind Childrens Center.

Chen, D., & Smith, J. (1992). Developing orientation and mobility skills in students who are multihandicapped and visually impaired. *RE:view, 24,* 133–139.

Committee to Develop Guidelines for Literacy. (1991). Selecting appropriate learning media for visually handicapped students. *RE:view, 23*(2), 64–66.

Council of Executives of American Residential Schools for the Visually Handicapped. (1990). Literacy for blind and visually impaired school-age students. *RE:view, 22*(3), 159–163.

Cress, P., Spellman, C., DeBriere, T., Sizemore, A., Northam, J., & Johnson, J. (1981). Vision screening for persons with severe handicaps. *Journal of the Association for Persons with Severe Handicaps, 6,* 41–50.

Curry, S., & Hatlen, P.H. (1988). Meeting the unique educational needs of visually impaired pupils through appropriate placement. *Journal of Visual Impairment & Blindness, 82,* 417–424.

Cutsforth, T.D. (1951). *The blind in school and society.* New York: American Foundation for the Blind.

Dote-Kwan, J. (1995). Impact of mothers' interactions on the development of their young visually impaired children. *Journal of Visual Impairment & Blindness, 89,* 47–58.

Dote-Kwan, J., & Hughes, M. (1994). The home environments of young blind children. *Journal of Visual Impairment & Blindness, 88,* 31–42.

Downing, J., & Bailey, B.R. (1990). Developing vision use within functional daily activities for students with visual and multiple disabilities. *RE:view, 21,* 209–220.

Downing, J., & Seigel-Causey, E. (1988). Enhancing the nonsymbolic communicative behavior of children with multiple impairments. *Language, Speech, and Hearing Services in Schools, 19,* 338–348.

Erhardt, R.P. (1987). Vision function in the student with multiple handicaps: An integrative transdisciplinary model for assessment and intervention. *Education of the Visually Handicapped, 19*, 87–98.

Erin, J.N. (1986). Teachers of the visually handicapped: How can they best serve children with profound handicaps? *Education of the Visually Handicapped, 18*, 15–25.

Erin, J.N., Daugherty, W., Dignan, K. & Pearson, N. (1990). Teachers of visually handicapped students with multiple disabilities: Perceptions of adequacy. *Journal of Visual Impairment & Blindness, 84*, 16–20.

Erin, J.N., Dignan, K., & Brown, P.A. (1991). Are social skills teachable? A review of the literature. *Journal of Visual Impairment & Blindness, 85*, 58–61

Erwin, E.J. (1993). Social participation of young children with visual impairments in integrated and specialized settings. *Journal of Visual Impairment & Blindness, 5*, 138–142.

Ferrell, K.A., Trief, E., Dietz, S.J., Bonner, M.A., Cruz, D., Ford, E., & Stratton, J.M. (1990). Visually impaired infants research consortium (VIIRC): First-year results. *Journal of Visual Impairment & Blindness, 84*, 404–410.

Fraiberg, S. (1977). *Insights from the blind.* New York: Basic Books

Goetz, L., & Gee, K. (1987a). Teaching visual attention in functional contexts: Acquisition and generalization of complex motor skills. *Journal of Visual Impairment & Blindness, 81*, 115–117.

Goetz, L., & Gee, K. (1987b). Functional vision programming. A model for teaching visual behaviors in natural contexts. In L. Goetz, D. Guess, K. Stremel-Campbell (Eds.), *Innovative program design for individuals with dual sensory impairments.* Baltimore, MD: Paul H. Brookes.

Good, W.V., Jan J.E., DeSa, L. Barkovich, A.J., Groenveld, M., & Hoyt, C.S. (1994). Cortical visual impairment in children. *Survey of Ophthalmology, 38*(4), 351–364.

Hall, A., & Bailey, I.L. (1989). A model for training vision functioning. *Journal of Visual Impairment & Blindness, 83*, 390–396.

Hazekamp, J., & Huebner, K.M. (Eds.). (1989). *Program planning and evaluation for blind and visually impaired students: National guidelines for educational excellence.* New York: American Foundation for the Blind.

Head, D.N. (1989). The future of low incidence training programs: A national problem. *RE:view, 21*, 145–152.

Head, D.N., Bradley, R.H. & Rock, S.L. (1990). Considerations for the use of home environment measures with visually handicapped children. *Journal of Visual Impairment & Blindness, 88*, 377–380.

Head, D.N. Maddock, J. Healey, W.C., & Griffing, B. L. (1993). A comparative study of residential schools for children with impairments: 1985–1990. *Journal of Visual Impairment & Blindness, 87*, 216–218.

Howze, Y.S. (1987). The use of social skills training to improve interview skills of visually impaired young adults: A pilot study. *Journal of Visual Impairment & Blindness, 81*, 251–255.

Hoyt, C. (1986). Optic nerve hypoplasia: A changing perspective. In *Pediatric ophthalmology and strabismus. Transactions of the New Orleans Academy of Ophthalmology.* New York: Raven Press.

Huebner, K.M. (1986). Social skills. In G.T. Scholl (Ed.), *Foundations of education for blind and visually handicapped children and youth: Theory and practice* (pp. 314–362). New York: American Foundation for the Blind.

Huebner, K.M., & Paige Strumwasser, K. (1987). State certification of teachers of blind and visually impaired students: A report of a national study. *Journal of Visual Impairment & Blindness, 81*, 244–250.

Jones, G. (1991). Recruitment efforts to save a teacher-preparation program. *Journal of Visual Impairment & Blindness, 85*, 29–30.

Kederis, C. Morris, J., & Nolan, C. (1964). *Training for increasing braille reading rates. Final report*. Louisville, KY: American Printing House for the Blind.

Kekelis, L.S., & Anderson, E.S. (1984). Family communication styles and language development. *Journal of Visual Impairment & Blindness, 78*, 54-65.

Kekelis, L.S., & Sacks, S.Z.(1992). The effects of visual impairment on children's social interactions in regular education programs. In S.Z. Sacks, L.S.Kekelis & R.J. Gaylord-Ross (Eds.). *The development of social skills by blind and visually impaired students*. New York: American Foundation for the Blind.

Kelley, P., Davidson, R., & Sanspree, M.J. (1993). Vision and orientation and mobility consultation for children with severe multiple disabilities. *Journal of Visual Impairment & Blindness, 87*, 397–404.

Kirchner, C. (1989). National estimates of children with visual impairments. In M.C. Wang, M.C. Reynolds, & H.J. Walberg (Eds.), *Handbook of special education: Research and practice: Vol 3. Low incidence conditions* (pp. 135–153). Oxford, England: Pergamon Press.

Krebs, B.M. (1993). *Transcriber's guide to English Braille*. Los Angeles: California Transcribers and Educators of the Visually Handicapped.

Kusajima, T. (1974). *Visual reading and braille reading: An experimental investigation of the physiology and psychology of visual and tactual reading*. New York: American Foundation for the Blind.

Lundervold, D., Lewin, L.M., & Irvin, L.K. (1987). Rehabilitation of visual impairments: A critical review. Clinical Psychology Review, 7, 169–185.

Miller, W.H. (1989). The role of residential schools for the blind in educating visually handicapped pupils. In J. Hazekamp & K.M. Huebner (Eds.), *Program planning and evaluation for blind and visually impaired students: National guidelines for educational excellence* (pp. 77–78). New York: American Foundation for the Blind.

Mosk, M., & Bucher, B. (1984). Prompting and stimulus shaping procedures for teaching visual motor skills to retarded children. *Journal of the Association for Persons with Severe Handicaps, 17*,23–34.

Mullen, E.A. (1990). Decreased braille literacy: A symptom of a system in need of reassessment. *RE:view, 22*(3), 164–169.

Murray-Branck, J., Udvari-Solner, A., & Bailey, B.R. (1991). Textured communication systems for individuals with severe intellectual and sensory impairments. *Language, Speech and Hearing Services in Schools, 22*, 260–268.

Neilsen, L. (1991). Spatial relations in congenitally blind infants: A study. *Journal of Visual Impairment & Blindness, 85*, 11–16.

Nelson, K.A., & Dimitrova, E. (1993). Severe visual impairment in the United States and in each state, 1990. *Journal of Visual Impairment & Blindness, 87*, 80–85.

Nolan, C., & Kederis, C. (1969). *Perceptual factors in braille word recognition*. New York: American Foundation for the Blind.

Olson, M. (1976). Faster braille reading: Preparation at the reading readiness level. *New Outlook for the Blind, 70*, 341–343.

Olson, M.R. (1983). A study of the exploratory behavior of legally blind and sighted preschoolers. *Exceptional Children, 50*, 130–138.

Parsons, S.A. (1986a). Function of play in low vision children (Part 1): A review of the research and literature. *Journal of Visual Impairment & Blindness, 80*, 627–630.

Parsons, S.A. (1986b). Function of play in low vision children. Part 2. Emerging patterns of behavior. *Journal of Visual Impairment & Blindness, 80*, 777–784.

Parsons, S.A. (1990). A model for distance delivery in personnel preparation. *Journal of Visual Impairment & Blindness, 84*, 445–450.

Patton, J.R., Payne, J.S., & Bierne-Smith, M.(1986). *Mental retardation* (2nd ed.). Columbus, OH: Charles E. Merrill.

Piaget, J. (1952). *Origins of intelligence in children*. New York: International University Press.

Poppe, K.J. (1991). *Distribution of quota registrants in 1990 by grade placement, visual acuity, reading medium, school or agency type, and age: A replication of Wright's 1988 study.* Louisville, KY: American Printing House for the Blind.

Preisler, G., & Palmer, C. (1989). Thoughts from Sweden: The blind child goes to nursery school with sighted children. *Child Care, Health and Development, 5,* 45–52

Rex, E.J. (1989). Issues related to literacy of legally blind learners. *Journal of Visual Impairment & Blindness, 83,* 306–313.

Rock, S.L., Head, D.N., Bradley, R.H., Whiteside, L., & Brisby, J. (1994). Use of the HOME Inventory with families of young visually impaired children. *Journal of Visual Impairment & Blindness, 88,* 140–151.

Rogers, S.J., & Puchalski, C.B. (1984). Social characteristics of visually impaired infants' play. *Topics in Early Childhood Special Education, 3,* 52–56.

Rogers, S.J., & Puchalski, C. B. (1988). Development of object permanence in visually impaired infants. *Journal of Visual Impairment & Blindness, 82,* 137–142.

Rowland, C. (1984). Preverbal communication of blind infants and their mothers. *Journal of Visual Impairment & Blindness, 78,* 297–302.

Sacks, S.Z., & Gaylord-Ross, R.J. (1989). Peer-mediated and teacher-directed social skills training for visually impaired students. *Behavior Therapy, 20,* 619–638.

Sacks, S.Z., & Pruett, K.M. (1992). Summer transition training project for professionals who work with adolescents and young adults. *Journal of Visual Impairment & Blindness, 86,* 211–214.

Schneekloth, L.H. (1989). Play environments for visually impaired children. *Journal of Visual Impairment & Blindness, 83,* 196–201.

Schroeder, F. (1989). Literacy: The key to opportunity. *Journal of Visual Impairment & Blindness, 83,* 290–293.

Schroeder, F. & Hatlen, P. (1993, March). *Braille Bills.* Paper presented at the annual conference of the California Transcribers and Educators of the Visually Handicapped. Sacramento, CA.

Seitz, J.A. (1994). Seeing through the isolation: A study of first-year teachers of the visually impaired. *Journal of Visual Impairment & Blindness, 88,* 299–309.

Silberman, R.K., Corn, A.L., & Sowell, V.M. (1989). A profile of teacher educators and the future of their personnel preparation programs for serving visually handicapped children and youth. *Journal of Visual Impairment & Blindness, 83,* 150–155.

Skellenger, A.C., & Hill, E.W. (1994). Effects of a shared teacher-child play intervention on the play skills of three young children who are blind. *Journal of Visual Impairment & Blindness, 88,* 433–445.

Sonksen, P.M., Petrie, A., & Drew, K.J. (1991). Promotion of visual development of severely impaired babies: Evaluation of a developmentally based programme. *Developmental Medicine and Child Neurology, 33,* 320–335.

Spungin, S.J. (1990). *Braille literacy: Issues for blind persons, families, professionals, and producers of braille.* New York: American Foundation for the Blind.

Stainback, S. & Stainback, W. (1990). Inclusive schooling. In W. Stainback & S. Stainback (Eds.), *Support networks for inclusive schooling: Interdependent integrated education* (pp. 3–23). Baltimore, MD: Paul H. Brookes

Stolarski, V.S. & Erwin, E.J. (1991). Course content and assignment of teacher preparation programs in vision. *Journal of Visual Impairment & Blindness, 85,* 125–128.

Tavernier, G.G.F. (1993). The improvement of vision by vision stimulation and training: A review of the literature. *Journal of Visual Impairment & Blindness, 87,* 143–148.

Trief, E., Duckman, R. Morse, A.R. & Silberman, R.K. (1989). Retinopathy of prematurity. *Journal of Visual Impairment & Blindness, 83,* 500–504.

Troster, H. & Brambring, M. (1994). The play behavior and play materials of blind and sighted infants and preschoolers. *Journal of Visual Impairment & Blindness, 88,* 421–432.

Urwin, C. (1984). Dialogue and cognitive functioning in the early language development of three blind children. In A.E. Mills (Ed.), *Language acquisition in the blind child* (pp. 142–161). San Diego, CA: College-Hill Press.

U.S. Department of Education (1993). *Fifteenth annual report to Congress on the implementation of the Individuals with Disabilities Education Act.* Washington, DC: Author

Utley, B., Duncan, D., Strain, P., & Scanlon, K. (1983). Effects of contingent and noncontingent vision stimulation on visual fixation in multiply handicapped children. *Journal of the Association for Persons with Severe Handicaps, 8,* 29–42.

Van Hasselt, V.B., Hersen, M., Kazdin, A. (1985). Assessment of social skills in visually handicapped adolescents. *Behavior Research and Therapy, 23,* 53–63.

Warren, D. (1994). Blindness and children. New York: Cambridge University Press.

Wittenstein, S.H. (1994). Braille literacy: Preservice training and teachers' attitudes. *Journal of Visual Impairment & Blindness, 88,* 516–524.

Workman, S.H. (1986). Teachers' verbalizations and the social interactions of blind preschoolers. *Journal of Visual Impairment & Blindness, 80,* 532–534.

Wormsley, D. (1981). Hand movement training in braille reading. *Journal of Visual Impairment & Blindness, 75,* 327–331.

8

Learners Who Are Deaf or Hard of Hearing

JOSEPH E. FISCHGRUND
Headmaster, Pennsylvania School for the Deaf, Philadelphia, Pennsylvania, USA

The education of learners who are deaf and hard of hearing is in a period of significant change both in terms of conceptualization and practice. These changes impact particularly upon issues of curriculum and instruction, including choice of language instruction and specially designed instructional materials, assumptions about placement and program organization, and most dramatically in the areas of language, literacy and communication. Likewise, recent trends in identification of children with hearing loss and, at the other end of the educational spectrum, transition and post-secondary planning are also of significance. These developments are described within the changing socio-cultural constructs of the education of learners who are deaf or hard of hearing and shifting paradigms for the provision of educational services and programs for this low incidence, communicatively unique population.

Introduction

On April 8, 1864 Gallaudet University, the world's only university for the deaf, was founded by Edward Miner Gallaudet. E. M. Gallaudet, the son of Thomas Hopkins Gallaudet, the person for whom the university is named, believed that higher education was possible for deaf individuals and that only through educational opportunity could deaf individuals realize their dreams. One hundred and twenty-four years later, on March 14, 1988, Dr. I. King Jordan was selected as Gallaudet's first Deaf President.

The events that led up to Dr. Jordan's selection forever changed the manner in which the world would view deaf and hard of hearing individuals. On March 7, 1988 Gallaudet's Board of Trustees choose as its seventh President Elizabeth Ann Zinser, an outstanding and well respected university administrator, but one who had no experience with deafness, sign language, or deaf culture. In a protest that drew world-wide attention, students from

Gallaudet closed the University and demanded that they have a Deaf President. The demonstration became known as the Deaf President Now movement, and in the words of Jack Gannon became "the week the world heard Gallaudet." One week later, Dr. Zinser resigned and Dr. Jordan was selected as the University's first Deaf President.

In his acceptance, Dr. Jordan stated what the world has now come to recognize; "the only thing a deaf person can't do is hear." Moreover, in the words of Edward C. Merrill (Gallaudet's 4th President), "Dr. I. King Jordan's advancement to the Presidency of Gallaudet University was a vindication of Edward Miner Gallaudet's beliefs." (adapted from Jack R. Gannon (1989))

Incidence and Prevalence

In the United States alone, a breakdown of hearing losses by age group and suggests that over one million preschool, elementary and secondary-aged children have a hearing loss (Punch, 1983; National Center for Health Statistics, 1988). Also in the U. S. almost 12 million working aged individuals (18–65) suffer hearing loss while roughly nine million retired and senior citizens must contend with hearing loss (National Center for Health Statistics, 1988). These estimates, which indicate that between eight and ten percent of the population of the U.S. experience a hearing loss may not be generalizable to other countries and settings. Reporting practices may vary tremendously depending on the nature of the health care system, the availability of reporting mechanisms, the existence of a social and educational service and intervention systems and a host of other variables. In the U.S., for example, 1989–90 Annual Survey of Children and Youth identified over 46,000 students in educational programs for the deaf and hearing impaired in the United States, only a small percentage of the estimated one million school age children who are estimated to have some degree of hearing loss. The implications of the low incidence of deafness or hearing loss that significantly impacts upon educational prospects are the possibility of lack of awareness on the part of regular and special educators, lack of appropriately trained personnel and the scattering and lack of accessibility of necessary support services. In addition, the inability to establish the necessary critical mass for meaningful communication interaction may negatively impact upon educational opportunities for deaf learners who depend upon sign communication for effective processing of linguistic information.

Definitions from a Functional Perspective

The following definitions will be utilized throughout this chapter. Note that these definitions represent a functional perspective, not only a medical or audiological view. Current practice utilizes this functional approach rather than one that specifies a specific decibel level of hearing loss for categorization purposes. This is because for educational purposes, categorization of hearing loss by audiometric criteria alone cannot be demonstrated to have a direct impact upon the delivery of educational services to children who are deaf or hard of hearing. An important distinction must be made between general terms "deafness" and other more general terms such as "hearing loss", "hearing impairment", and "hard of hearing." Deafness denotes a hearing loss so great that hearing cannot be used to develop oral language, whereas the other terms are used to describe any deviation from normal hearing, regardless of its severity. There are also philosophical, socio-cultural and educational implications of the above terms as they also reflect how deaf and hard of hearing individuals are perceived. For example, "hearing impairment," once the accepted neutral term used to identify an wide range of individuals with hearing loss, is now felt to be a negative term by many Deaf and hard of hearing individuals who do not wish to be identified as hearing individuals who are impaired.

These definitions are as follows:

Hearing Impairment—A generic term indicating a hearing disability that may range in severity from mild to profound. For educational purposes, this generic term consists of two groups of learners, those who are *deaf* and those who are *hard of hearing*:

An audiologically *deaf* person is one whose hearing loss precludes successful processing of linguistic information through audition, with or without a hearing aid. This category may include individuals who are congenitally deaf(those who were born deaf) and the adventitiously deaf (those who were born with normal hearing but who now function with a hearing loss which has a significant impact upon their communication style, modality or choice of communication partners or community).

A *hard of hearing* person is one who, generally with the use of a hearing aid, has residual hearing sufficient to enable successful processing of linguistic information through audition (Report of the Ad Hoc Committee to Define Deaf and Hard of Hearing, 1975). In general, hard of hearing persons utilize the dominant oral language of their society for the purpose of communication.

A *culturally Deaf* person (note that the word Deaf is capitalized to denote a specific cultural orientation and identification) are those individuals who affiliate and identify themselves with the Deaf community in their place of residence. In general, culturally Deaf individuals communicate primarily in the indigenous sign language of their society.

Identification and Evaluation of Hearing Loss

Special educators and educators of learners who are deaf or hard of hearing are generally well versed in issues relating to audiometric testing as a means of identification and evaluation of hearing loss. Thorough discussions of audiometric evaluation procedures are prevalent in the literature, and recent advances relate in large part to the more sophisticated computer technology available. Well-defined methods of formal audiometric testing necessary to determine the precise extent of the hearing loss and to determine the best approach to auditory rehabilitation are described extensively throughout the literature, with pure tone audiological assessment remaining as the single most prevalent form of audiological assessment.

Of critical importance for the future are programs that provide early detection and screening of infants who may have a hearing loss. Practices in this area vary throughout the world, and often are a function of the particular approach to health care in general which exists in that location or to the availability of resources. According to the Joint Committee of ASHA and the Council on Education of the Deaf;

> Children who are deaf or hard of hearing and their families/caregivers constitute a unique group whose need differ from those of their families. The variables that set children with hearing loss apart from those with other disabilities are related to the lack of full access to communication. This can have long-term effects on the child's cognitive, speech, language and social-emotional development, as well as affect the family system. Early identification, assessment, and management should: (a) be conducted by professional who have the qualification to meet the need os children who are deaf or hard of hearing, particularly infants, toddles, and their families; (b) be designed to meet the unique needs of the child and family; and (c) include families in an active, collaborative role with professionals in the planning and provision of early intervention services.

The value of early identification has long been recognized, particularly among high risk groups. As a result in countries with universal health care systems and in some other countries, high risk registries exist and are relatively effective in identifying hearing losses in infants born under identifiable high risk conditions. In a recent study by White, Behrens and Strickland, (1995) the efficacy of different types of infant screening, the authors note that even following a widely accepted protocol, (ASHA, 1991) "at least half and probably substantially more of the infants and young children with significant hearing loss would be missed by using this protocol." (p.10). They suggest that a procedure using transient evoked otacoustic emissions is a cost effective and reliable method for screening and identification and can be utilized in a variety of settings. Screening procedures themselves, however, do not guarantee appropriate follow-up or intervention where necessary and much remains to be understood about the critical next steps after screening.

Curriculum and Instruction

Choosing Communication Methods for Deaf Children

For many, many years, a fundamental debate about what communication methodology should be used to educate deaf children had raged throughout the world. In 1880 The International Congress of Educators of the Deaf passed the now infamous Milan Decree which expressed the superiority of speech over sign language. This set the stage for almost a century of exclusively oral education in the United States. In this philosophy or approach, deaf children were required to gain an education entirely through the use of residual hearing, speech reading, and print representations. However, sign languages of the Deaf continued to be used for the purposes of communication among Deaf people throughout the world.

The issue of how deaf children should be taught continues to revolve around the fundamental issues of language and communication, especially in relationship to the form of the language of instruction. Moores (1991) notes, "In a bewilderingly short period of time, in the education of the deaf in most countries have gone from oral-only, or predominately oral-only, to oral-manual." This oral-manual trend, often known as Total Communication "usually consists of a combination of speech and a sign system designed to be used in conjunction with the spoken language." Moores (1991) suggests that there is some evidence that the use of Total Communication "contributed to improved academic and linguistic performance in deaf children," but also notes that their performance, as a group, still lags significantly behind their hearing peers in educational achievement. While the "oral vs. manual" debate continues vigorously in some academic and educational forums it is less of an issue now than it once was. Programs that espouse the use of oral/aural methods are (with a few notable exceptions) less likely to criticize programs that use Total Communication or other manual approaches and simply focus on improving their own approach. Programs that use sign language are now the majority, and are rarely defensive about their use of a sign system. In fact, the major focus among programs using sign language in the instructional process is not whether to, but how to and which sign system to use.

In 1989, sparked by the publication of *Unlocking the Curriculum: Principles for Achieving Access to Deaf Education*, educators began to seriously consider the sign languages used by Deaf adults in the instructional process. For example, in the United States, Johnson, Liddell, and Erting (1989) argued for the use of American Sign Language (ASL) as the language of instruction in

234 Joseph E. Fischgrund

American schools for deaf children and youth. At the same time, a mandated program of the use of Swedish Sign Language as the primary language of instruction in schools for the deaf in Sweden attempted to put this philosophy in practice, as so-called "bilingual" programs were implemented in both Denmark and Sweden (cf. Davies, 1991). Much of the debate has focused on *which* language to use, but there unfortunately has been very little about the content and quality of the language used in the instruction of deaf students. It should be emphasized that choosing the most appropriate language or modality alone does not guarantee full access to curriculum. As important as the form of language we use is *what* is communicated in instructional settings, *how* interactions with deaf and hard of hearing learners take place and what is *expected* of language interactions with children with a range of hearing losses. Thus, it is not only the form, but the *content and function of language in the classroom* that is also an accessibility to the curriculum issue. While the field has chosen to focus on the form issue–perhaps since it is more clear cut–it is the more complex linguistic and communicative issues of how language functions in the classroom, that will determine whether full accessibility to the curriculum will be achieved.

Specially Designed Instruction

While communication continues to remain as a central issue in the education of learners who are deaf or hard of hearing, the design of instruction itself is also critical.

Much of the discussion around the structuring of instructional environments centers around group size. The trend over the last to decades has been towards more individualized, service provision driven educational programs. This has led to smaller classes, increased pull-out programs to provide support or related services such as speech, occupational or physical therapy and increased numbers of paraprofessionals in classrooms. Noted Deaf researcher Carol Padden, commenting on this trend from a cultural rather than medical/pathological perspective on learners who are deaf, comments as follows (interpreted into English from sign):

> . . . I'd like to introduce yet another, related model, which I'll call the 'psychological model.' This model espouses that we need to control how much information we give to a deaf student. . . . This model also holds that there must be a one-to-one, teacher-student relationship in which the teacher carefully controls each child's input. . . . I think this whole idea is basically flawed. . . . The key would be to encourage more interaction and information sharing among everyone in the class. . . . There would be much more interaction, more talking back and forth, and more sharing among the students. The teacher would ultimately be in control of

everything, of course, but the students would also be learning from each other. And the reason that we need to *share* control is because it gives more opportunity to the students. (p. 26–27).

The implementation of a model that reflects Padden's assumptions about deaf learners (see specific suggestions for implementing just such a model in Johnson, Liddell and Erting, 1989) will require changes in staffing patterns, recruitment, training and retention of teachers and placement practices as well as the local and national regulatory processes that govern those three areas. It will also require a far greater number of qualified deaf professionals than currently exists and a far greater involvement of the adult Deaf community in the educational process than is currently the case in most educational systems.

A second major area of instructional design relates to how learners who are deaf or hard of hearing acquire language and literacy skills in the language of their national origin. Acquisition of the oral language of national origin and its accompanying written code (where formalized) has always presented unique challenges for the individual who does not fully hear that language. In the case of deaf children, who hear so little of that language and function through a visual language, this a particularly problematical situation. Perhaps growing out of the assumption that this process needed to be managed in qualitatively different ways, specifically designed structural methods for "teaching language" have dominated language and literacy instruction for deaf individuals in most countries. Bowe (1995), citing the overwhelming weight of studies in language acquisition, calls for an abandonment of the "teaching language" paradigm and a focus on the establish environments where deaf and hard of hearing learners can *acquire* language through more natural processes.

In recent years, however, the approach known as "whole language" has been introduced into the education of learners who are deaf or hard of hearing as a promising practice in the area of literacy development. (See Abrams, 1991 for a bibliography of articles on this subject). In this approach there is more focus on "top-down" reading that incorporates prior experience, context and world knowledge as opposed to the focus on "bottom-up" reading which emphasizes decoding and the building of smaller phonological, morphological and syntactic elements into meaning. While there has been an increase in the use of whole language based approaches, questions about the efficacy of this method have been recently raised by some researchers (cf. Dolman, 1992; Shapiro, 1992; and Kelly, 1995). This criticism centers around the notion that "whole language's 'as-needed' approach to bottom up skills threatens to leave gaps in the learner's basic skill repertoire and, second, because indirect instruction is considered inadequate for teaching certain basic skills." (Kelly, 1995, p.320). This discussion will no doubt dominate most of the field's discussion of language and literacy skills acquisition in the next decade.

Program Organization

Organization of programs for learners who are deaf or hard of hearing is varied, but several primary options are identifiable. First, there are students with educationally significant hearing losses who are *individually placed* in regular education settings. Support services for these learners may consist itinerant teacher services (for either individual work or consultation to the classroom teacher), audiological support, speech/language direct intervention or consultation and, in case of students whose primary mode of communication is a signed language, interpreters. This option is most frequently utilized by learners with less severe hearing losses and those who, despite their hearing loss, communicate primarily through oral language for both social and academic purposes.

Often, where a significant number of students is present in a particular locality, classes are organized into a *local or regional day program* where there is a concentration of services to address the needs of the group. While these arrangements may require the movement of a student from their neighborhood or local school, they do maintain the student's placement in a primarily hearing environment. Depending upon the organization of the program and, even more important, the receptivity of the larger school to hosting the program, meaningful interaction with the larger school community may be available and occurring on a regular basis. In the United States, approximately 70% of all students with hearing losses reported in a 1988 survey were educated in just such local or regional day programs.

A third option for program organization is the *special school*, either day or residential. These programs, whose origins date back to the early 1800s in France, were once the most utilized placement for students who are profoundly deaf but now have become less prevalent and serve a more narrowly defined population. Most of the programs serve learners who communicate primarily through sign language, although there are those special schools which do maintain the auditory/oral approach. Often too special schools, and particularly those with residential facilities, serve children with significant special needs in addition to those required by their hearing loss.

The question of whether deaf children should be taught in a special school for deaf children (residential or day) or in a variety of other "regular" classrooms or school programs remains a controversial one throughout the world. Moores (1991) notes that "very little is currently known about the relative benefits of the different options for the educational placement of deaf children." (p.46) This is consistent with the findings of Lowenbraun and Thompson (1989) whose synthesis of studies of the influence of school placement on social, emotional, and/or academic development suggests results that are at best, equivocal. For example, the findings from a review of seven studies (Lowenbraun and Thompson, 1989) suggested overall academic advantages to mainstream placements and disadvantages relative to social and emotional growth, and findings from other studies suggest that physical integration does not necessarily promote either the use of oral language or social interaction. Educators of the deaf in the U.S., where the "mainstreaming" or "integration" movement is now nearly two decades old, are finding that the extensive support services necessary to provide full inclusion are often impossible to implement and extremely costly. In addition, even the provision of these support services does not overcome the social and cultural isolation often experienced by the deaf student.

When considering placement for a student with hearing loss, especially a placement on a primarily regular educational setting, the following are important considerations:

A fully inclusive community for children who are deaf requires full access to all communication in the environment which enables the student to feel and function as a fully participating member of the community. Establishing such an environment is complicated by the fact that children who are deaf do not hear the language of the majority community directly from its speakers. While it may be possible for a deaf student to feel a sense of belonging in a predominately hearing community, it should also be recognized that the deaf student might experience dual membership in both hearing and Deaf communities or a primary sense of identity with a Deaf community.

When considering inclusive placement options for a student who is deaf, the following processes and supports, over time, are necessary and are consistent with the NASDSE (1994) Educational Service Guidelines for deaf and hard of hearing students:

—assessment and evaluation practices utilized in developing the IEP and arriving at a placement decision must include evaluators knowledgeable about the educational ramifications of deafness/hearing loss and who are proficient in the student's communication mode, style, or language (cf. NASDSE, p. 36).

—an array of placement options in regular classes, special classes and special schools for children who are deaf must be considered for each deaf child (cf. NASDSE, p. 49).

—specially designed instruction which is communicatively accessible must be available on a consistent and ongoing basis. This instruction must be provided by qualified professionals and must take into account the academic impact of the deaf student's unique communication and language needs and the cognitive demands of functioning visually (and to the extent possible auditorily) in an oral language educational environment. (cf. NASDSE, p. 8–9, 22)

—related services which are necessary for the deaf student to benefit from both special and regular education services must be communicatively accessible to the student. (cf. NASDSE, p. 55).

—educational interpreting services, where specified in the IEP, must be effective in transmitting academic information and facilitating social interaction and must be provided for all school activities and throughout the school community. Educational interpreters should demonstrate the qualifications recommended in recognized national guidelines (cf. NASDSE, p. 52).

—necessary physical modifications of the regular school environment which promote full accessibility to all information, including captioning of all film/tape resources, acoustic treatments where appropriate, visual displays of safety and other information, lighting necessary for visual learners, etc., should be in place prior to the initiation of the placement. Appropriate assistive technology, both auditory and visual, should be readily available and staff trained in their operation and use. (cf. NASDSE. p. 26–28).

—comprehensive training of regular education personnel in communication techniques, use of assistive technology, cultural perspectives on deafness, appropriate use of interpreters, etc. must be provided on an ongoing basis. (cf. NASDSE. p. 58).

—recognition, acceptance and support for the unique cultural, communication and community needs of a deaf student which impact upon his or her social, emotional and personal development in a school setting. These needs require opportunities for communicatively appropriate peer support and interaction and the availability of deaf adult role models. (cf. NASDSE, p. 53–54).

—parents and family are the child's first community and their preference for the child's school community must be respected, considered and meaningfully included in the placement decision. (cf. NASDSE, p. 50–51).

A fundamental principle is that the choice of an educational placement for a deaf child should be made on an individual basis for each child and should not be based on prior assumptions about the relative merits of a type of setting. As Brownley (1987) noted,

> The placement of a deaf child in a public school does not assure quality. Neither does placement in a special school. . . . Quality happens where committed groups of knowledgeable people working together make individual decisions based on objective evidence for each child's present and future. (p. 341)

Transition and Post-Secondary Opportunities

A final major area of concern for educators of deaf children and youth is the area of transition to either post secondary education or to the world of work. In the area of vocational preparation, testimony to the United States Commission on Education of the Deaf (1987) indicated that if intensive specialized training does not become available, a seventy percent rate of labor force non-participation or unemployment could be predicted for deaf individuals as technological advances reduce the number and kinds of jobs they have traditionally filled. The days of technology, including complete and thorough familiarity with computers, are here. Although technology in itself will not create jobs, jobs will be there for those who have the ability to use technological applications, such as word processing, spread sheet usage, programming, and simulation models. Data entry itself is limited, and opportunities in that area may be reduced for deaf individuals as we move into an era of optical scanners and voice recognition programs. In addition, the Commission findings also indicated that about 60% of deaf high school students who graduate or drop out of school every year go directly into the labor market in semi- or unskilled jobs or remain unemployed. These rather depressing figures underscore the importance of planning, collaboration, and new approaches in the area of transition. Once again the issues of literacy must we woven into these discussions, as well as the on-going problems of appropriate post secondary training and educational opportunities, accessibility, and interpreter training and availability.

In a survey of programs in the United States for the transition from school to work, Bull and Bullis (1991) noted that "the value scale was always rated higher than the implementation scale" (p. 342) and concluded that while educators have an awareness of the importance of transition programs, "they may lack the time, methods, and resources needed to fully implement transition options." (p. 342). Interestingly, they also found that special schools, especially residential institutions "have higher implementation rates than do mainstream and other groups," and suggest that consolidating transition services for planning and referral purposes is an option that needs further investigation. A further area of concern is that of transition of deaf individuals with developmental or other disabilities. In a review of the empirical literature in this area, Davis and Bullis (1990) conclude that little empirical data exist which would lead to effective program design and thus "limits [this population's] opportunities to become contributing members of society." Within the general population of deaf and hard of hearing learners there are several specially defined groups—individuals with additional disabilities, individuals of color, late-deafened learners—who appear to be particularly under-served in the transition area.

Summary

The education of learners who are deaf or hard of hearing entered the 1990s is a state of ferment, exchange of new ideas, and intense self and re-evaluation. Much of this can be attributed to the Deaf President Now (DPN) movement at Gallaudet University in 1987. The events of that period not only had a significant effect upon the university itself, but also upon educational programs for school-age deaf children. The DPN also provided impetus for the renewed interest in the use of the language of the Deaf community in school and a greater role for Deaf persons in leading and guiding the educational systems for deaf students. This was clearly in evidence at the 17th International Congress of Educators of the Deaf (ICED) where Deaf persons assumed unprecedented primary leadership roles in the planning and implementation of the Congress. This greater participation by Deaf people themselves in the decision making process within the education of deaf children and youth signals a de-emphasis of the medical/pathological or deficit model of the education of the deaf and an emphasis upon developmental and sociocultural based approaches that "tap in the wealth of the language and culture of deaf communities throughout the world to enhance the education of all deaf children and to give them the pride and confidence they need to become members of the world community." (Hicks and Stuckless, 1990). McChord (1990) provides a fitting a challenge for the remainder of this century and the next:

> We have challenged the deaf child's right to choose his or her language; we have impeded the opportunities for deaf adult role models to work in his or her school; we have prescribed where and with whom he or she must be educated; we have restricted the o pportunities for deaf adults to serve as administrators in educational programs; and inadvertently, we have conveyed to that perceptive and tolerant child that he or she does not stand on equal footing with the child who has normal hearing. (p. 55)

McChord further challenges educators to reaffirm the "right of every deaf human being to a quality education, and the competence of deaf people to learn, to work, to lead and have premier voice in defining and attaining their individual destinies." (p. 56) It is within this changing social construct that the issues related to learners who are deaf or hard of hearing has been examined. The strengthening identity of Deaf communities, the questioning of the century old intervention paradigm, the emergence of socio-cultural based models for educating learners and the far greater participation of deaf and hard of hearing individuals themselves in the educational systems at all levels will indeed provide the contexts for research and practice in the next century.

References

Abrams, M. (Ed.). (1991). *Whole language: A folio of articles from Perspectives in Education and Deafness.* Washington, DC: Pre-College Programs, Gallaudet University.

American Speech-Language-Hearing Association (ASHA), Committee on Infant Hearing. (1989). Audiologic sreening of newborn infants who are at risk for hearing impairment. *ASHA, 31,* 89–92.

Bowe, F. (1995). *COED report: Commission perspective*. Address to CAID/CEASD national conference, Bloomington, MN.

Bull, B., & Bullis, M. (1991). A national profile of school-based transition programs for deaf adolescents. *American Annals of the Deaf, 136*(4).

Brownley, J. (1987). Quality education for all deaf children: An achievable goal. *American Annals of the Deaf, 135*(5), 340–343.

Commission on the Education of the Deaf. (1988) *Toward equality: Education of the deaf*. Washington, DC: U.S. Government Printing Office.

Davies, S.N. (1991). *The transition toward bilingual education of deaf children in Sweden and Denmark: Perspectives of language*. Washington, DC: Gallaudet University, Gallaudet Research Institute, Occasional Paper 91–1.

Davis, C., & Bullis, M. (1990). The school to community transition of hearing-impaired persons with developmental disabilities. *American Annals of the Deaf, 135*(5).

National Association of State Directors of Special Education. (1994). *Deaf and hard of hearing students: Educational service guidelines*. Alexandria, VA: Author.

Dolman, D. (1993). Some concerns about using whole language approaches with deaf children. *American Annals of the Deaf, 137*(3), 278–282.

Gannon, J. (1989). *The week the world heard Gallaudet*. Washington, DC: Gallaudet University Press.

Joint Committee of ASHA and the Council on Education of the Deaf Regarding Service Provision Under the Individuals with Disabilities Education Act–Part H, as Amended(IDEA-Part H) to Children who are Deaf and Hard of Hearing Ages Birth to 36 Months. (1993).

Hicks, D., & Stuckless, E. (Eds.) (1991). *1990 International Congress of Education of the Deaf: proceedings II: Topical addresses*. Rochester, NY: National Technical Institute for the Deaf, Rochester Institute of Technology.

Johnson, R.E., Liddell, S.K., & Erting, C.J. (1989) *Unlocking the curriculum: Principles for achieving access in deaf education*. Washington, DC: Gallaudet University, Gallaudet Research Institute Working Paper 89–3.

Kelly, L. (1995). Processing of bottom-up and top-down information by skilled and average deaf readers and implications for whole language. *Exceptional Children, 61*(4), 318–334.

Lowenbraun, S., & Thompson, M. (1989). Environments and strategies for learning and teaching. In M.C. Wang, M.C. Reynolds, & H.J. Walberg (Eds.), *Handbook of special education: research and practice. Vol, 3: Low incidence conditions*. Pergamon, Oxford.

McChord, W. (1991). Organization and administration of schools. In D. Hicks & E. Stuckless (Eds.), *1990 International Congress of Education of the Deaf: Proceedings II: Topical addresses*. Rochester, NY: National Technical Institute for the Deaf, Rochester Institute of Technology.

Moores, D. (1991). Educational policies and services. In D. Hicks & E. Stuckless (Eds.), *1990 International Congress of Education of the Deaf: Proceedings II: Topical addresses*. Rochester, NY: National Technical Institute for the Deaf, Rochester Institute of Technology.

National Center for Health Statistics. (1988). *Report of the 1985 National Health Survey*.

Padden, C. (1989). Panel response. In *Access: Language in deaf education*. Proceedings of a seminar sponsored by the Gallaudet Research Institute concerning "Unlocking the Curriculum: Principles for Achieving Access in Deaf Education." Robert C. Johnson, ed. Washington, DC: Gallaudet Research Institute Occasional Paper 90–1, Gallaudet University.

Punch, J. (1983). The prevalence of hearing impairment. *ASHA, 25*, 27.

Shapiro, H. (1992). Debatable issues underlying whole-language philosophy: A speech-language pathologists perspective. *Language, Speech, and Hearing Services in Schools, 23*, 308–311.

White, K., Behrens, T., & Strickland, B. (1995). Practicality, validity, and cost-efficiency of universal newborn hearing screening using transient evoked otacoustic emissions. *Journal of Childhood Communication Disorders, 17*(1), 9–14.

9

Learners with Emotional or Behavioral Difficulties

REECE L. PETERSON

Associate Professor, Department of Special Education and Communication Disorders,
University of Nebraska-Lincoln, Lincoln, NE 68583, USA

Students with behavioral disorders are arguably worse off today than they were 10 years ago. Controversy and disagreement regarding how these students should be defined and identified continue. No matter what one's definition, the numbers of these students unserved or underserved is large, and is increasing. The seriousness and complexity of the behavioral disorders among those students who are served have significantly increased. The social context of students with emotional or behavioral disorders continues to worsen, with students evidencing more and more environmental risk factors. With the possible exception of "service coordination," there are few new services or service delivery models that have been created to serve this population during the past 10 years. While there may be certain elements of programming or services for behaviorally disordered students that have developed or improved during the past decade (e.g., social skills instruction), on balance services and programming for these students has not significantly improved or changed from the previous decade. This may be due in part to serious shortages of trained personnel prepared to serve this population, and a decline, therefore, in the availability of expertise to treat these students.

Strong interest in service integration and coordination is one area where progress has been made, although widespread implementation of service coordination has not yet occurred. The apparent emergence of much stronger parent advocacy organizations is another area of optimism for the field, and may provide the political pressure necessary to increase the resources available for these students. Finally, the increasing recognition that earlier identification and intervention are essential to providing better outcomes for students with emotional and behavioral disorders is significant.

As was so starkly described by Schorr in her book Within Our Reach on how to break the cycle of disadvantage for children, we do know what needs to be done in order to serve students with emotional or behavioral disorders effectively. The question remains whether taxpayers, our schools, and our society have the will to do so. We must identify students with special emotional or behavioral needs earlier, we much obtain the resources and expertise to use effective interventions, and we must support the growth of resiliency factors we know can help these students, while we attack and diminish the risk factors that permeate these students' lives.

Introduction

The purpose of this chapter is to provide a brief status report of programs and services for students with emotional and behavioral disorders, and to provide an outlook for progress in this area. Focusing primarily on changes during the past 10 years, this paper will address some of the traditional topics of concern in special education related to students with behavioral disorders, such as definition, identification, and programming. It would be well beyond the scope of this chapter to attempt to summarize developments in all of these various areas, let alone actually review research across this field. Instead, this chapter will identify some of the key topics which have received attention, and will attempt to note some examples of recent developments, if any, or areas in need of additional work. It will, of necessity, be more of a brief survey than a thorough analysis. At the end, some interpretations will be made, and conclusions drawn about the current status of programs and services for students with emotional or behavioral disorders.

Definition

Given the desire of most of the professional community serving students who are considered "emotionally or behaviorally disordered" (as described below), throughout this chapter, the terminology "emotional or behavioral disorders" (EBD) will be used to refer to students who fit the federal special education category of "Seriously Emotionally Disturbed" (SED) or other varying but equivalent state terminology. All of these terms refer to students whose school progress is disabled as a result of their behavior.

Definition of this category of special education students has been a focus of intense parent and professional attention during the past 10 years. Almost from the adoption of the Bower definition in 1957 of social maladjustment as the federal definition of "serious emotional disturbance" in the 1960s, this definition was controversial (Cline, 1990; Bower, 1982). The definition, which was adopted in federal regulations even before the passage of the Education of All Handicapped Children Act of 1975, and which has remained virtually the same since that time, defines "Serious Emotional Disturbance" as:

(i) "...a condition exhibiting one or more of the following characteristics over a long period of time and to a marked degree, which adversely affects school performance: (a) an inability to learn which cannot be explained by intellectual, sensory, or health factors; (b) an inability to build or maintain satisfactory relationships with peers and teachers; (c) inappropriate types of behavior or feelings under normal circumstances; (d) a general pervasive mood of unhappiness or depression; or (e) a tendency to develop physical symptoms or fears associated with personal or school problems." (ii) "The term includes children who are schizophrenic [or autistic]. The term does not include children who are socially maladjusted, unless it is determined that they are seriously emotionally disturbed." (45 C.F.R. 121a.5(b).

In 1990, this definition was modified to make autism a separate category of disability for purposes of special education.

The definition is not an example of administrative clarity. Bower himself was one of the strongest critics of the use of his definition by federal authorities, and the subsequent interpretation and use of that definition by state and local policymakers (Bower, 1982). He objected to the terminology that was used (serious emotional disturbance) when the definition was created as a definition of "social maladjustment" for a research project concerning mental health needs of children in California in the 1950s. Bower strenuously objected to the absurdity of the clause that was added to his definition by the U.S. Congress, which specifically excluded from the definition students who are "socially maladjusted." This was the very group he had intended the definition to identify.

The Council for Children with Behavioral Disorders has, as its own name states, advocated for the use of "Behavioral Disorders" as the appropriate educational terminology for this group of students (CCBD, 1984; 1985). This was viewed as being more educationally meaningful, less stigmatizing, and more likely to be objective. Controversy about the terminology triggered a federal study of terminology (Tallmadge, Gamel, Munson, & Hanley, 1985), which concluded that a change in terminology would have relatively little impact on the field. Supporters of a change of terminology have taken exception to the study (CCBD, 1985), and have continued to advocate for new terminology along with definition revisions (CCBD, 1987).

Exclusion Clause

Recently, the use of parts of the definition to specifically exclude students with certain characteristics from special education services, has also become a major point of controversy. In particular, the clause that excludes "social maladjustment" from being considered "emotionally disturbed" was used to provide a rationale for schools to not exclude from special education services students with "Conduct Disorders," a category of the *Diagnostic and Statistical Manual*, Third Edition Revised (DSM-IIIR) (American Psychiatric Association [APA], 1987; Kelly, 1990; Slenkovich, 1983). Such an interpretation was viewed by many as inappropriate for both philosophical and practical reasons (CCBD, 1987; CCBD, 1990a; Cheney & Sampson, 1990). Many states and local districts continue this practice, while others do not. The controversy has triggered a great amount of attention about the use of DSM-IIIR (and now DSM-IV [APA, 1994]) as a diagnostic tool in determining eligibility for special education services. Some states have adopted a requirement for a DSM diagnosis in order for students to be eligible for services, often with limitations regarding which specific DSM-IIIR categories are permissible for eligibility. Others states have completely avoided using the DSM-IIIR as a component of special education diagnosis.

Emergence of Related Categories

During the late 1980s, Attention Deficit Hyperactive Disorders (ADHD), another DSM-IIIR category, also received much attention. While ADHD had been identified as a diagnostic category for some time, interest in this category mushroomed, particularly among parents, who presumed that once their child had received a diagnosis of ADHD that their child should receive services in the schools based on this diagnosis. Parents pushed very hard to have ADHD become a new category of special education, with school administrators resisting this addition in part due to two concerns. First, the large numbers of students potentially identified as ADHD would overwhelm the available resources, and, second, because they felt that the most needed ADHD students could be served in existing categories of special education. Final decision on this issue was delayed pending a federal study of ADHD, with the understanding that in the meantime these students might be served if otherwise eligible, under other categories of special education including EBD. The issue will likely be raised again in the future. In the meantime, literature on this topic has also mushroomed in the special education journals, and has included philosophical discussions as well as practical treatment options (Maag & Reid, 1994; Swanson et al., 1992).

At the same time that this controversy was occurring, Congress did add two other categories to the list of disabilities included under federal special education legislation. These were autism and traumatic brain injury (TBI). Although autism had been originally included as a part of the definition of "Serious Emotional Disturbance," it had been previously differentiated from the SED definition, and was now to become its own category. TBI was itself a rather new and vaguely defined category, which clearly had potential to overlap with some students within the SED category. Although conceptually no more clear than ADHD, both of these categories may have been success-fully added to federal special education law because it was clear that the students in these categories were already being mostly served in special education programs, whereas ADHD was not added because it had the potential to bring a large number of new students into special education.

Proposed Replacement Definition

In about 1988, the Council for Children with Behavioral Disorders began drafting a proposed substitute for the present terminology and definition. This draft became the primary focus of discussion by a newly formed group of organizations, the National Mental Health and Special Education Coalition (Forness & Knitzer, 1992). The coalition consisted of over 20 professional and parent organizations that had an interest in this topic. The coalition, by 1993, had thoroughly discussed and refined the draft until it appeared that there was near consensus on a proposed substitute for the present terminology and definition. The new terminology would be a compromise; this group of students would be identified as having "Emotional or Behavioral Disorders." Although minor modifications of the definition continue, the proposed definition is as follows:

Proposed Definition of "Emotional or Behavioral Disorder":

(i) The term emotional or behavioral disorders means a disability char-acterized by behavioral or emotional responses in school so different from appropriate age, cultural, or ethnic norms that they adversely affect educational performance. Educational performance includes academic, social, vocational, and personal skills. Such a disability (a) is more than a temporary, expected response to stressful events in the environment; (b) is consistently exhibited in two different settings, at least one of which is school-related; and, (c) is unresponsive to direct intervention in general education or the child's condition is such that general education interventions would be insufficient.
(ii) Emotional and behavioral disorders can co-exist with other disabilities.
(iii) This category may include children or youth with schizophrenic dis-orders, affective disorders, anxiety disorders, or other sustained disorders of conduct or adjustment when they adversely affect educational performance, in accordance with section (i). (Forness, 1992).

Importance of Definition

While some advocates continue to believe that a definition is not really an important issue, many on the Coalition seemed to feel that this definition could be significant. It presumably would end the controversy about terminology. It would also diminish or end the inappropriate exclusion of "conduct disordered" students from this category. It had the potential to make more objective the identification criteria, while directly adding the element of "ethnic norms" to the definition, which might address the overrepresentation of African Americans in this category. Perhaps the most significant aspect of this effort was that it represented the first instance when these broad groups of rather diverse organizations, which included parents, administrators, and mental health professionals as well as special educators, had come together to arrive at compromise and consensus regarding a substantive issue in this field. If this broad group of organizations could agree on a definition, perhaps they could continue to address other important issues and needs in the field.

The new proposed terminology and definition were supposed to be included in the next reauthorization of the Individuals with Disabilities Education Act (IDEA); however, the unity of the coalition cracked, and other events began to overtake the effort. Even after participating in the development of the proposals, the National Association of School Boards later withdrew its support, and indicated that it would oppose the changes. In addition, the Council for Administrators of Special Education, whose support had been shaky all along, also indicated that it could not support the proposals. It became unclear whether the coalition's proposals would be able to overcome the opposition of these groups, and the U.S. Office of Education attempted to mediate a resolution to this dispute. The major stumbling block appeared to be concerns that the new definition would permit too many students to be identified, with concomitant concerns for increased special education budgets to accommodate these students. As a result of this development it is unclear whether these ideas will even be proposed in the next reauthorization of IDEA, let alone win passage.

Decategorization

Other events and issues may make this issue moot. In the context of much criticism of special education labeling, a very powerful movement towards full inclusion of all students with disabilities into regular education programs was launched. Simultaneously, intense efforts to limit special education funding, and concern for the paperwork, time, and cost of special education categorization have occurred. It seems possible now that the U.S. Department of Education may propose the elimination of all of the special education eligibility categories to be replaced with one definition similar to the one currently in the 1990 Americans with Disabilities Act and the 1973 Rehabilitation Act (Section 504). Such a definition might simply be that any student with "a physical or mental impairment which substantially limits the major life activity of learning, and who by reason thereof requires special education" would be entitled to IDEA services (Capitol Publications, 1995). If this occurs, specific reference to both terminology and definition of SED or EBD would appear moot at the federal level, and it is unclear what the result of this would be for individual states.

Conclusion

Aside from the obvious potential for loss of the huge amount of time and energy that have been spent on terminology and definition of EBD students, this controversy well illustrates the general difficulties of establishing eligibility for special education, and the very clouded and confused situation of EBD within that larger picture. It would seem to illustrate well the incapacitating effect of definitional indeterminacy that affects the politics of special education for emotionally or behaviorally disordered students.

Identification

Little recent attention in the field has been directed to general procedures and processes for identifying EBD students, with the exception that some specific attention has been directed to the implications for identification procedures of the controversies discussed above regarding "social maladjustment," "conduct disorders," "attention-deficit hyperactive disorders," and the proposed new terminology and definition. As mentioned earlier, the very use of DSM-IIIR diagnoses as a part of special education identification procedures has been controversial.

Best professional practice continues to suggest that a multisource, multimethod type of comprehensive assessment procedure should be employed (Wood, Smith, & Grimes, 1985). Such procedures should obtain information across many domains (behavior, academic, intellectual, social skills, vocational, etc.), should be obtained from multiple sources (parents, various teachers, psychologists, other service providers, etc.), and should use multiple methods of assessment (interviews, questionnaires, standardized and informal tests, curriculum-based assessment, behavior rating scales and checklists, and direct observations). While not all of these domains, sources, or methods of data gathering would be used in each case, a broad-based assessment that intentionally gathers data about a student is less likely to be mistaken or biased, and it will be able to encompass a more complete understanding of that child's functioning. A multidisciplinary assessment based on this broad base of information is not only valuable for making an accurate identification or verification of the student's eligibility as a student with EBD, but is also most likely to lead to an effective plan to address the student's educational needs in his or her IEP (Rizzo & Zabel, 1988).

One issue that has been re-emphasized and that cuts across the various domains of an individual child evaluation is the need for "functional" assessment. A functional assessment approach focuses on the specific deficits or excesses in performance that are associated with the problem or problems which have lead to a student being referred, and/or to the ecological variables that affect the student's functioning in a particular environment (Maag, 1989). It focuses less on comparisons of one student to a norm group, and more on the meaning, value, or purpose of the behavior. In so doing, a functional assessment, like a curriculum-based assessment, is viewed as more likely to directly lead to interventions that will be effective in addressing the problems (Cessna, 1993). Calls for more functional assessment and programming have been made related to specific populations such as autism (Donnellan, 1984), and attention deficit hyperactivity disorder (Maag & Reid, 1994). This has led to a focus on "needs based programming" as an approach in serving students with behavioral disorders (Cessna, 1993).

Low Incidence but Increasing Prevalence

One consistent theme in discussions of students with EBD is that it is an underidentified population. The number of SED students served with educational programming according to the U.S. Department of Education (1993) was approximately 400,000 children during the 1991–92 school year. This represents about 8.4% of all of the students with disabilities, and is a 2.6% increase from the previous year. Nevertheless, it represents only just less than 1% of all children of school age, making it a "low incidence" disability. The number of students with EBD is substantially less than the three very large groups of students with disabilities (learning disabilities, speech or language impairments, and mental retardation), and yet substantially larger than the other traditional "low incidence" disabilities (deafness, visual impairments, etc.). While there has been a very slow growth both in the numbers of students identified as EBD, and in the proportion of students with EBD among all students with disabilities, the increases have been quite small, particularly in comparison to the large growth in the learning disabilities and speech/language disabilities categories.

By contrast, estimates of the prevalence of EBD have universally expected many more students than the number actually served. These estimates have ranged from 2% to 25% of the total school aged population, with a conservative estimate of the number of students with serious mental health needs that significantly interfere with school functioning to be in the 2% to 5% range (Rizzo & Zabel, 1988; Kauffman, 1989; Walker, Colvin, & Ramsey, 1995). A recent report (Zill & Schoenborn, 1990) that asked parents to indicate whether their child had emotional or behavior problems indicated that 18.5% of the children had experienced these problems by age 17, and that the percentage of children who had received psychological treatment for emotional or behavioral problems had nearly doubled from a study conducted just eight years earlier. This study tends to confirm once again that a substantially larger number of children seem to have serious and significant emotional or behavioral disorders than are presently served in special education programs. Probably well more than a million students could, or perhaps should be, identified, but only about 400,000 are presently served.

There are many potential explanations for this underidentification. Certainly there is confusion about definitions and varying criteria for identifying EBD students from state to state, and from school district to school district. Given the confusion and lack of agreement about definitions, prevalence rates may be viewed as arbitrary. A relative lack of expertise and training in this area tends to make schools reluctant to identify students. Concern for the stigma of special education, and the EBD or SED label in particular, may also be preventing some students from being identified and receiving services. Since EBD is an "invisible" disability, and is often related to intangible and changing environmental factors, it may vary in the degree to which it interferes with school functioning. It is then possible that school personnel simply do not "see" many EBD children as having these disorders. Possibly some students with emotional or behavioral disorders are being served under other categories of special education, particularly "learning disabilities," for many of the reasons just discussed; many of these students with EBD also have learning and academic problems, and are simply labeled as learning disabled. Prevalence estimates may also have very large errors and be subject to financial and political pressures. Kauffman (1989) has also noted social policy and economic factors, such as the costs of serving all students with EBD, which likely limit the number of students in this category (see also Moynihan, 1993). The lack of quick, easy, or effective interventions may also make it psychologically more difficult to accept this label for a student by parents or by educators. Many school personnel have stated that they would resist categorization of students as EBD because to do so would limit the use of disciplinary options such as suspension or expulsion with these students. While all of these explanations are plausible, at least in part, the fact is we simply do not know why there is such a discrepancy between the numbers of children who we believe could or should be identified as EBD and the actual numbers of children identified as EBD who are receiving special education services.

To the extent that EBD in children is associated with factors of poverty, poor health, nontraditional family status, and other sociological factors (discussed as risk factors in more detail below), the actual prevalence of this disability has likely continued to increase during the past 10 years (Zill & Schoenborn, 1990). The prevalence of these risk factors has also risen dramatically in the past 10 years, and continues to increase. Anecdotal evidence suggests that the number of children experiencing behavioral difficulties is increasing, probably much faster than the minor increases in identified EBD students during the past 10 years. This increase is not reflected in traditional estimates of prevalence and is very difficult to quantify or identify specifically in light of definitional problems.

The Nature of EBD Students

Anecdotal evidence also suggests that the nature of the EBD student served in special education has significantly changed during the past 10 years. Teachers who have worked with EBD students for an extended period of time indicate that the severity and complexity of students served in these special education programs have increased dramatically. Students who had serious problems of adjustment to school and who were serious classroom management problems were in programs for EBD students 10 or 15 years ago. Teachers report that today such students would not even be considered eligible for EBD programs, and that the nature and complexity of the problems the current students have are much more severe today (Peterson, 1993b). These teachers were not implying that the students 10 years ago should not have been served. They were genuinely in difficulty and deserving of services. The students today simply have far more serious and complex problems than did these earlier students.

One recent study indicated that all 181 of the EBD students in a sample of students in public school EBD programs were elevated more that two standard deviations on one or more of the broad band scores (total problem, internalizing, or externalizing scores) of the Child Behavior Checklist (Peterson & Mattison, 1995). Each and every one of these EBD students received a DSM-IIIR diagnosis based on ratings of their behavior, while in the same study comparison groups of 138 at-risk nonlabeled students and 162 students with learning disabilities received these highly elevated scores and DSM-IIIR diagnoses only in a small percentage of cases (Peterson, Conoley, & Lawson, 1995). It would certainly appear that the EBD category is serving only the most severe behaviorally disordered students.

There is virtually no discussion today of false positives, students incorrectly placed in this category, either in the literature on behavioral disorders or in programs serving these students. There is no longer discussion of the possibility that these students were inappropriately placed in this category, even though such discussion does occur in the other larger categories of disability. Conversely, there seems to be a general recognition that there are numerous false negatives for EBD in the schools, students who are not identified, perhaps because they do not "act out" their aggression, or because their level of aggression has not yet reached a crisis threshold. Underidentification seems acknowledged and widespread.

The EBD students served in special education today most likely have serious aggressive behavior that is literally threatening to others or to the students themselves. The vast majority of students with EBD evidence antisocial and aggressive behavior (Walker et al., 1995), and more and more frequently these students are capable of violence against themselves or others. These violent behaviors are indeed likely to receive attention. While many of the aggressive students are likely to also be clinically depressed, a wide range of diagnoses are possible among those who are identified EBD (Peterson et al., 1995). The most common feature continues to be aggressive or acting out behavior. Of course, aggressive behavior is not restricted to identified EBD students. Schools have again become very concerned about the apparently increasing levels of violence committed by children and adolescents.

Risk Factors

There are, however, other environmental or situational factors that are associated with increased risk of emotional or behavioral disorders. These risk factors include poverty; family status; family violence, abuse, or neglect; drug or alcohol use; poor health; ethnicity; and gender.

Poverty

By far the most extensive factor that is associated with emotional or behavioral disorders is poverty. Of the 408 clients and their families involved in the Kentucky IMPACT study (Nelson & Pearson, 1992), over 52.4% had incomes below the poverty level. Schorr (1989) has detailed the debilitating and compounding impact of poverty on children that spills over into all aspects of their lives. Poverty also probably plays a role in lessening the inability of the parents and family to be able to respond to their children's difficulties.

Family Status

Family status, particularly when a child is in a situation other than with both natural parents, is nearly equal to poverty as a correlate of EBD. Many have noted that single parent families are much more likely than other families to be in poverty. In one study, the Kentucky IMPACT study, 54.2% of the families had divorce between natural parents. Evidence is mounting that dissolution of intact two-parent families is harmful to large numbers of children (Whitehead, 1993).

Family Violence, Abuse, or Neglect

A very large number of students with EBD have experienced abuse. Kentucky IMPACT data indicated that 40.3% of students with EBD had at least one report of physical abuse in their files, 36.3% had documented sexual abuse in their files, and 58.1% were reported as having a verified history of family violence (Nelson & Pearson, 1992). Farber and Egeland (1987) in their review of invulnerability among abused and neglected children stated, "To our knowledge, no one has presented data indicating that there are children who function competently despite an ongoing exposure to abuse." (p. 283) Clearly abuse, particularly sexual abuse, may be an antecedent of emotional or behavioral disorders.

Drug or Alcohol Use or Abuse

A study in Kentucky (Nelson & Pearson, 1992) indicated that 48.1% of a sample of 408 clients ages birth to 20 and their families had a history of chemical dependence or abuse. 8.7% of the youngsters themselves were drug or alcohol dependent.

Poor Health

Although perhaps related to poverty and the lack of appropriate medical treatment, students with EBD seem also to have histories of serious health problems (Schorr, 1989). Chronic nutrition problems, infections, fatigue, and other health problems exacerbate if not trigger some behavioral disorders (Peterson, 1993a). Social supports need to be in place to foster good health (Brunswick, 1985), and often these supports are not in place for EBD students, resulting in poor health.

Ethnicity

African Americans (males in particular) are represented in programs for EBD students at more than twice the expected rate. Although good data are not available, it appears that Hispanic and Native American groups are approximately represented in proportion to their population, while Asian groups are underrepresented. While much speculation has occurred, the reason for the overrepresentation of African-American males remains mysterious. It is also not clear whether the other groups are inappropriately underrepresented, or whether cultural factors serve to insulate some groups from identification in this category.

Gender

Boys have been overrepresented in programs for EBD students by more than four or five to one. This differential propensity for boys to be more likely to develop mental health difficulties in stressful situations has been observed across numerous situations (Garmezy & Rutter, 1983). Boys have been noted to be more likely than girls to be antisocial, aggressive, or violent (Walker et al., 1995). Therefore, boys are much more likely than girls to be sent to the office for school discipline.

Other Risk Factors

While some other factors have also been identified such as a history of family mental illness, or a history of parent criminal activity, many of these factors begin to overlap with and become intertwined with those listed above. Others have looked at similar factors, but have formulated them in terms of life events that are stressors (Johnson, 1986). Many such factors can be identified, and are associated with other related problems such as delinquency (Wright & Wright, 1994).

Interaction of Risk Factors

Seifer, Sameroff, Baldwin, and Baldwin (1992, as reported in Conoley, 1995) have shown that children experiencing more than 3 risk factors (from a list of more than 30) are likely to have significant maladjustment. These risks make them less able to manage the stresses within their lives and to deal with other environmental demands such as learning and adapting to school environments. Conoley (1995) reports that when a group of about 40 EBD students who were served in a public school run day-treatment program were rated on the presence or absence of 31 risk factors, the students had a mean of 10.6 factors and a median of 9.5 factors present. Secondary students had a mean of 12 risk factors, while elementary students had a mean of 9. These data support the fact that identified EBD students are at an extremely high level of risk according to these kinds of factors, and probably the more the risk the more severe the problems among EBD students.

Resiliency Factors

Although much slower in being recognized, resiliency factors that might have a prophylactic effect related to the development of mental health problems have also been identified (Brooks, 1994; Grizenko & Pawliuk, 1994). Although formulated differently in various studies, these usually include temperament, caring adult relationships, community support, and feelings of competence.

Temperament

Some children due to innate personality or temperament seem to have better coping resources than others. Temperament probably has both biological and environmental components, and includes many within child characteristics (Constantino, 1992). Temperament can at an early age define very positive as well as very difficult characteristics, and the degree to which temperament can be affected or changed by the environmental components is not known. There is little doubt that genetic predispositions to various types of problems (e.g., dyslexia [Omaha researchers aid discovery of dyslexia gene, 1994]), as well as to resiliency factors will likely be discovered in the future (Kagen, 1994). These discoveries may be very important to the ways that emotional and behavioral problems are conceptualized, and to the ways that interventions for EBD are designed.

Caring Adult Relationships

The role of a significant adult relationship (typically a parental relationship) is one that is being viewed as being more and more important to the outcome for children with mental health needs (Anthony & Cohler, 1987; Garmezy & Rutter, 1983). A great deal of attention has focused on the disintegration of the traditional two parent family, and how this can affect the availability of adult caring relationships. Some have even begun to characterize EBD students as children who have impairments in adult relationships (Brendtro, Brokenleg, & Van Bockern, 1990) or as children who are intensely lonely. The role of caring among teachers has also begun to be emphasized (Morse, 1994) and will likely be a larger issue. Can teachers systematically develop important relationships with the children who are most in need? How can teachers foster caring adult relationships for children, regardless of who the adult is?

Community Support

Close personal relationships may also result from social networks and the community (Garmezy & Rutter, 1983). It is possible that relationships to adults or others may serve to increase the ability of some children to resist stress and build internal coping resources. Although research on peer networks (Leone & Trickett, 1990) and friendships is not well developed, it is commonly believed that even peer support can be a useful boost to coping resources. Gangs are hypothesized as filling this type of need in an antisocial form (Kodluboy, 1994). The work of Coleman (1987) to conceptualize this support as "social capital" may be very useful and applies at the community support level (Flora, 1994), as well as within an individual family. Many new initiatives in schools, such as mentoring programs, business partnerships, foster grandparent programs, and many others, might directly or indirectly help to begin to provide this kind of community support or "social capital."

Feelings of Competence

Teachers have long recognized the apparent value and desirability of having students have strong self-esteem. Such self-esteem may also be a factor in a child's ability to weather adversity. Creative competence may be an important way that children adapt (Anthony, 1987). For children, both academic and social competence are important (Scott-Jones, 1991). Many school programs for EBD students focus on building both academic and social competence.

Other Resiliency Factors

As with risk factors, many other resiliency factors could be identified that may be important. Differing formulations of these factors overlap and interact with each other. However, research on resiliency factors has not occurred as frequently as on risk factors, and therefore the impact of these resiliency factors seems more difficult to trace (Luthar & Zigler, 1991). Resiliency factors need additional research, but like other preventive interventions, they have been slow to capture attention or resources.

Interaction of Resiliency Factors

As Schorr (1989) describes eloquently, it may not be the existence of any one risk factor that is most important. Instead it is likely the existence of multiple risk factors that creates the increased likelihood of negative outcomes. The more risk factors present, the more likelihood of negative outcomes. To some degree the resiliency factors may work the same way. The likelihood that resiliency factors will insulate or inoculate children from negative outcomes probably increases with a larger number of these kinds of positive factors available to an individual child.

Early Identification

While students with EBD have tended to enter special education in the largest numbers at adolescence, there is increasing evidence that these students have had serious problems throughout their school careers. The vast majority of EBD students enter special education programs while they are in fifth to ninth grade or from ages about 10–13 (U.S. Department of Education, 1992; Peterson, 1993b). Given the extreme complexity of these students, combined with the onset of "normal" problems of adolescence, the likelihood is high that EBD students have by that time exhausted the resources of both their families and the school, and their behaviors have worsened to the point that they many times cannot be maintained in the classroom setting. It only makes sense that these students be identified and treated at an earlier time when their inappropriate behavior is less ingrained, and more effective resources can be employed to bring about change.

The situation would be analogous if students with preexisting serious hearing impairments were not identified until ages 10–13. By that time the hearing loss would be compounded by serious language and academic difficulties, social and adjustment problems, frustration, and loss of self-esteem. At that point, the chances of effective remediation of language or other deficits would be difficult if not impossible, even given maximal resources. Yet this appears to be the current situation for most students with EBD. Problems that could be identified early are not identified until they are so severe that they are almost impossible to overcome.

Fortunately, researchers have begun to develop the tools for earlier identification, have begun to prove the effectiveness of early identification, and have begun to advocate more strenuously for it (Constantino, 1992; Walker et al., 1995). According to Walker, Colvin, and Ramsey (1995):

> Using three measures of school-related adjustment in grade 5, the arrest status of a high-risk sample can be correctly predicted in 80% of cases five years later. These measures are (1) a 5-minute teach rating of social skills, (2) two 20-minute observations of negative-aggressive behavior on the playground involving peers, and (3) the number of discipline contacts with the principal's office that are written up and place in the child's permanent school record. (p. 6)

The predictive value of a behavior rating checklist, the Child Behavior Checklist (Achenbach & Edelbrock, 1979), has been examined and found to very likely predict negative outcomes (McConaughy & Achenbach, 1989). The risk factor discussed earlier also could be useful in identifying children for intervention at an earlier age. Probably the best known and most comprehensive such screening procedure is the multiple gating model proposed by Walker and colleagues (Walker et al., 1988). It involves three steps or gates that screen a group of youngsters who are highly probable to experience later behavioral difficulties in school. The first step or gate consists of teachers simply being asked to identify the three highest ranked children on internalized and externalized behavioral criteria. The second gate takes these children and determines which if any exceed normative criteria on critical event checklists and on a frequency of behavior index. The final screen or gate consists of normative criteria on direct observations and parent questionnaires regarding behavior in free-play and structured activities. Walker and his colleagues have just recently published a preschool-aged version of a screening system for students with emotional or behavioral disorders. These screening instruments and procedures are based on measurements of the child's behavior in the school and home setting. The instruments are based on the concept that the best predictor of a child's behavior is his or her previous behavior.

In order for screening to be of any value, it will be crucial that intervention is provided in order to bring about change in the students who are identified in the screening process. Unfortunately, there are no clear or obvious interventions that can occur strictly within the school setting which guarantee a changed trajectory for students who are identified. The most promising programs are the ones that strive to build competence in the students to face difficult situations. These include conflict resolution, mediation strategies, problem solving skills, social skills, and efforts to build confidence and self-esteem. In short, the strategy would be to identify and enhance any resiliency factors that exist in that child. Unfortunately, there are no data as to whether these efforts can in fact overcome the risk factors that press against EBD children. Urgent research is needed in this area.

Programming

The most common way of discussing programming for EBD students has been to examine program "placement." School placement options available to students with behavioral disorders today are identical to what they were more than 25 years ago when Deno conceptualized the Cascade of Services Model. These range from psychiatric hospitalization or institutionalization through a series of steps to full integration into regular classes in schools. In mental health, this is better known as the "Continuum of Services Model" and ranges from informal counseling and outpatient care to psychiatric hospitalization (Friedman, 1985). To be sure, there continue to be many places where some levels of service on the cascade do not exist (e.g., day school or day-treatment program options in certain areas), and efforts have been under way to fill in these gaps with varying degrees of success. Rural areas continue to experience difficulty in making complete arrays of service or placement options available. Other service options seem to have excess capacity from time to time (e.g., psychiatric hospitalization, where availability of funding from private and public insurance has sometimes been easily accessible). Many times, the effort to build missing services is framed as a vehicle to save the costs of more expensive and intensive services such as psychiatric hospitalization or residential treatment, particularly when those costs would be borne by the public schools.

Most recently, there has been a tremendous amount of attention devoted to the "inclusion movement" in special education. There have also been much discussion and perhaps elaboration of some components of the cascade (e.g., much literature on mainstreaming/integration/inclusion of students with EBD), but no new conceptualizations of this array of services have emerged. Much of the inclusion debate relies on teacher tactics similar to those advocated when federal special education law was just being passed. Some have stated that the system of special education is flawed, others have stated that the basic principles are correct but that implementation of them has been flawed, others have framed the issue in other ways (see generally Lloyd, Singh, & Repp, 1990). Some have cautioned that inclusion may be difficult or impossible for some EBD students (Braaten, Kauffman, Braaten, Polsgrove, & Nelson, 1988; Walker & Bullis, 1990; CCBD, 1989a).

Classroom Models for EBD Programs

Perhaps this is why there have been virtually no new models for classrooms or model roles for teachers who serve students with emotional or behavioral disorders during the past 20 or more years. Two basic models continue. Probably the best known and the one that directly or indirectly represents the largest number of classrooms exclusively serving EBD students is the "engineered classroom model" described by Hewett in 1968. While this term is only infrequently used, the engineered classroom is still one of the best descriptions of a prototypic self-contained EBD classroom that exists. The second most influential model, not originally specifically designed for students with EBD but serving more EBD students than any other, is the "resource room" model described by Hammill and Wiederholt in 1972 (Wiederholt, Hammill, & Brown, 1978). While some other models have been proposed such as the "crisis teacher model" (Morse, 1980) or the "consulting teacher model," and have been adopted in localities for some periods, they have not been widely adopted. As schools move toward more "inclusion" of students with disabilities, the "consulting teacher" or "behavior specialist" model is likely to become more widespread.

Most EBD classrooms, and thus teachers, probably represent an amalgam of the elements of these two main program models. From the engineered classroom model most EBD classrooms have attempted to implement a structure for behavior management, typically including a token economy system of some kind, along with reinforcers for appropriate behavior and sanctions for inappropriate behavior. These measures have received some serious criticism for not attending to the need to bring about change in overall student behavior and instead focusing on merely controlling student behavior while in these programs. When used inappropriately, they have been termed the "curriculum of control" (Knitzer, Steinberg, & Fleisch, 1990; Nichols, 1992).

The classrooms offer schedules designed to meet the attention spans of students and the demands for efforts to maintain students in the school's academic curriculum. The teacher is the primary instructor for all subjects for students in the program. In addition, the teacher is the manager of the behavioral change programs and classroom management program for the students. These teachers are also the intermediaries for EBD students with the rest of the school, often serving as advocates for the EBD students. Only one recent text (Morgan & Jensen, 1988) has discussed the differences between serving EBD students in resource as opposed to self-contained settings. Although they did not address the combination of these two models, they distinguished them by the mix of roles of the teachers, not by strong differences in the organization, curriculum, or classroom management techniques that should be employed in these situations.

In the resource model, students come and go to regular classrooms where they are integrated for portions of their day. While the students are in the EBD resource classroom, the teacher assists and supports them with their work from these mainstream classes, often consulting with these teachers to motivate and assist them to make both academic and behavioral accommodations for EBD students. The teacher has a major role in assessment of students, in assisting in the preparation of academic materials both for the resource and for the regular classroom, and in consulting with all other school personnel regarding the success of the student in the other classes and school environment.

Since most EBD programs in public school settings attempt to integrate students as soon as they are ready, the self-contained EBD classroom is really a mixture of both the self-contained and the resource models, with the teacher attempting to accomplish both roles for the various students in the program depending on the degree to which they are integrated into regular classrooms. The teacher serves as an advocate for the student in the building, a troubleshooter for academic or social difficulties, and a crisis specialist providing mediation and counseling to staff and students when students get into trouble.

Day programs (special school programs) for EBD students have tended to be of two types. Admission of day students to the educational programs of residential treatment centers or psychiatric hospitals is common. Obviously these programs are identical to those for residential students in those settings. Other special day school programs that serve students are run by public schools and occasionally by mental health centers or other agencies. The outcomes of day programs are also not well researched (Zimet & Farley, 1985) and in many cases are driven by economic issues (forestalling more expensive residential treatment). While some innovative day programs may already be in development (Paige, Caudle, & Hart, 1995), the classrooms and structure of these programs are not particularly distinguishable from those in residential settings, or those in self-contained classrooms in regular public school settings (except of course that the "resource" component is diminished). Many of these programs have added special individual or group therapy programs, but otherwise look similar to other "self-contained" programs.

The same is true for residential treatment. Programs in residential treatment centers or psychiatric hospitals have primarily continued to focus on the overall creation of a therapeutic milieu based on the work of Redl (1959) in order to have all aspects of the child's life contribute to treatment goals while in treatment. These programs tend to provide individual and group therapy along with other services. Hospitals tend to provide medical, particularly psychopharmacological, interventions as well. Classrooms in these environments, with the possible exception of the participation of the teacher in the larger program case management process, are not particularly different from classrooms for EBD students in other settings. Research on the placement or effectiveness of residential treatment is lacking (Wells, 1991), and results are difficult to interpret (Curry, 1991). Instead of classroom-based token economy systems or level systems, in residential settings these approaches tend to be carried over into the entire waking day for the child.

The one level of programming for EBD students that has not yet been addressed is that of the "regular classroom." Have regular classrooms changed significantly in the past several years? What is the capacity of these classrooms to accommodate students with emotional or behavioral disabilities? There has been a kind of revolutionary change that has occurred in all of special education during the past several years. That is the strong emphasis on "inclusion" of students with disabilities into regular schools and classrooms. Students with all categories of disability including students with EBD who might have previously been in more segregated settings, especially in self-contained settings, have begun to be placed in "regular" classrooms and programs for substantial periods of time. What is not clear is the degree to which regular classrooms will be able to develop structured behavior management systems capable of dealing with serious aggressive and disruptive behavior of EBD students as they are included in regular classrooms.

Teachers of EBD Students

Some states offer an undergraduate program in EBD, but preparation of specialized EBD teachers has primarily been on a graduate level. Typically this includes an approximately 36-semester-hours program where the teacher also completes a master's degree (Zabel, 1988). These programs have typically included courses or components dealing with characteristics, identification procedures, and programming. Most have courses in teaching interventions and behavior management, as well as courses addressing controversial and emerging topics.

As a result of the inclusion movement and efforts to de-emphasize categorical programming in special education, many states have moved or are moving to more noncategorical preparation of special education teachers. It is not clear to what extent these programs will provide specialized training in managing behavior of exceptional students. In the past, noncategorical programs have tended to emphasize academic rather than behavioral interventions.

For a considerable period of time, teachers who have special expertise in working with students with EBD have been highly sought by schools (Zabel, 1988). This is one of the highest areas of personnel shortage in all of education, let alone special education. A large number of positions serving EBD students go unfilled. In addition, the trend has been to fill these positions with persons with little or no training. In some cases this is done by special state certification procedures that permit quick access to endorsements by circumventing traditional college or university course work. In other cases, provisional endorsements are used to hire persons into EBD positions who then are expected to complete university EBD preparation programs within a period of time in order for them to obtain normal credentials so that they can continue to be employed. In either case, most of the immediate skills needed for that teacher's survival or effectiveness are learned "on the job," many times under very difficult circumstances where students had not been effectively managed for some time.

Teachers are of course not the only providers of services to students with EBD. School counselors and psychologists in particular have significant roles. As more and more EBD students are to receive education in "regular" classrooms and schools, the classroom teacher and building administrators have increased roles in serving this group of students. Certainly special education resource teachers, many of whom will now be in more consulting and teaming roles within general education classrooms, will have a larger and larger role in serving EBD students. Historically, however, almost all of these personnel have had minimal training in addressing students with serious behavioral difficulties.

A growing concern related to EBD students is the apparent decline of expertise in the schools. A serious teacher shortage has meant that an increasing number of untrained, inadequately trained, or poorly trained teachers have been hired to serve EBD students. In addition, other teachers who have good training and substantial successful experience are being reassigned to positions where they will serve as consultants to classroom teachers who may have some EBD students in their classes (this is not in itself a problem, but it may exacerbate the decline in expertise among teachers having direct contact with EBD students). One question is, Who will provide the expertise needed to effectively educate these students, once the specialists in this area have been incorporated into more generic roles within the school?

Increasingly there has been a recognition that the schools will not be able to effectively meet the needs of EBD students unless educational services are an integrated and coordinated part of a much larger system of care (Stroul & Friedman, 1986). (See also the discussion of service integration and coordination under that heading later in this paper.) Many have called on schools to become one-stop centers for a variety of community agencies, and some have even suggested that schools become brokers for these services (Heath

& McLaughlin, 1987; Melaville, & Blank, 1991). Although EBD self-contained and resource teachers have to some degree served as the liaison between the schools and other agencies, there does not appear to be anyone within the typical school situation who would be in a good position to serve this coordination or case management function with community agencies. Schools may need to develop new roles for people to serve in these capacities. Currently few training programs prepare educators for this type of role or responsibility, although some such programs are being developed (Donner, 1987).

Interventions to Change EBD Behavior

It is hard to know what might constitute a new intervention for students with behavioral disorders. Very few, if any, new interventions have appeared during the past 10 years, with one possible exception. While the idea of social skills instruction has existed for more than 10 years, this is one major intervention that has clearly blossomed during the past 10 years. Social skills instruction was not addressed in the classic "engineered classroom model" of Hewett, although it could be argued that it was implied in many of the underlying concepts. Nevertheless, social skills instruction is an essential component of any program for EBD students today. These programs, which focus broadly on social competence, can be divided into three interrelated but distinct components (Coleman, 1992): social skills instruction (social interaction skills, conflict resolution, etc.); affective/emotional education (anger control training, relaxation training, etc.); and self-management (including self-control, self-monitoring, and other related cognitive behavior change approaches). Each of these components has spawned specific strategies for teachers to use, as well as ongoing research.

Other behaviorist interventions based on reinforcers and punishers continue to be a mainstay of EBD programs, with more psychodynamic interventions (counseling, various activity therapies, etc.) also being a part of almost all programs for EBD students. While these interventions have tended to be combined in programs under an ecological framework, the interventions themselves have changed very little during the past 10 years. Research has continued to attempt to document their effectiveness, and to elaborate the applications of the principles, but few if any interventions could be viewed as new.

Service Integration and Coordination

The "System of Care" Model proposed by Stroul and Friedman (1986) is a significant restatement of a broad ecological conceptualization of services for students with EBD, and has received widespread attention. This model illustrates the belief that to be effective in serving students with EBD, the student and his or her family must have available a comprehensive array of services that address the child's physical, emotional, social, and educational needs. By placing not just the child, but the child *and his/her family* at the center of the model, it also illustrates the belief that families must be full participants in all aspects of the planning and delivery of services. A series of major programming elements must be in place for successful care to surround the family. These include mental health services, social services, educational services, health services, vocational services, recreational services, and operational services (services such as case coordination, transportation, communication, etc).

One element that is not emphasized in the System of Care Model to the extent that it was in the earlier Study of Child Variance is the importance of religion and religious institutions in providing caring and services (Summerlin, 1980). Religious beliefs may be more important than many human service professionals understand in a "System of Care" (Wood, 1994).

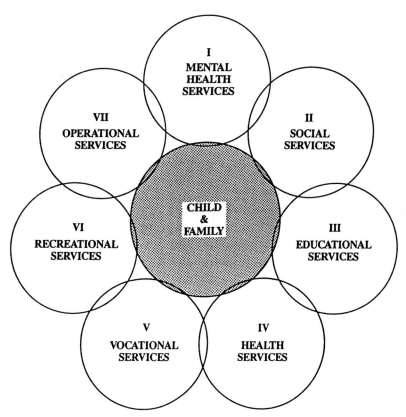

Figure 9.1 System of care framework. Stroul, B.A. & Friedman, R.M. (1986). *A System of Care for Severely Emotionally Disturbed Children and Youth.* Washington, D.C.: CASSP Technical Assistance Center, Georgetown University Child Development Centre.

While the System of Care Model has focused attention on the issue of the interplay among the various service delivery systems, such an awareness has grown out of the work of Rhodes and Tracy (1974) and Hobbs (1975).

While not entirely different from the conceptualization of service delivery systems developed as a part of the Study of Child Variance (Rhodes & Head, 1974), the System of Care Model includes a component called "operational services," which includes case management, self-help and support groups, advocacy, transportation, legal services, and volunteer programs. Of these the concept of "case management" has received the most attention. This is probably the result of recognition by many service providing agencies at all levels that the array of services provided by various agencies is bewildering to clients, can be difficult to access, and has not been coordinated. There was also recognition that the families of many or even most EBD students were typically involved with numerous human service agencies, or at least were eligible to be involved with these agencies. One person from a particular agency would serve as a "case coordinator" for the family, planning and coordinating services delivered by the various agencies, inviting key staff members to meetings, and also initiating contacts with "new" agencies that might have programs or benefits which could meet the needs of a family, but that the family might otherwise not be aware were available (Epstein, Quinn, Nelson, Polsgrove, & Cumblad, 1993; Burchard & Clarke, 1990).

Case management has been recognized as an important missing element and the need to integrate and coordinate services makes case management crucial (Nelson & Pearson, 1992). As an outgrowth of the problems of trying to create appropriate levels of services for EBD students within states (rather than sending students to out-of-state residential treatment), a series of service coordination projects has emerged. The purpose of these efforts has been to build ongoing mechanisms to better coordinate services for families with children who have EBD, to also build service capacity in weak areas, and to eliminate costly duplicative services. The Alaska Youth Initiative, the Kentucky Bluegrass IMPACT Project, the Venture Model (Nelson & Pearson, 1992), and the Vermont Wrap-Around Model are four examples that have received considerable attention. In each case the projects involved multiple agencies, created mechanisms for service coordination, and attempted to use creativity and ingenuity to build and provide needed services that were unavailable due to distance or simply did not exist. The Annie E. Casey Foundation has taken the lead in numerous communities to build cooperative and innovative state and local systems of services for children and youth (White & Wehlage, 1995). In many cases, public school personnel have taken the lead in these efforts to coordinate services, and the public schools have been suggested as the logical location for "one-stop" social service centers that could house the offices of various agencies likely to be involved with families of EBD students (Cohen, 1995, p. 22).

In spite of the strong interest these projects have created, and the widespread attempts to imitate parts of them in other states, these ideas have not yet been systematically implemented in other states. Even the concept of interagency cooperation is not new. A federally funded project in 1979 had the prophetic title, "Interagency Collaboration of Full Services for Handicapped Children and Youth" (Regional Resource Center, 1979). Varying state organizational patterns, agency turf battles in periods of declining governmental resources, and lack of effective leadership have likely been part of the problem. Certainly those states that have been most successful seem to be those which have received extraordinary resources and guidance like those participants in the various programs funded by the Casey Foundation (White & Wehlage, 1995; Melton, 1995), and yet even those programs have encountered difficulty (Cohen, 1995; White & Wehlage, 1995). The adoption of collaborative, coordinated service systems, like other educational innovations, depends heavily on the perspectives of the individual participants, and these elements are often underestimated even if other resources are available (McLaughlin, 1990).

Other Programming Issues

A variety of other programming issues may also be important, although an analysis is beyond our scope at this time. Briefly these include the following.

Technology

While technology has had huge impacts on some types of disability, it seems to have had relatively little impact on programs for EBD. There is limited literature in this area, and given the relatively little attention being paid to this issue in EBD, it seem unlikely that technology will affect EBD programming in the future. To the extent that EBD students are impaired in their relationships with adults and peers, most technological advances seem at best only tangentially relevant to the basic problems of EBD students.

Discipline

The seeming unfairness of a separate system of school discipline for students with disabilities has caused much concern in the schools and in communities. This was created as a result of court ruling that interpreted IDEA to preclude students with disabilities from being expelled (if that meant receiving no education). It interpreted suspension or expulsion for more than 10 cumulative days as being a change of educational placement requiring multidisciplinary team approval, and placing key decision making for these kinds of issues in the hands of the multidisciplinary team rather than administrators. Because of the likelihood that EBD students will behave in ways that would normally engage the school discipline procedures, this has been a major concern for many administrators. It is alleged that many potentially EBD students are not identified for special education in order to maintain the traditional disciplinary options. However, the value of these disciplinary procedures has been challenged for any student. The Council for Children with Behavioral Disorders has developed a position paper on this topic (1989b), and there have been proposals to change the federal legislation in order to permit disabled students to come under traditional school discipline procedures. Current intense concern to limit violence and gang influence in school may also result in new legislation that might affect EBD students' role in school discipline.

Aversive Procedures/Punishment

Ethical and moral concerns have emerged over the use of aversive procedures and "punishment" with children. The procedures in question range from using electric shock to verbal reprimands. Much attention has focused on the use of time-out procedures. These concerns have applied to all children, not just students with EBD. There has also been some controversy about the effectiveness of these procedures. As a result some states have enacted legislation limiting the use of aversive procedures or attempting to regulate their use in publicly supported institutions and programs, including schools. The Council for Children with Behavioral Disorders has opposed most of these limitations (1990b), indicating that there may be a place for some or all of these procedures if used appropriately and ethically. Related controversies have surfaced over the effectiveness of and ethical use of procedures that remove students from instruction such as time-out or in-school suspension.

School Reform

Most proposals to reform schools have not mentioned students with disabilities, let alone students with EBD. Therefore it is difficult to anticipate what effect, if any, these efforts will have on programs for EBD students. It is probably safe to assume that most EBD students have not even participated in the schoolwide assessments of academic progress (Shriner, 1995), let alone other reforms. Leone, McLaughlin, and Meisel (1992) have examined some of the proposals with implications for EBD students in mind, and have maintained some optimism that the reforms could benefit EBD students.

Antisocial and Violent Behavior

Violence has become a very big issue in schools. While clearly not all EBD students are violent, many of these students are aggressive and are capable of violence. Additionally, many EBD students have been the perpetrators of violence in schools. As a result, it is has become difficult to separate discussions of violence in schools from discussions of how to serve EBD students. Virtually all of the programming elements of EBD programs (structured classroom instruction, social skills instruction, behavior de-escalation, etc.) are similar to programming elements of programs intended to reduce violence. Although being concerned with the problems of effectively serving EBD students in schools is not particularly fashionable, discussions of violence, gangs and crime prevention are extremely fashionable in the current times of intense public interest and concern for these topics. As a result, many new materials and activities have attempted to capitalize on this interest. Examples are new books including *Antisocial Behavior in School: Strategies and Best Practices* by Walker et al., (1995), *Delinquent Gangs: A Psychological Perspective* by Goldstein (1991), *The Gang Intervention Handbook* by Goldstein and Huff (1993), *Special Education in the Criminal Justice System* by Nelson, Rutherford, and Wolford (1987), and *School Violence* by Conoley & Goldstein (1995). Most of the time these materials apply the same intervention strategies that have historically been a part of EBD programs without using the EBD category, or placing the programs in the context of special education. It is also possible, however, that this new focus on violence and related issues will permit new interventions to be developed or new perspectives to be appreciated that may move our understanding in this area along (Eron, Gentry, & Schlegel, 1995), particularly as it may relate to the complexity and interconnectedness of violence in our lives (Gelles & Loseke, 1993).

Parents, Families, and Advocacy Organizations

One genuinely positive development during the past few years has been the emergence of parent advocacy organizations for families of students with emotional or behavioral disorders. Two of these deserve particular mention. The newest organization that has had very strong growth and leadership, particularly at the national level, is the Federation of Families for Children's Mental Health. It appears to have been developed in part as a result of a coalition of state parent groups, and as a fortuitous grouping of key parents in the Washington, D.C., area that has permitted the group to overcome organizational hurdles. Its leadership and newsletter, *Claiming Children*, have provided current information and innovative ideas to its constituents and to parents. Perhaps in part in competition with the Federation, the Alliance for the Mentally Ill (AMI) has also seemed to have undergone a revitalization and has widened its tendency to serve primarily families of institutionalized mentally ill individuals to an increased interest in families with all types of mental illness on all levels. AMI has also undertaken an effort to devote more attention to children and children's issues. As could be expected, the state and local units of these organizations vary in their development; many of these appear to be very vital if small. In at least some cases, local and state organizations serve as affiliates of both of these national organizations.

Historically EBD has been an area where effective parent and advocacy organizations have been lacking (Smith & Wood, 1986). The arrival on the scene of these two organizations has been a welcome addition to the advocacy efforts on behalf of EBD children. These organizations have worked hard to become credible lobbyists at the national level and to some degree the national organizations have also had an impact on state levels as well. These organizations have found effective financial support among foundations (such as the Annie E. Casey and Robert Wood Foundations), which have been anxious to support children's mental health issues and have in many cases sponsored key state reform efforts related to children's policy.

While lobbying and advocacy have also occurred at the local level, these organizations have probably served primarily as parent information and support networks. They have offered parents and families the chance to talk with others who have experienced similar devastating mental health problems with their children. In many cases, the local organizations are able to provide information on how to access services and financial assistance, and have been working to develop services such as respite care and day care capable of handling the unusual and often difficult needs of these children.

While there has been intensified recognition of the important role of parents in effectively treating students with EBD, there has not been much tangible progress in more effectively involving parents in the school programs of their EBD children. The social and economic costs and depletion of social capital for these families is devastating. Stigma and blaming of parents remain common. Schools and other agencies have not acted sufficiently to overcome the obstacles to better family involvement. Much work remains to be done in this area (Harry, 1994).

Cultural Contexts for EBD

As stated previously, African Americans are overrepresented in programs for EBD students. Other ethnic groups are not, in fact Asians are underrepresented. The behavior problems of new immigrant groups from Eastern Europe, particularly from war zones, have also raised questions about the relationship between cultural issues and behavior. This chronic issue has begun to trigger a closer examination of specifically how cultural issues affect families with children identified as having EBD children.

Peoples of differing cultural backgrounds may interpret the meaning of "emotional or behavioral disorders" in quite different ways. The culturally appropriate responses may also vary dramatically. The contexts of differing families, religion, and language may be as important as race and specific ethnicity. Immigration, refugee, and generational status may also be very important in examining psychological issues of children in school. Previous exposure of children to violence, and traumatic exposure to physical or psychological injuries (such as that of war refugees), may also be important. All of these issues and many more will be important to a more sophisticated understanding of how cultural issues affect student behavior and mental health in school (Peterson & Ishii-Jordan, 1994).

While our awareness and understanding of these issues is only now beginning to emerge, these issues may be very useful in helping to design or redesign programs that serve EBD and other students whose behavior is problematic in schools (Grossman, 1995a; 1995b). Some of our most common behavior management strategies may not adapt well to the values of some cultures, and in fact may be offensive.

Outcomes for EBD Students

Outcome measures for EBD students are particularly poor. They have poor academic skills, have the highest dropout rate of any identified group in school, and are likely to be involved with the juvenile justice system and other community services even before they leave school (Chesapeake Institute, 1994; Neel, Meadows, Levine, & Edgar, 1988). These students continue to be disruptive within the school environment, are involved with incidents of aggression and violence in school, have poor independent living skills, and often are engaged with numerous social agencies once they leave school. There has been little indication that these outcomes are improving during the past 10 years.

Mood of Mean Spiritedness

Although difficult to document, most educators have acknowledged a perception that the public has taken on an attitude of mean spiritedness during the past several years. This has resulted in less toleration and more blaming of persons whose behavior is judged to be inappropriate. An example of this is the slogan "zero tolerance" that is often applied in the context of school discipline procedures to students' inappropriate behavior in school. Certainly this attitude may be a reflection of concern for apparent increases in community and school violence and crime, frustration with children and schools for insufficient levels of achievement, and frustration at the cost and slow degree of positive change apparently created by social programs. It has seemed to result in diminished support for school programs for students with EBD, and in a willingness to see these problems taken to the criminal justice system. Even educators can show this frustration and apparent mean spiritedness (see Carlin, 1988). The long-term effect of these attitudes is not known, but does not appear positive.

Conclusions

What Is the State of Affairs for Students with EBD?

For the most part, the level of progress in serving EBD students in special education during the past decade is discouraging. EBD students as a group by just about any standard are worse off today than they were 10 years ago. EBD students who have been receiving special education services have not made significant improvements in their status or outcomes. More and more potentially EBD students are not being served at all, and therefore are getting worse. The likelihood that the number of children with EBD is increasing, the likelihood that only between one half to one third or less of the existing EBD eligible students are receiving special education services, and the increasing severity and complexity of the students in these special education programs constitute by almost any standard a serious and out-of-control epidemic.

A struggle to define who is emotionally or behaviorally disordered continues. Although initiated as a way to move toward a common vision among all concerned with children and youth who have EBD, the effort has apparently been stalemated over concerns about the costs of providing services to too many children (regardless of how they are identified). No matter how one chooses to designate this group of students, it is clear that there are more and more of these students, and that the complexity and severity of problems among those students who are identified continue to increase. More and more of students with EBD will not be receiving special services or programs of any kind. Even the students who are identified and who are receiving services have more serious and complex problems than ever before, to the extent that the programs that serve them may be near to loosing hope of making any real difference in the lives of the students they serve. Most of these students are served only when they have reached adolescence, and after almost all of the potential resiliency factors have been exhausted. Only when family, school, and community resources are depleted and overwhelmed by problems is the student made eligible for special education services. At this point the student behavior is so out of control or dangerous that interventions with strong elements of behavioral control are required (curriculum of control—Knitzer et al., 1990), and more restrictive placements are required. While these programs are employing generally useful strategies to manage behavior and teach social skills, there have been few innovative ideas for new interventions or services delivery.

Instead of the traditional triage concept of devoting treatment resources to those for whom the treatment is most likely to be successful, the educational system is devoting its resources to EBD students for whom there is precious little hope left for significant improvement. The number of programs is inadequate. Personnel are often inadequately trained. Good intentions are not sufficient to overcome the difficult problems of the students. Programs have inadequate resources to accomplish significant improvement against over-whelming odds. While it may be important to manage the seriously EBD students in order to prevent these students from further aggression and school disruption, the programs become holding tanks until the students can drop out of school. The programs are likened to hospice programs. *Even with the most complete, thorough, and effective programming*, the severity of behavioral disorders faced makes a positive outcome doubtful.

At the same time, virtually no school resources are being provided for primary or secondary prevention. While the research has begun to emerge which suggests that we can identify many students who have or will develop EBD at a much earlier period in their school careers, it is not yet clear whether schools will have the will to do so. It is also not clear whether the related interventions will be developed and implemented to be able to make such early identification an advantage. Early identification programs are still experi-mental, and the support for creating these programs, like support for any preventive program, will be difficult to secure. They also potentially are at odds with some efforts to diminish labeling. Nevertheless, this remains one of the few places in this field that offers hope for significant improvement during the next decade.

Recognizing the Problems and Charting a Course

Several major national studies and reports have chronicled these problems (Knitzer et al., 1990; Hill, 1989; Koyanagi & Gaines, 1993; Annie E. Casey Foundation, 1992; Steinberg, 1991). Knitzer's report found only small improvements since her earlier report (Knitzer, 1982). A whole host of other reports at the state level have documented the problems of students with EBD (for a list, see Peterson, 1993b), and many, many more have documented closely related problems of foster care, homelessness, mental health, child abuse, and neglect that affect children and families (see list in Peterson, 1993). A "Draft National Agenda for Achieving Better Results for Children and Youth with Serious Emotional Disturbance" (Chesapeake Institute, 1994; Osher, Osher, & Smith, 1994), has also been created that identifies seven targets for improvement:

1. Expand Positive Learning Opportunities and Results
2. Strengthen School and Community Capacity
3. Value and Address Diversity
4. Collaborate with Families
5. Promote Appropriate Assessment
6. Provide Ongoing Skill Development and Support
7. Create Comprehensive and Collaborative Systems.

The problems have been recognized and actions to remediate them identified.

Are There Any Hopeful Signs?

Programs and services for students with EBD have changed little in the past decade. Service delivery models have for the most part remained unchanged, with relatively few new interventions discovered. The emphasis on social skills instruction and cognitive behavior modification during the past several years may be one exception, and certainly these have been added to the repertoire that teachers can use to address behavioral needs of students in school and classroom settings. These are particularly useful because they can be applied effectively in various service delivery options from regular classroom to residential treatment settings. Certainly, too, there have been examples of useful innovative concepts demonstrated such as "case management" and "wrap around" services which suggest the possibility that reorganization of services to better meet the needs of children and their families is possible. Nevertheless it is not clear whether the capacity of schools to implement these kinds of ideas will increase given the decline in the expertise of teachers, and the teacher shortages in this area of special education, let alone the declining resources available to special education (or any special programs), or education generally.

The emergence of parent advocacy organizations has been one of the other positive developments of the last decade. The support these organizations can provide to families and the advocacy efforts they provide are clearly needed and welcomed. It is too early to determine their long-term impact.

Conclusion

As was so eloquently described by Schorr in her book *Within Our Reach* on breaking the cycle of disadvantage, we do know what needs to be done to serve students with EBD effectively. The question remains whether the public, our schools, and our society have the will to do so. We must identify EBD students earlier, we must obtain the resources and expertise to use effective interventions, and we must support the growth of resiliency factors we know can help these students, while we attack and diminish the risk factors that permeate these students' lives.

References

Achenbach, T.M., & Edelbrock, C.S. (1979). The Child Behavior Checklist: Boys aged 12–16 and girls aged 6–12 and 12–16. *Journal of Consulting and Clinical Psychology, 47,* 223–233.

Anthony, E.J., & Cohler, B.J. (Eds.) (1987). *The invulnerable child.* New York: Guilford Press.

American Psychiatric Association. (1987). *Diagnostic and statistical manual of mental disorders* (third edition–revised). Washington, DC: Author.

American Psychiatric Association. (1994). *Diagnostic and statistical manual of mental disorders* (fourth edition). Washington, DC: Author.

Annie E. Casey Foundation. (1992). *Kids count data book–state profiles of child well-being.* Washington, DC: Center for the Study of Social Policy.

Bower, E.M. (1982). Defining emotional disturbances: Public policy and research. *Psychology in the Schools, 19,* 55–60.

Braaten, S., Kauffman, J., Braaten, B., Polsgrove, L., & Nelson, M. (1988). The regular education initiative: Patent medicine for behavioral disorders. *Exceptional Children, 55,* 21–28.

Brendtro, R.K., Brokenleg, M., & Van Bockern, S. (1990). *Reclaiming youth at risk: Our hope for the future.* Bloomington, IN: National Educational Service.

Brooks, R.B. (1994). Children at risk: Fostering resilience and hope. *American Journal of Orthopsychiatry, 64*(4), 545–553.

Brunswick, A.F. (1985). Health services for adolescents with impairment, disability, and/or handicap. *Journal of Adolescent Health Care, 6*(2), 141–151.

Burchard, J.D., & Clarke, R.T. (1990). The role of individualized care in a service delivery system for children and adolescents with severely maladjusted behavior. *Journal of Mental Health Administration, 17*(19), 48–60.

Capitol Publications. (1995). ED's proposals would change IDEA eligibility, funding. *Special Education Report, 21*(2), 1–2

Carlin, P.M. (1988). Modest proposal: Shoot troublesome kids. *American School Board Journal,* 36.

Cessna, K. (Ed.). (1993). *Instructionally differentiated programming: A needs-based approach for students with behavior disorders.* Denver: Colorado Department of Education.

Cheney, C., & Sampson, K. (1990). Issues in identification and service delivery for students with conduct disorders: The "Nevada Solution." *Behavioral Disorders, 15*(3), 174–179.

Chesapeake Institute. (August 1, 1994). *National agenda for achieving better results for children and youth with serious emotional disturbance* (pp. 1–14). Washington, DC: U.S. Department of Education, Office of Special Education and Rehabilitative Services, Office of Special Education Programs.

Cline, D. (May, 1990). A legal analysis of policy initiatives to exclude handicapped/disruptive students from special education. *Behavioral Disorders, 15*(3), 159–173.

Cohen, D.L. (1995, January 18). A concerted effort. *Education Week,* 21–25.

Coleman, F.S. (1987). Families and schools. *Educational Researcher, 16*(6), 32–38.

Coleman, M.C. (1992). *Behavior disorders: Theory and practice,* second edition. Boston, MA: Allyn & Bacon.

Conoley, J.C. (1995, January 27). *Evaluation report: Behavior skills program.* Lincoln, NE: Special Education, Lincoln Public Schools.

Conoley, J., & Goldstein, A.P. (1995). *School violence.* New York: Prentice Hall.

Constantino, J. (1992, October). On the prevention of conduct disorder: A rationale for initiating preventative efforts in infancy. *Infants and Young Children,* 29–41.

Council for Children with Behavioral Disorders. (May, 1984). Position paper on substituting "behaviorally disordered" for "seriously emotionally disturbed" as a descriptor term for children and youth handicapped by behavior. *Behavioral Disorders, 10*(8), 167–174.

Council for Children with Behavioral Disorders. (April, 1985). *Response to: The Department of Education's 1985 "Special Study on Terminology–Comprehensive Review and Evaluation Report"* (unpublished paper). Reston, VA: Author.

Council for Children with Behavioral Disorders. (November, 1987). Definition and identification of students with behavioral disorders. *Behavioral Disorders, 13*(1), 9–19.

Council for Children with Behavioral Disorders. (1989a). Position statement of the regular education initiative. *Behavioral Disorders, 14*(3), 201–208.

Council for Children with Behavioral Disorders. (November, 1989b). School discipline policies for students with significantly disruptive behavior. *Behavioral Disorders, 15*(1), 57–61.

Council for Children with Behavioral Disorders. (1990a, May). Provision of service to children with conduct disorders. *Behavioral Disorders, 15*(3), 180–189.

Council for Children with Behavioral Disorders. (1990b, August). Use of behavior reduction strategies with children with behavioral disorders. *Behavioral Disorders, 15*(4), 243–260.

Curry, J.F. (1991). Outcome research on residential treatment: Implications and suggested directions. *American Journal of Orthopsychiatry, 63*(3), 348–357.

Deno, E. (1970). Special education as developmental capital. *Exceptional Children, 37,* 229–237.

Donnellan, A. (1984). The criterion of the least dangerous assumption. *Behavioral Disorders, 9,* 141–150.

Donner, R. (1987, May). *Resource training manual for case management with adolescents with emotional problems and their families.* Topeka, KS: Division of Mental Health Services, Social and Rehabilitation Services, State of Kansas.

Edgar, E. (1990). Quality of life for persons with disabilities: A time to change how we view the world. In R. Rutherford & S. DiGangi (Eds.) *Monograph in behavioral disorders, severe behavior disorders of children and youth.* Reston, VA: Council for Children with Behavioral Disorders.

Eron, L.D., Gentry, J.H., & Schlegel, P. (Eds.). (1995). *Reason to hope: A psychosocial perspective on violence & youth.* Washington, DC: American Psychological Association.

Epstein, M.H., Quinn, K., Nelson, C.M., Polsgrove, L., & Cumblad, C. (1993, Winter). Saving students with emotional and behavioral disorders through a comprehensive community based approach. *OSERS News in Print,* 19–23.

Farber, E., & Egeland, B. (1987). Invulnerability among abused and neglected children. In E. Anthony & B.J. Cohler (Eds.), *The invulnerable child* (pp. 253–288). New York: Guilford Press.

Flora, C. (1994, December 1). *Social capital, the family and community in times of change.* Lincoln, NE: Unpublished lecture, Center for Children, Families and the Law, University of Nebraska-Lincoln.

Forness, S.R., & Knitzer, J. (1992). A new proposed definition and terminology to replace "serious emotional disturbance" in Individuals with Disabilities Education Act. *School Psychology Review, 21*(1), 12–20.

Friedman, R.M. (1985, December). *Serving seriously emotionally disturbed children: An overview of major issues.* Tampa, FL: Research and Training Center for Improved Services for SED Children, Florida Mental Health Institute, University of South Florida.

Garmezy, N., & Rutter, M. (1983). *Stress, coping, and development in children.* New York: McGraw-Hill.

Gelles, R.J., & Loseke, D.R. (Eds.) (1993). *Current controversies on family violence.* Newbury Park, CA: Sage Publications.

Goldstein, A.P. (1991). *Delinquent gangs: A psychological perspective.* Champaign, IL: Research Press.

Goldstein, A.P., & Huff, C.R. (Eds.). (1993). *The gana intervention handbook*. Champaign, IL: Research Press.

Grizenko, N., & Pawliuk, N. (1994). Risk and protective factors for disruptive behavior disorders in children. *American Journal of Orthopsychiatry, 64*(4), 534–544.

Grossman, H. (1995a). *Classroom behavior management in a diverse society*, second edition. Mountain View, CA: Mayfield Publishing Company.

Grossman, H. (1995b). *Special education in a diverse society*. Boston: Allyn & Bacon.

Harry, B. (1994). School behavior disorders in the context of families. In R. Peterson & S. Ishii-Jordan *Multicultural Issues in the education of students with behavioral disorders*. Cambridge, MA: Brookline Books.

Heath, S.B., & McLaughlin, M.W. (1987, April). A child resource policy: Moving beyond dependence on school and family. *Phi Delta Kappan*, 576–580.

Hewett, F. (1968). *The emotionally disturbed child in the classroom*. Boston: Allyn & Bacon.

Hill, A.S. (1989). *Children with serious emotional disturbance: Bridging the gap between what we know and what we do*. Washington, DC: National Governors' Association.

Hobbs, N. (1975). *The futures of children*. San Francisco: Jossey-Bass.

Johnson, J.H. (1986). *Life events as stressors in childhood and adolescence*. Newbury Park, CA: Sage Publications.

Kagen, J. (1994, October 5). The realistic view of biology and behavior. *Chronicle of Higher Education, 41*(6), A64.

Kauffman, J.M. (1989). *Characteristics of behavior disorders of children and youth*, fourth edition. Columbus, OH: Merrill.

Kelly, E.J. (1990). *The differential test of conduct and emotional problems*. Aurora, IL: Slossen.

Knitzer, J. (1982). *Unclaimed children: The failure of public responsibility to children and adolescents in need of mental health services*. Washington, DC: Children's Defense Fund.

Knitzer, J., Steinberg, Z., & Fleisch, B. (1990). *At the schoolhouse door: An examination of programs and policies for children with behavioral and emotional problems*. New York: Bank Street College of Education.

Kodluboy, D. (1994). Behavioral disorders and the culture of street gangs. In R. Peterson & S. Ishii-Jordan, *Multicultural issues in the education of students with behavioral disorders*. Cambridge, MA: Brookline Books.

Koyanagi, C., & Gaines, S. (1993). *All systems failure: An examination of the results of neglecting the needs of children with serious emotional disturbance*. Alexandria, VA: National Mental Health Association and the Federation of Families for Children's Mental Health.

Leone, P.E., McLaughlin, M.J., & Meisel, S.M. (1992, August). School reform and adolescents with behavior disorders. *Focus on Exceptional Children, 25*(1).

Leone, P.E., & Trickett, E.J. (1990). Social networks of students in special education programs: Contrasts with non-special education students and correlates of school adjustment. In R. Rutherford & S. DiGangi (Eds.), *Monograph in Behavioral Disorders, Severe Behavior Disorders of Children and Youth*. Reston, VA: Council for Children with Behavioral Disorders.

Lloyd, J., Singh, N., & Repp, A. (1990). *The regular education initiative: Alternative perspectives on concepts, issues, and models*. Sycamore, IL: Sycamore Publishing Co.

Luthar, S., & Zigler, E. (1991). Vulnerability and competence: A review of research on resilience in childhood. *American Journal of Orthopsychiatry, 61*(1), 6–22.

Maag, J.W. (1989). Assessment in social skills training: Methodological and conceptual issues for research and practice. *Remedial and Special Education, 10*(4), 7–17.

Maag, J.W., & Reid R. (1994). Attention-deficit hyperactivity disorder: A functional approach to assessment and treatment. *Behavioral Disorders, 20*(1), 5–23.

McConaughy, S.M., & Achenbach, T.M. (1989). Empirically based assessment of serious emotional disturbance. *Journal of School Psychology, 27*, 91–117.

McLaughlin, M.W. (December, 1990). The Rand Change Agent Study revisited: Macro perspectives and micro realities. *Educational Researcher, 19*(9), 11–16.

Melaville, A.I., & Blank, M.J. (1991). *What it takes: Structuring interagency partnerships to connect children and families with comprehensive services.* Washington, DC: Education and Human Services Consortium.

Melton, G. (1995, March). *Beyond symbolism: Family policy for a renaissance* Lincoln, NE: Unpublished Roberta A. Morris Memorial Lecture, Law and Psychology Program, University of Nebraska-Lincoln.

Morgan, D.P., & Jensen, W.R. (1988). *Teaching behaviorally disordered students: Preferred practices.* Columbus, OH: Merrill.

Morse, W.C. (1980). The crisis or helping teacher. In N. Long, W. Morse, & R. Newman (Eds.), *Conflict in the classroom.* Belmont, CA: Wadsworth Publishing Co.

Morse, W.C. (1994, Spring). The role of caring in teaching children with behavior problems. *Contemporary Education, 65*(3), 132–136.

Moynihan, D.P. (1993, Winter). Defining deviancy down. *American Scholar, 62*(1), 17–30.

National Advisory Mental Health Council. (1990). *National Plan for Research on Child and Adolescent Mental Disorders.* Rockville, MD: U.S. Department of Health and Human Services, National Institute of Mental Health.

Neel, R., Meadows, N., Levine, P., & Edgar, E. (1988). What happens after special education: A statewide follow-up study of secondary students who have behavior disorders. *Behavioral Disorders, 13*(3), 209–216.

Nelson, C.M., & Pearson, C.A. (1992). *Integrating services for children and youth with emotional or behavioral disorders.* Reston, VA: Council for Exceptional Children.

Nelson, C.M., Rutherford, R.B., & Wolford, B.I. (Eds.). (1987). *Special education in the criminal justice system.* Columbus, OH: Merrill.

Nichols, P. (1992, Winter). The curriculum of control: Twelve reasons for it; Some arguments against it. *Beyond Behavior, 3*(2), 5–11.

Omaha researchers aid discovery of dyslexia gene. (1994, October 14). *Lincoln Journal.*

Osher, D., Osher, T., & Smith, C. (1994, Fall). Toward a national perspective in emotional and behavioral disorders: A developmental agenda. *Beyond Behavior, 6*(1), 6–17.

Paige, R., Caudle, M., & Hart, C. (1995, Winter). Educational intensive care. *School Safety,* 7–10.

Peterson, R.L. (July/August, 1993a). Behind the behavior. *Instructor Magazine, 103*(1), 78–79.

Peterson, R.L. (1993b). *Dealing with the broken human spirit: Students with behavioral disorders in Nebraska, an action plan report.* Lincoln, NE: Nebraska Department of Education.

Peterson, R.L., Conoley J.C., & Lawson, S. (1995). *DSMIII-R diagnoses as criteria for the Identification of behaviorally disordered students.* Unpublished paper.

Peterson, R.L., & Ishii-Jordan, S. (Eds.). (1994). *Multicultural issues in the education of students with behavioral disorders.* Cambridge, MA: Brookline Books.

Peterson, R.L., & Mattison, R. (1995). *Behavior rating scales and checklists for behaviorally disordered students.* Presentation at the Midwest Symposium for Leadership in Behavioral Disorders, February, Kansas City, MO.

Redl, F. (1959). The concept of therapeutic milieu. *American Journal of Orthopsychiatry, 29,* 721–734. (Reprinted in P. Newcomer (Ed.). (1994). *Readings in emotional disturbance.* Austin, TX: Pro-Ed.)

Regional Resource Center Task Force on Interagency Collaboration. (1979). *Interagency collaboration on full services for handicapped children and youth, volumes 1–4.* Washington, DC: Bureau of Education for the Handicapped, Department of Health, Education and Welfare.

Rhodes, W.C., & Head, S. (1974). *A study of child variance. Volume 3: Service delivery systems.* Ann Arbor, MI: Conceptual Project in Emotional Disturbance, University of Michigan.

Rhodes, W.C., & Tracy, M.L. (Eds.). (1974). *A study of child variance, vols. 1–4.* Ann Arbor, MI: University of Michigan Press.

Rizzo, J.V., & Zabel, R.H. (1988). *Educating children and adolescents with behavioral disorders: An integrative approach.* Boston: Allyn & Bacon.

Scott-Jones, D. (1991). Families and academic achievement: Risks and resilience. In M.C. Wang, M. C. Reynolds, & H.J. Walberg (Eds.), *Handbook of special education: Research and practice. Volume 4: Emerging programs.* Oxford: Pergamon Press.

Schorr, L.B. (1989). *Within our reach: Breaking the cycle of disadvantage.* New York: Anchor Doubleday.

Shriner, I. (February 1995). *State assessments of student outcomes: Changing patterns of inclusion and implications for students with E/BD.* Presentation at the Midwest Symposium for Leadership in Behavior Disorders, Kansas City, MO.

Slenkovich, J. (1983). *P.L. 94–142 as applied to DSMIII diagnoses: An analysis of DSMIII diagnoses vis-a-vis special education law.* Cupertino, CA: Kinghorn Press.

Smith, C.R., & Wood, F.H. (July, 1986). *Education of behaviorally disordered students: Accomplishments, problems and prospects.* Reston, VA: Council for Children with Behavioral Disorders, Council for Exceptional Children (unpublished paper).

Stroul, B.A., & Friedman, R.M. (1986). *A system of care for severely emotionally disturbed children and youth.* Washington, DC: CASSP Technical Assistance Center, Georgetown University Child Development Center.

Summerlin, F.A. (1980). *Religion and mental health: A bibliography.* Rockville, MD: U.S. Department of Health and Human Services, National Institute of Mental Health.

Swanson, J.M., Cantwell, D., Lerner, M., McBurnett, K., Pfiffner, L., & Kotkin, R. (1992). Treatment of ADHD: Beyond medication. *Beyond Behavior, 4*(1), 13–22.

Tallmadge, G.K., Gamel, N.N., Munson, R.G., & Hanley, T.V. (1985). *Special study on terminology–Comprehensive review and evaluation report.* Mountain View, CA: SRA Technologies, Inc.

U.S. Department of Education. (1992). *To assure the free appropriate public education of all children with disabilities: Fourteenth annual report to Congress on the implementation of the Individuals with Disabilities Education Act.* Washington, DC: Office of Special Education Programs, U.S. Department of Education.

U.S. Department of Education. (1993). *To assure the free appropriate public education of all children with disabilities: Fifteenth annual report to Congress on the implementation of the Individuals with Disabilities Education Act.* Washington, DC: Office of Special Education Programs, U.S. Department of Education.

Walker, H.M., & Bullis, M. (1990). Behavior disorders and the social context of regular class integration: A conceptual dilemma? In J. Lloyd, N. Singh, & A. Repp. *The regular education initiative: Alternative perspectives on concepts, issues, and models.* Sycamore, IL: Sycamore Publishing Co.

Walker, H.M., Colvin, G., & Ramsey, E. (1995). *Antisocial behavior in school: Strategies and best practices.* Pacific Grove, CA: Brooks/Cole Publishing Co.

Walker, H.M., Severson, H., Stiller, B., Williams, G., Haring, N.G., Shinn, M.R., & Todis, B. (1988). Systematic screening of pupils in the elementary age range at risk for behavior disorders: Development and trial testing of a multiple gating model. *Remedial and Special Education, 9*(3), 8–14.

Wells, K. (1991). Placement of emotionally disturbed children in residential treatment: A review of placement criteria. *American Journal of Orthopsychiatry, 61*(3), 339–347.

White, J.A., & Wehlage, G. (1995). Community collaboration: If it is such a good idea, why is it so hard to do? *Educational Evaluation and Policy Analysis, 17*(1), 23–38.

Whitehead, B.D. (1993, April). Dan Quayle was right. *Atlantic Monthly,* 47–84.

Wiederholt, J.L., Hammill, D.D., & Brown, V. (1978). *The resource teacher: A guide to effective practices.* Boston: Allyn & Bacon.

Wood, F. (1994). Religion and mental health. In R.L. Peterson & S. Ishii-Jordan (Eds.), *Multicultrual issues in the education of students with behavioral disorders.* Cambridge, MA: Brookline Books.

Wood, F., Smith, C.R., & Grimes, J. (1985). *The Iowa Assessment Model.* Des Moines, IA: Iowa Department of Public Instruction.

Wright, K.N., & Wright, K.E. (1994). *Family life, delinquency, and crime: A policy maker's guide–research summary*. Washington, DC: Office of Juvenile Justice and Delinquency Prevention, U.S. Department of Justice.

Zabel, R.H. (1988). Preparation of teachers for behaviorally disordered students: A review of the literature. In M.C. Wang, M.C. Reynolds, & H.J. Walberg (Eds.), *Handbook of special education: Research and practice. Volume 2: Mildly handicapped conditions*. Oxford: Pergamon Press.

Zill, N., & Schoenborn, C.A. (1990). Developmental, learning, and emotional problems, Health of our nation's children, United States, 1988. *Advance Data*, #190, November 16, From vital and health statistics of the National Center for Health Statistics, Public Health Service, U.S. Department of Health and Human Services.

Zimet, S.G., & Farley, G.K. (1985). Day treatment for children in the United States. *Journal of the American Academy of Child Psychiatry*, 24(6), 732–738.

10

Learners with Severe Intellectual Disabilities

DIANE M. BROWDER and KARENA COOPER
Lehigh University, Australia

LEVAN LIM
Charles Sturt University, New South Wales, Australia

This chapter focuses on the developments in services for individuals with severe intellectual disabilities in the early 1990s. The specific developments discussed are inclusion, person-centered planning, choice, multicultural issues, and innovations in behavioral interventions. Facilitated communication is examined as an issue that reflects the themes of the early 1990s. Recommendations are offered for future research.

Introduction

The last two decades have been an era of rapid development in methods and services for individuals with severe disabilities. The passage of PL 94-142, the Education for All Handicapped Children Act, helped to initiate this period of development by fostering the creation of new educational services for this underserved population. During the decade from 1975–1985, major advances occurred in determining what and how to teach individuals with severe disabilities. Systematic instruction of functional life skills, with teaching occurring in both school and community, became the hallmark of services for students with severe disabilities (Snell, 1983). During this era, major advances were also made in analyzing the function of problem behavior and teaching alternative skills (Carr & Durand, 1985). The themes of systematic instruction of age appropriate life skills and positive approaches for problem behavior were evident in early intervention (Vincent, Salisbury, Walter, Brown, Gruenewald, & Powers, 1980), school age services (Snell, 1983), and employment training (Bellamy, Rose, Wilson, & Clark, 1982).

The themes of the second decade, 1985–1995, stemmed from an important shift in thinking about individuals with severe disabilities. Whereas professionals had focused on improving the behavior and skills of individuals with severe disabilities to increase opportunities, the focus shifted to changing systems to create fuller inclusion of all individuals regardless of current skills or deficits. The goal was to increase physical integration in schools and communities and, more importantly, to enhance social integration and overall quality of life. Professionals such as Stainback, Stainback, and Forest (1989) provided a vision for including all students in the mainstream of education. Others focused on inclusion in job settings (Moon, Inge, Wehman, Brooke, & Barcus, 1990) or leisure activities (Schleien & Ray, 1988). By the early 1990s, rapid growth had occurred in research and practice related to the inclusion of individuals with severe disabilities. By contrast, some professionals began to question the push toward full inclusion in regular classes for school aged children (Fuchs & Fuchs, 1994).

Concurrent with the emphasis on inclusion, professionals began to focus more on person-centered planning and natural support systems (O'Brien, 1987). Person-centered planning provided the format for envisioning new options for individuals with severe disabilities. Social and other forms of nonprofessional support provided a means to make these dreams attainable whether at work, home, or school (Nisbet, 1992).

Consistent with person-centered planning, professionals became increasingly committed to recognizing and honoring the choices and preferences of individuals with severe disabilities (Brown et al., 1991). This commitment to understanding the perspective of the person with disabilities renewed interest in augmentative and alternative forms of communication (Warren & Reichle, 1992). The focus on person- centered planning also generated interest in understanding perspectives and issues in diversity, especially cultural diversity (Lim & Browder, 1994).

Although the shift in thinking among professionals towards empowering people with severe disabilities focused attention on service delivery, many professionals continued to build on the instructional and behavioral technology of the first decade. Advanced instructional procedures were developed to be more effective and efficient (Wolery, Ault, & Doyle, 1992). Functional analysis became more informative and advances were made in positive procedures for problem behavior (Horner, O'Neill, & Flannery, 1993).

This chapter focuses on services for individuals with severe intellectual disabilities in the latter part of the second decade since the passage of PL 94-142, the years 1990–1995. In 1990, this legislation was reauthorized as the Individuals with Disabilities Education Act (IDEA) PL 101-476. The 1990 act added two categories of disability that are eligible for funding—autism and traumatic brain injury. The IDEA also added assistive technology as a related service and required Individualized Transition Plans for students 16 years of age and older. Also in 1990, the Americans with Disabilities Act was passed

(PL 101-336) which requires equal access and reasonable accommodations in employment and other areas of public and private services.

In this era of the early 1990s, as important legislation was enacted, new directions also emerged in service delivery. This chapter will discuss the primary innovations from this era: inclusion, person-centered planning, choice, multicultural issues, and behavioral interventions. Facilitated communication will be examined as an issue that reflects the themes of the early 1990s. Because other chapters in this handbook address early intervention, severe behavior disorders, and transition, these topics will not be emphasized here.

Definition and Prevalence

What is the 1990 understanding of the population of individuals who have severe cognitive impairments? This can be inferred by considering terminology, definitions, and the methods used to determine prevalence.

Terminology

The term "mental retardation" has been professionally accepted since the 1950s (Heber, 1959). Despite debate, professionals have retained this term in subsequent classifications. This is especially evident in the classification guidelines of the American Association on Mental Retardation and in legislation for school aged children.

In 1983, Grossman updated the guidelines of the American Association on Mental Retardation for *Classification in Mental Retardation.* This revision retained much of the 1973 perspective that mental retardation was manifested by deficits in intellectual functioning and adaptive behavior. Levels of mental retardation were defined based on these criteria (i.e., mild, moderate, severe, and profound.)

Similarly, updating of the Education of All Handicapped Children Act PL 94-142 as the IDEA, PL 101-476, maintained consistent terminology for individuals with intellectual disabilities. In the IDEA, 12 categories of individuals with disabilities were described. Although the new categories of autism and traumatic brain injury were added, 10 categories were retained including mental retardation.

In 1992, the American Association on Mental Retardation (AAMR) made substantial changes in their classification guidelines (Luckasson et al., 1992). Before considering the shift in thinking reflected in these guidelines, it is important to recognize that the AAMR Classification Committee retained the term "mental retardation" after considerable debate. In the preface to the classification manual, the committee provided their rationale:

"The manual retains the term **mental retardation.** Many individuals with this disability urge elimination of the term because it is stigmatizing and is frequently mistakenly used as a global summary about complex human beings. After considerable deliberation, we concluded that we were unable at this time to eliminate the term, despite its acknowledged shortcomings" (p. xi).

Similarly, the subdivision of the Council of Exceptional Children (CEC) retained the term "mental retardation" but added "developmental disabilities" (Division on Mental Retardation and Developmental Disabilities). The AAMR Classification Committee considered the term developmental disabilities, but chose the more specific term mental retardation (Luckasson et al., 1992, p.7).

In legislation (the IDEA), professional organizations (e.g., CEC), and classification manuals (e.g., AAMR's manual), the term "mental retardation" has been retained in the early 1990s because of professional preference. As acknowledged by the AAMR Classification Committee, this term has not been chosen by people with disabilities. Also, this "professional" preference may be of limited scope beyond the United States of America. Mittler and Serpell (1985) have advocated that "intellectual disability" is a more universally accepted term. Recent professional movements in Great Britain have chosen to use the term "learning disability" more broadly to encompass individuals once referred to as mentally retarded based on the preferences of individuals with this disability.

Because of the international perspective of this *Handbook of Special Education*, the term "intellectual disability" is used in this chapter. The term "severe intellectual disability" is used to refer to the subpopulation of individuals who need more intensive support in daily living. Serpell (1988) found international consensus for the concept of severe intellectual disability based on children's deficits in communication, ability to learn a novel task, social skills, self help, and domestic responsibility.

Definition

Although the AAMR Classification Committee retained the most widely accepted term among American professionals—mental retardation—the committee provided some important new directions in understanding this disability.

The 1992 AAMR definition reads as follows:

Mental retardation refers to substantial limitations in present functioning. It is characterized by significantly subaverage intellectual functioning, existing concurrently with related limitations in two or more of the following applicable adaptive skill areas: communication, self-care, home living, social skills, community use, self-direction, health and safety, functional academics, leisure, and work. Mental retardation manifests itself before age 18.

The following four assumptions are essential to the application of the definition:

1. Valid assessment considers cultural and linguistic diversity as well as differences in communication and behavioral factors.
2. The existence of limitations in adaptive skills occurs within the context of community environments typical of the individual's age peers and is indexed to the person's individualized needs for supports.
3. Specific adaptive limitations often coexist with strengths in other adaptive skills or other personal capabilities; and
4. With appropriate supports over a sustained period, the life functioning of the person with mental retardation will generally improve" (Luckasson et al., 1992, p. 5).

Several important changes are reflected in this new definition. These modifications reflect the evolution in professional thinking in recent years about the concept of intellectual disability. First and foremost, the definition recognizes a paradigm shift from viewing mental retardation[1] as an absolute trait of an individual to viewing it as a state created by an interaction between the individual and the environment. Second, it extends the concept of adaptive behavior by describing more specific adaptive behaviors. This specification increases the focus on adaptive behavior and is concurrent with recent deemphasis of IQ testing (Evans, 1991). Third the definition eliminates the old descriptors that implied levels of a trait (mild, moderate, severe, profound) and instead focuses on the level of support the individual needs. Finally, the definition extends the understanding of mental retardation to strengths and personal capacities. This is consistent with the movement towards searching for capacity instead of listing deficits in describing individuals with disabilities (O'Brien & Mount, 1991).

Prevalence

Methods used to determine prevalence reflect the difficulties in terminology and definition. Typically researchers have had to rely on reported IQ scores or classification of individuals as mentally retarded to determine prevalence figures. Based on studies using an IQ of 50 or below, McLaren and Bryson (1987) found the rate of individuals with severe intellectual disability in the United States to be about 3 to 4 per 1,000 persons which is higher than would be expected from a normal distribution of IQ scores. They also found differences across age groups with prevalence being slightly lower in the preschool years (1.2 to 2.5 per 1000), an increase in rates between 10 and

---

[1]The term "mental retardation" is used here in describing the AAMR classification manual (Luckasson et al., 1992).

20 years of age, and stability of the rate after age 20. More males than females were identified as severely intellectually disabled. Wolery and Haring (1994) theorize that these gender differences may be due to biological differences or gender role expectations.

Research on prevalence reports given by state departments of education in the United States conducted by Ysseldyke, Algozzine, and Thurlow (1992) indicated a decline in the number of individuals reported with mental retardation across time. This may be due to increased awareness of culturally fair assessment procedures, the importance of adaptive behavior, or the use of other categories in classification (e.g., learning disabilities). While the U.S. Department of Education data do not specify level of mental retardation, their most recent count of students with mental retardation was 484,871 which is 10.9% of all students with disabilities served by Part B of IDEA (U.S. Department of Education, 1994). This figure showed a continuation of the decreasing number of students classified as mentally retarded (2.98% decrease from prior year). These data also revealed that the large majority of students with mental retardation (312,402 or 64%) were served in separate classes with comparatively few served in regular classes (about 5%) and resource rooms (about 28%). The remainder were served in segregated facilities or through homebound instruction.

Current Themes in Research and Practice

Full Inclusion

Full inclusion has been defined as "full participation of all students with disabilities in the social milieu of the regular classroom" (Sailor, Gerry, & Wilson, 1991). Sailor et al. (1991) defined the characteristics of full inclusion as including: a) a "home school" placement, b) natural proportion of disability at the school site, c) zero-rejection/heterogeneous grouping, d) age-and grade-appropriate school and classroom placements, e) strong site-based coordination and management, and f) use of cooperative learning and peer instruction models in the regular educational instructional systems. Sailor et al. (1991) also provided a rationale for full inclusion by discussing the potentially negative effects of educational isolation such as the loss of opportunity to learn from peers, reduction of educational time on task when pull out occurs, lowered teacher expectations, and poor transition from school to gainful employment.

Since Sailor et al.'s (1991) overview of full inclusion, more information has emerged on models for inclusion and its impact. For example, a program evaluation conducted by Hunt, Farron-Davis, Beckstead, Curtis, and Goetz (1994) found that fully included students initiated more social interactions with peers, spent less time alone, and had more IEP objectives related to integration. Hollowood, Salisbury, Rainforth, and Palombaro (1995) found that the presence of students with severe disabilities did not negatively impact the instructional time of students without disabilities. Students with severe disabilities actually had higher levels of instructional time in the inclusive setting compared to their nondisabled classmates. The use of paraprofessionals in the inclusive setting contributed to this result. These two studies provide important examples of the type of research being conducted on the impact of full inclusion of students with severe disabilities and the beneficial results found to date in innovative programs in California (Hunt et al., 1994) and Johnson City Schools (Hollowood, et al. 1995). Future research is needed to determine optimal models for inclusion (e.g., training and use of paraprofessionals) and its impact on achievement of students with severe disabilities.

The debate on the inclusion of students with severe disabilities has focused on how much time students should spend in regular classes. Brown et al. (1991) advocated for students attending their home school, but urged professionals to use individual planning in determining time in regular classes. Others, such as Stainback and Stainback (1992), proposed eliminating all separate special services and instead, serving students fulltime in regular classes. It was this proposal to end all separate special services that created resistance from critics such as Fuchs and Fuchs (1994). Fuchs and Fuchs (1994) noted that the elimination of all alternatives to regular education was not supported by experts in vision and hearing impairments. They also questioned the deemphasis of standard curriculum proposed by some inclusionists and their priority of social versus academic competence.

The benefit of a "radical" stance in a human service field such as special education, is that it can create the tension necessary to transform the status quo. Because some took a radical stance on inclusion, benefits such as those described by Hunt et al. (1994) have occurred in integrated services for students with severe intellectual disabilities. By contrast, social change in America has often been followed by "backlash" energy to reinstate the status quo (e.g., comments at the 1993 Council for Exceptional Children Convention in favor of segregation). In a democracy, a debate of extreme positions may be one of the most important vehicles to formulate clear thinking. The current debate and research on how much time students with severe intellectual disabilities should spend in regular classes is needed for professional guidance. Such debate cannot replace child-centered planning on how to develop an appropriate education for an individual student.

Within America, the debate on full inclusion must intersect with overall school reform efforts to be relevant and effective. The debate also needs to be expanded to assume a multicultural and international perspective. The term "inclusive education" is beginning to be used in the field of special education across the world (Mittler, Brouillette, & Harris, 1993). The United Nations has made a greater commitment to the inclusion of children with intellectual disabilities in the regular education system as reflected in the enactment of rules for equal opportunity (UNESCO, 1990). The research and debate on inclusion in the United States can be important to the international movement towards inclusion. Similarly, the international discussions on inclusion, can help broaden the cultural relevance of these practices throughout the world as well as within the diverse population of the United States.

The concept of inclusion at the international level must be considered in the broader context of a nation's educational policies for all children. For example, in most developing countries, less than one percent of children with significant intellectual disabilities attend any form of school (UNICEF, 1994). By contrast, highly developed special education services in countries that rely on government funded, formal systems, such as the United States, sometimes have created exclusive services that must be dismantled to meet the individual needs of children. As Parmenter (1993) notes, Western countries would benefit from closer examination of the informal systems of support used by some developing countries. Professionals also need to remember that the concept of inclusion may hold different meanings across international contexts. As Calvez (1993) describes, cultural diversity exists in people's understanding of social structures and roles. This understanding will impact on what is meant by "inclusion." International discussions can help broaden the understanding of the impact of culture on the goals of inclusion. International partnerships in special education are needed to consider inclusion and other issues critical to the lives of students with disabilities. To reap the full benefit of these partnerships, participation must be egalitarian (Davidson, Goode, & Kendig, 1992).

Leisure Inclusion

To encourage individuals with severe intellectual disabilities to be fully included in their communities, efforts have extended beyond school and work to leisure activities. The three primary ways researchers and practitioners have facilitated inclusion in leisure activities are to: a) assess the leisure skills and preferences of the individual, b) identify or create cooperatively structured leisure activities, and c) provide any instructional or other forms of support needed for participation. The identification of leisure preferences often involved sampling of potential activities or interviewing staff. Newton, Ard, and Horner (1993) demonstrated that consensus could be obtained between preferences identified by staff and the choices of participants with disabilities. Specifying these preferences in an individualized habilitation plan has also been shown to encourage staff to provide greater access to these activities (Newton, Horner, & Lund, 1991).

Besides preferences, professionals also developed ways to create or identify inclusive leisure activities (Moon, 1994). Cooperative activities within these contexts were essential to encouraging interaction between students with and without disabilities. Rynders, Schleien, and Meyer (1993) reviewed the literature on cooperatively structured recreation activities and reached several conclusions. First, Rynders et al. (1993) found that a wide variety of recreational contexts enhanced cooperative activities (e.g., bowling, art, sociodrama.) Within these contexts, a cooperative goal structure enhanced social interaction more than a competitive or independent goal structure. Also, social turn taking, opportunities to make choices, and the ratio of individuals with disabilities to those who were nondisabled all influenced cooperative activities. Slightly older peers (1 to 2 years) were also beneficial in facilitating social interactions for children with severe intellectual disabilities in these leisure activities. In addition to enhancing cooperative activities, professionals also created access to leisure inclusion by adapting the activity for the person with disabilities. For example, Banks and Aveno (1986) utilized adaptive devices for miniature golf and Bernabe and Block (1994) modified the rules in a softball league.

The third way professionals focused on encouraging leisure inclusion was to provide whatever support was necessary to the individual with disabilities to participate in the activity. This involved either a leisure "coach" who encouraged participation during the activity or instruction in skills required by the activity. In Datillo's (1991) review of research on teaching recreational skills to individuals with disabilities, he found many examples of how to teach leisure tasks, but few conducted outside the home and in integrated contexts. One concern in teaching leisure activities in integrated contexts is that the instruction may be counterproductive to goals of social inclusion. Rynders et al. (1993) posited that a focus on peer socialization may create less negative results in attitudes of peers who are nondisabled than a tutorial role. More research is needed on ways to teach participation in leisure activities that do not stigmatize the individual nor discourage social interaction with peers.

In the last five years, professional attention to the growing population of older adults with intellectual disabilities has also increased (Factor, 1993). For older adults, paid work and formal education often diminish in importance. Leisure activities become the primary means of continued socialization. Browder and Cooper (1994) described ways to encourage leisure inclusion of older adults with severe intellectual disabilities including: a) assessing preferences of participants, b) respecting the challenges of aging, c) creating access to community opportunities, and d) teaching new skills. Research is needed on ways to meet the unique leisure needs of older adults with severe intellectual disabilities.

Person-Centered Planning and Choice

The last decade's trend to include individuals with severe disabilities stemmed from a broader commitment to enhancing quality of life for all people with disabilities (Dennis, Williams, Giangreco, & Cloninger, 1993).

Equally important to this lifestyle enhancement is person-centered planning (O'Brien, 1987). Person-centered planning is both a philosophy and a methodology. Its philosophy is to identify an individual's personal goals and life dreams and create ways to realize these goals and dreams. Its methodology is to utilize a planning process for goal setting and action. Several models for this planning were developed in recent years such as Lifestyle Planning (O'Brien & Lyle, 1987), the McGill Action Planning System (Vandercook, York, & Forest, 1989) and the Lifestyle Development Process (Malette, Mirenda, Jones, Bunz, & Rogow, 1992). These models emphasized the importance of inclusion with nondisabled peers, community access, choice, and autonomy.

In these models, the planning process typically incorporated three phases: a) describing the person (e.g., preferences, social networks), b) brainstorming future goals, and c) creating an action plan to reach the goals. This planning differs from traditional assessment in that person-centered planning identifies the gifts and capacities of the person, whereas traditional assessment diagnoses weaknesses. Also, the implementation of traditional assessment is usually the responsibility of professionals, whereas, a person-centered approach relies more on a network of family and friends. Perhaps it is because person-centered planning has focused on informal networks of support that little research exists on the impact of this approach on individuals with severe intellectual disabilities.

The trend in lifestyle enhancement also fostered increased interest in choice making by individuals with severe intellectual disabilities. Research in the 1980s suggested that individuals with severe disabilities had inadequate opportunities for expression of preference and to make choices (Houghton, Bronicki, & Guess, 1987; Kishi, Teelucksingh, Zollers, & Meyer, 1988). This lack of opportunity might have been due to either professionals and caregivers not knowing how to present choices or to the communicative difficulties of individuals with severe intellectual disabilities. Research in the early 1990s, such as that by Parsons, McCarn, and Reid, (1993), demonstrated how to increase choice making through both expanding opportunities and teaching choice making skills. Recently, the need for increasing diversity of choices has been described (Brown, Belz, Corsi, & Wenig, 1993). More research is needed on increasing opportunities for choice and teaching choice making that address a wide range of daily activities.

Multicultural Competence

To conduct person-centered planning and respect the preferences of individuals with severe intellectual disabilities, cultural awareness and sensitivity is required. Multicultural competence is a term used to describe the skills and awareness needed to work with families and individuals from a variety of ethnic cultures (Lynch & Hanson, 1992). Although resources on multicultural special education have been increasing in the last decade (e.g., Baca & Cervantes, 1989; Plata & Chinn, 1989), considerations of cultural perspective for individuals with severe intellectual disabilities is more recent (Lim & Browder, 1994; Rueda & Martinez, 1992). As the field moves towards viewing severe intellectual disabilities in terms of the concept of social competence (Greenspan & Granfield, 1992), it becomes important to consider the "goodness of fit" between individuals and their primary community culture. "Goodness of fit" with other cultural communities (e.g., the majority culture) may also be important, but cannot supplant social competence within the primary culture.

Recent work in multicultural education illustrates how the goal of "goodness of fit" can be achieved through culturally sensitive assessment and instruction, curricula, and personnel preparation. For example, Rueda and Martinez (1992) described how the acceptability of service delivery was increased by considering language and cultural values in designing parent training programs for families from Latino families. Lim and Browder (1994) outlined methods to make lifeskills assessment more multicultural through gaining cultural awareness and applying it to the assessment process. More work is needed to extend these ideas and to evaluate their impact on individuals with severe intellectual disabilities from culturally diverse backgrounds.

More work is also needed in applying multicultural awareness to curriculum development for individuals with severe intellectual disabilities. In recent years, the curriculum focus of programs for individuals with severe intellectual disabilities has been life skills (e.g., skills of daily living for work, home, and the community). Curriculum guides (Ford et al. 1989, Neel & Billingsley, 1989, and Wilcox & Bellamy, 1987) and assessment models (Browder, 1991) have emphasized the need to select life skills that are relevant to the individual's current and future environments. Attention also needs to be given to selecting and teaching skills with cultural relevance (Lim & Browder, 1994). Future research is also needed on how cultural diversity influences this curriculum selection process. For example, Hamre-Nietupski, Nietupski, and Strathe (1992) surveyed parents about their curricular priorities given a choice between academics, life skills, and socialization. If this study were replicated comparing cultural groups, differing sets of priorities might be revealed.

In addition to making curriculum relevant to students' home culture, more research is also needed on how to help individuals with severe disabilities function across cultural contexts. For example, the student may need to learn styles of social exchange typical of the work context (e.g., shaking hands), as well as of the home (e.g., kissing the cheek). Sometimes a culture may devalue an individual with disabilities. Cross cultural advocacy, including self advocacy for people with disabilities, may be important in challenging negative attitudes.

To achieve these goals, personnel preparation is needed as well as research on multicultural issues. Although literature has been written about preparing personnel to work with individuals with severe disabilities (e.g., Kaiser & McWhorter, 1990), little has been described on the skills needed for working with culturally diverse individuals and their families. By contrast, Lynch & Hanson (1992) and Harry (1992) provide examples of recent resources that can help achieve this goal. Future research is also needed on the impact of culture on the overall lives of individuals with severe disabilities. For example, Heller and Factor (1988) found that African American (black) families were less likely than European American (white) families to seek residential placement outside the home for their relative with intellectual disabilities. Most research studies do not state the cultural background of participants so it is difficult to know how representative current literature is of the diversity of individuals in the general population. Extending this research to consider issues of cultural diversity is a critical step for the next decade.

Research and personnel preparation programs also need to acquire an international focus. International dialogue and learning are increasing because of (a) improved technologies in the international exchange of information, (b) increased recognition of the benefits of cross-cultural research, and (c) deeper appreciation of the international applicability of solutions to disability concerns (Davidson, Goode, & Kendig, 1992). Rowitz (1990) has urged professionals to maintain interest in this dialogue so that the benefits of international exchange can be achieved. One way this dialogue has been evident in recent years is through descriptions of a nation's special services that are written for an international audience (e.g., Ballard, 1990; Befring, 1990; Hrnjica, 1990). A second example of this dialogue are the discussions of family support such as the resource provided by Gartner, Lipsky, and Turnbull (1991). More resources such as these can help increase the international perspective on disabilities.

Instructional Technology

During the 1980s, important gains were made in understanding what and how to teach individuals with severe intellectual disabilities. Several methods of prompting to teach discrete or chained responses were developed such as time delay, the system of least prompts, and graduated guidance (Demchak, 1990). Ault, Wolery, Doyle, and Gast (1989) synthesized this literature and highlighted the importance of considering efficiency of learning, as well as the effectiveness of the technique. In the 1990s, the dissemination of these techniques increased through guidelines targeted for teacher training (e.g., Wolery, Ault, & Doyle, 1992).

Research guidelines also emerged for establishing effective small group instruction (Collins, Gast, Ault, & Wolery, 1991), cooperative learning (Cosden & Haring, 1992) and encouraging social relationships (Haring, 1991). While these strategies were important for school contexts, other researchers demonstrated how to apply instructional strategies to community contexts (e.g., McDonnell & Laughlin, 1989). New directions also emerged for teaching individuals to self-manage and solve problems in contexts such as work (Hughes & Agran, 1993; Hughes & Rusch, 1989). Teachers were also encouraged to become better problem solvers with new information on how to use data to increase instructional effectiveness (Farlow & Snell, 1994). Table 10.1 provides a synthesis of current literature with 10 guidelines for planning effective instruction.

Table 10.1 Summary of Effective Instruction for Individuals with Severe Intellectual Disabilities

Strategy	For More Information
1. Use life skills assessment to select curriculum that is relevant to person and family. Consider cultural relevance.	Browder (1991) Lim & Browder (1994)
2. Write instructional objectives that target both acquisition & generalization.	Billingsley, Burgess, Lynch, & Matlock (1991)
3. Select inclusionary teaching contexts-school, work, community, leisure while planning for individual needs. Teach across contexts and invivo for generalization.	Snell & Brown (1993)
4. Encourage natural supports for individual in these settings.	Nisbet (1992)
5. Plan for effective and efficient instruction. Use least intrusive procedures possible.	Wolery, Ault, & Doyle (1992)
6. Use self-direct learning whenever possible and effective. Teach learners to solve their own problems.	Hughes & Agran (1993) Hughes & Rusch (1989)
7. Encourage social inclusion. Grenot-Scheyer (1994)	Haring (1991)
8. Evaluate instruction through review of student performance data.	Farlow & Snell (1994)
9. Teach for generalization and fluency as well as acquisition.	Horner, Dunlap & Koegel (1988)
10. Consider social validity of instructional methods and outcomes.	Billingsley & Kelley (1994) Wolfe (1994)

Strategies for Challenging Behavior

In the 1980s, significant progress occurred in understanding the functions of behavior (Durand, 1990). This understanding led to practical guidelines for assessing the function of challenging behavior (e.g., O'Neill, Horner, Albin, Storey, & Sprague, 1991). By understanding the functions of problem behavior, alternative behavior could be taught. One set of alternative behaviors that proved especially powerful in reducing challenging behavior was functional communication (Durand & Carr 1991). Other positive approaches to challenging behavior emerged during the era of the early 1990s. For example, choice making offered one such positive strategy to reduce challenging behavior (Dyer, Dunlap, & Winterling, 1990). Planning curriculum to meet the unique support needs of individuals with problem behavior were also described (Horner, Sprague, & Flannery, 1993). Future directions for the work on challenging behavior of individuals with severe intellectual disabilities include continuing to develop functional analysis methodology (Mace, 1994) and behavior support plans that are conducive to inclusionary settings (Horner, O'Neill, & Flannery, 1993).

Communication and Controversy

Quality of life for individuals with severe intellectual disabilities can be enhanced as they gain skill in communication to make their needs known and to socialize with others. Important advances have occurred in methods to teach both verbal and nonverbal communication (Warren & Reichle, 1992). Communicative competence is relevant to all of the focal points of current research. Communication provides a tool for socialization within inclusive settings and for stating priorities in person centered planning. Communication often reflects cultural identification. For example, students with severe intellectual disabilities who come from homes where the primary language (e.g., Spanish) differs from that of society at large (e.g., English) may communicate with some words or gestures common to each language. Communicative competence can also be pivotal in minimizing challenging behavior and may enhance skill acquisition through self-instruction.

Perhaps because of the potential power of communication, professionals in the early 1990s were especially receptive to the concept of facilitated communication. Facilitated communication is a controversial technique that uses hand over hand guidance to facilitate the typing of a message by an individual with a disability. More specifically, facilitated communication is "an alternative to speech that involves providing physical and emotional support to individuals with severe communication impairments as they type or point to letters or pictures" (Biklen, 1993). The controversy arises from the question of who is the author of the communication- the person with disabilities or the facilitator who guides responding.

Facilitated communication was developed in Australia about 20 years ago by Crossley who found that through this method children with cerebral palsy, who were once thought to be mentally retarded, demonstrated the ability to communicate and read (Goode, 1994). Biklen (1990) introduced this procedure to the United States and quickly developed training procedures in response to demands for its dissemination. One of the surprising outcomes claimed by users of facilitated communication was that individuals with severe disabilities communicated concepts and skills in written language typically associated with higher levels of intelligence and demonstrated more advanced reading skills (Biklen, 1993).

These claims were not substantiated in most of the research that followed. Researchers using quantitative methods of analyses found little to no evidence to support the validity of the procedure (see review by Green & Shane, 1994). Biklen and Duchan (1994) criticized reliance on this positivist tradition and argued that their opponents who accepted this research viewed mental retardation as residing within the individual. With the move towards a social competence view of intellectual disabilities and the growing interest in qualitative research, these were important counterarguments. Green and Shane (1994) considered the validity of the descriptive evidence offered from the qualitative research of Biklen and others, but concluded that it failed to demonstrate authorship of the messages. Whitehurst and Crone (1994) considered the criticism that the research relied on a positivist perspective but noted that "even though the process of constructing scientific knowledge is strongly affected by human social, emotional, and cognitive processes, it also involves matters of fact that cannot be ignored" (p. 191). Continued use of facilitated communication has raised ethical concerns including: (a) use of the procedure without parental consent, (b) discontinuation of other methods of communication instruction, and (c) reports of abuse during facilitation and subsequent litigation regarding these allegations (Green & Shane, 1994).

Conclusions and Recommendations for Future Research

Facilitated communication has been a "lightening rod" for a field in transition. In the 1980s, the key to access new opportunities for individuals with severe disabilities seemed to lie in acquisition of new skills and deceleration of problem behavior. The positivist perspective, especially as reflected in the tradition of applied behavior analysis, provided a strong technology for the goal of gaining access through improving the person with disabilities. The paradigm shift of the late 1980s and early 1990s redirected the focus from changing the person with disabilities to changing social institutions to create access for the person with disabilities (Bradley & Knoll, 1990). With this shift in philosophy, professionals became interested in other paradigms of knowledge (Skrtic, 1986). During this era of transition, facilitated communication seemed to offer a powerful tool to help gain access to opportunities. For example, the person might attend college if the facilitated message could be accepted for academic responding. The evaluation of the technique tapped alternative paradigms of knowledge. But the lesson learned from the experience of facilitated communication was the need for professionals across paradigms to debate and research any technique that claimed such dramatic results.

Based on the lessons learned from the facilitated communication controversy and the overall current research on education for individuals with severe intellectual disabilities, the following recommendations for future research and practice are offered.

1. Multiparadigmatic knowledge construction is needed to achieve the ideals of inclusion of individuals with severe intellectual disabilities in school and society. The complexities of transforming educational and other social institutions to be accessible to individuals with disabilities requires knowledge beyond ways to change the individual (Skrtic, 1986). Diverse research perspectives can enrich overall knowledge about a new trend within the field. For example, quantitative (Hunt et al., 1994; Grenot-Scheyer, 1994) and qualitative (Staub, Schwartz, Gallucci, & Peck, 1994) methods of analyses are offering important information on the impact of inclusion. These various methodologies are beginning to uncover techniques that achieve positive effects on students with and without disabilities. By contrast, when methodologies seem to be yielding opposite findings as with facilitated communication, the solution is not to resort to monoparadigmatic knowledge even if this knowledge stems from a "new" perspective for special education such as constructivism. Rather, evidence from each perspective should be considered together (e.g., triangulation). As Horner (1994) notes, "We need more careful, rigorous research (both experimental and phenomenological) to understand if, when, and why FC is effective." The same is true for other new trends within the field (e.g., person-centered planning, full inclusion.) As Whitehurst and Crone (1994) have noted, belief systems have differential consequences on the technological changes valued by people with severe disabilities. Through evaluation these consequences are discerned.

2. The goals of education must be directed towards both the learner and his or her environments. Individuals with severe intellectual disabilities have benefitted from the continued research focus on instructional strategies and functional analysis of problem behavior as well as from new efforts to create opportunities and models of service delivery. One of the criticisms of the full inclusion movement is that educational achievement of the individual with severe disabilities sometimes seems lost in the goals of socialization and presence within the mainstream. School aged children, in particular, need the opportunity to gain new skills and knowledge through the most effective instructional procedures available. Research on both enhancing the skills of the learner and making the school environment more accessible for all students should continue as complementary priorities.

3. To achieve the goals of person-centered planning, special educators must gain the multicultural competence to support the diverse population of individuals with severe disabilities. Much more research and development is needed to understand and respond to the cultural diversity of individuals with severe intellectual disabilities. For example, the importance of belonging within the mainstream of a school for some families may depend on how responsive that school has been to their cultural values. Or, for some parents the presence of a disability may have religious or other cultural meaning that professionals must understand before proposing interventions. Special educators can also benefit from acquiring an international perspective on disability. Such a perspective may enhance understanding of diversity in one's own nation, lead to discovery of the cross cultural aspects of disability, and provide important advances in knowledge for the field.

4. Longitudinal research is needed to determine the impact of current methodologies on the long term community inclusion of individuals with severe disabilities. In the last two decades, there has been a proliferation of research and writing on ways to achieve the criterion of community inclusion for individuals with severe disabilities. Because the focus on community inclusion has a recent history, little information exists on the long term impact of current models of service delivery. For example, what types of supports are needed as individuals experience stages of adult development (e.g., middle age, early retirement, advanced aging)? How enduring are informal sources of support? What resources do individuals need to cope with changes in society and advances in technology? Longitudinal research may provide important insights about designing service delivery.

References

Ault, M.J., Wolery, M., Doyle, P.M., & Gast, D.L. (1989). Review of comparative studies in instruction of students with moderate and severe handicaps. *Exceptional Children, 55,* 346–356.

Baca, L., & Cervantes, H. (1989). *The bilingual special education interface.* Columbus, Ohio: Charles E. Merrill.

Ballard, K.D. (1990). Special education in New Zealand: Disability, politics, and empowerment. *International Journal of Disability, Development, and Education, 37,* 109–124.

Banks, R., & Aveno, A. (1986). Adapted miniature golf: A community leisure program for students with severe physical disabilities. *Journal of the Association for Persons with Severe Handicaps, 11,* 209–215.

Befring, E. (1990). Special education in Norway. *International Journal of Disability, Development, and Education, 37,* 125–137.

Bellamy, G.T., Rose, H., Wilson, D.J., & Clarke, J.Y. III. (1982). Strategies for vocational preparation. In B. Wilcox, & G. T. Bellamy (Eds.), *Design of high school programs for severely handicapped students* (p. 139–152). Baltimore, Md: Paul H. Brookes.

Bernabe, E.A., & Block, M.E. (1994). Modifying rules of a regular girls' softball league to facilitate the inclusion of a child with severe disabilities. *Journal of the Association for Persons with Severe Handicaps, 19,* 24–31.

Biklen, D. (1990). Communication unbound: Autism and praxis. *Harvard Educational Review, 60,* 291–315.

Biklen, D. (1993). *Communication unbound: How facilitated communication is challenging traditional views of autism and ability-disability.* New York: Teachers College Press.

Biklen, D., & Duchan, J.F. (1994). "I am intelligent": The social construction of mental retardation. *Journal of the Association for Persons with Severe Handicaps, 19,* 173–184.

Billingsley, F.F., Burgess, D., Lynch, V.W., & Matlock, B.L. (1991). Towards generalized outcomes: Considerations and guidelines for writing instructional objectives. *Education and Training in Mental Retardation, 4,* 351–360.

Billingsley, F.F., & Kelley, B. (1994). An examination of the acceptability of instructional practices for students with severe disabilities in general education settings. *Journal of the Association for Persons with Severe Handicaps, 19,* 75–83.

Bradley, V., & Knoll, J. (1990). *Shifting paradigms in services for people with developmental disabilities.* Cambridge, MA: Human Services Research Institute.

Browder, D. (1991). *Assessment of individuals with severe disabilities.* 2nd Edition. Baltimore, Md: Paul H. Brookes.

Browder, D.M., & Cooper, K. (1994). Inclusion of older adults with mental retardation. *Mental Retardation, 32,* 91–99.

Brown, F., Belz, P., Corsi, L., & Wenig, B. (1993). Choice diversity for people with severe disabilities. *Education and Training in Mental Retardation, 28,* 318–326.

Brown, L., Schwarz, P., Udvari-Solner, A., Kampschroer, E., Johnson, F., Jorgensen, J., & Gruenewald, L. (1991). How much time should students with severe intellectual disabilities spend in regular education classrooms and elsewhere? *Journal of the Association for Persons with Severe Handicaps, 16,* 39–47.

Calvez, M. (1993). Social interactions in the neighborhood: Cultural approach to social integration of individuals with mental retardation. *Mental Retardation, 31,* 418–423.

Carr, E.G., & Durand, V.M. (1985). Reducing behavior problems through functional communication training. *Journal of Applied Behavior Analysis, 18,* 111–126.

Collins, B.C., Gast, D.L., Ault, M.J., & Wolery, M. (1991). Small group instruction: Guidelines for teachers of students with moderate to severe handicaps. *Education and Training in Mental Retardation, 26,* 18–32.

Cosden, M.A., & Haring, T.G. (1992). Cooperative learning in the classroom: Contingencies, group interactions, and students with special needs. *Journal of Behavior Education, 2,* 53–71.

Datillo, J. (1991). Recreation and leisure: A review of the literature and recommendations for future directions. In L. Meyers, C. Peck, & L. Brown (Eds.), *Critical issues in the lives of people with severe disabilities* (pp. 171–194). Baltimore, Md: Paul H. Brooks.

Davidson, P.W., Goode, D.A., & Kendig, J.W. (1992). Developmental disabilities related education, technical assistance, and research activities in developing nations. *Mental Retardation, 30,* 269–275.

Demchak, M. (1990). Response prompting and fading methods: A review. *American Journal on Mental Deficiency, 94,* 603–615.

Dennis, R.E., Williams, W., Giangreco, M.F., & Cloninger, C.J. (1993). Quality of life as context for planning and evaluation of services for people with disabilities. *Mental Retardation, 30,* 269–275.

Durand, V.M. (1990). *Severe behavior problems: A functional communication training approach.* New York: Guilford Press.

Durand, V.M., & Carr, E.G. (1991). Functional Communication Training to reduce challenging behavior: Maintenance and application in new settings. *Journal of Applied Behavior Analysis, 24,* 251–264.

Dyer, K., Dunlap, G., & Winterling, V. (1990). Effects of choice making on the serious problem behaviors of students with severe handicaps. *Journal of Applied Behavior Analysis, 23,* 515–524.

Evans, I.M. (1991). Testing and diagnosis: a review and evaluation. In L.H. Meyer, C.A. Peck, & L. Brown (Eds.), *Critical issues in the lives of people with severe disabilities* (p. 25–44). Baltimore, Md: Paul H. Brookes.

Factor, A.R. (1993). Translating policy into practice. In Sutton, A.R. Factor, B.A. Hawkins, T. Heller, & G.B. Seltzer (Eds.), *Older adults with developmental disabilities* (pp. 257–276). Baltimore, Md: Paul H. Brookes.

Farlow, L., & Snell, M.E. (1994). *Effective and efficient use of student performance data.* AAMR Research to Practice Series. Washington, DC: AAMR.

Ford, A., Schnorr, R., Meyer, L., Davern, L., Black, J., & Dempsey, P. (Eds.). (1989). *The Syracuse community-referenced curriculum guide for students with moderate and severe disabilities.* Baltimore, Md: Paul H. Brookes.

Fuchs, D. & Fuchs, L. (1994). Inclusive schools movement and the radicalization of special education. *Exceptional Children, 60,* 294–309.

Gartner, A., Lipsky, D.K., & Turnbull, A.P. (1991). *Supporting families with a child with a disability: An international outlook.* Baltimore, Md: Paul H. Brookes.

Goode, D. (1994). Defining facilitated communication in and out of existence: Role of science in the facilitated communication controversy. *Mental Retardation, 32,* 307–311.

Green, G., & Shane, H.C. (1994). Science, reason, and facilitated communication. *Journal of the Association for Persons with Severe Handicaps, 19,* 151–172.

Greenspan, S., & Granfield, J.M. (1992). Reconsidering the construct of mental retardation: Implications of a model of social competence. *American Journal on Mental Retardation, 96,* 442–453.

Grenot-Scheyer, M. (1994). The nature of interactions between students with severe disabilities and their friends and acquaintances without disabilities. *Journal of the Association for Persons with Severe Handicaps, 19,* 253–262.

Grossman, H.J. (Ed.) (1983). *Classification in mental retardation.* Washington, DC: American Association on Mental Deficiency.

Hamre-Nietupski, S., Nietupski, J., & Strathe, M. (1992). Functional skills, academic skills, and friendship/social relationship development: What do parents of students with moderate/severe/profound disabilities value? *Journal of the Association for Persons with Severe Handicaps, 17,* 53–59.

Haring, T. (1991). Social relationships. In L.H. Meyer, C.A. Peck, & L. Brown (Eds.), *Critical issues in the lives of people with severe disabilities* (pp. 195–217). Baltimore, Md: Paul H. Brookes.

Harry, B. (1992). *Cultural diversity, families, and the special education system: Communication and empowerment.* New York: Teachers College.

Heber, R. (1959). A manual on terminology and classification in mental retardation. *American Journal of Mental Deficiency, 64,* Monograph Supplement.

Heller, T., & Factor, A. (1988). Permanency planning among black and white family caregivers of older adults with mental retardation. *Mental Retardation, 26,* 203–208.

Hollowood, R.M., Salisbury, C.L., Rainforth, B., & Palombaro, M.M. (1995). Use of instructional time in classrooms serving students with and without severe disabilities. *Exceptional Children, 61,* 242–253.

Horner, R.H. (1994). Facilitated communication: Keeping it practical. *Journal of the Association for Persons with Severe Handicaps, 19,* 185–186.

Horner, R.H., Dunlap, G., & Koegel, R.L. (Eds.). (1988). *Generalization and maintenance: Lifestyle changes in applied settings.* Baltimore, Md: Paul H. Brookes.

Horner, R.H., O'Neill, R.E., & Flannery, K.B. (1993). Effective behavior support plans. In M.E. Snell (Ed.), *Instruction of students with severe disabilities* (p. 184–208). New York: Macmillan.

Horner, R.H., Sprague, J.R., Flannery, K. (1993). Building functional curricula for students with severe intellectual disabilities and severe problem behaviors. In R.V. Houten & S. Axelrod (Eds.), *Behavior analysis and treatment* (pp. 47–71). New York: Plenum Press.

Houghton, J., Bronicki, G.J.B., & Guess, D. (1987). Opportunities to express preferences and make choices among students with severe disabilities in classroom settings. *Journal of the Association for Persons with Severe Handicaps, 12*, 18–27.

Hrnjica, S. (1990). Special education in Yugoslavia. *International Journal of Disability, Development, and Education, 37*, 169–178.

Hughes, C., & Agran, M. (1993). Teaching persons with severe disabilities to use self-instruction in community settings: An analysis of applications. *Journal of the Association for Persons with Severe Handicaps, 18*, 261–274.

Hughes, C., & Rusch, F. (1989). Teaching supported employee employees with severe mental retardation to solve problems. *Journal of Applied Behavior Analysis, 22*, 365–372.

Hunt, P., Farron-Davis, F.,Beckstead, S., Curtis, D., & Goetz, L. (1994). Evaluating the effects of placement of students with severe disabilities in general education versus special classes. *Journal of the Association for Persons with Severe Handicaps, 19*, 200–214.

Kaiser, A.P., & McWhorter, C.M. (1990). *Preparing personnel to work with persons with severe disabilities.* Baltimore, MD: Paul H. Brookes.

Kishi, G., Teelucksingh, B., Zollers, S.P., & Meyer, L. (1988). Daily decision making in community residences: A social comparison of adults with and without mental retardation. *American Journal on Mental Retardation, 92*, 430–435.

Lim, L.H.F., & Browder, D.M. (1994). Multicultural life skills assessment of individuals with severe disabilities. *Journal of the Association for Persons with Severe Handicaps, 19*, 130–138.

Lynch, E.W., & Hanson, M.J. (1992). *Developing cross cultural competence: A guide for working with young children and their families.* Baltimore, Md: Paul H. Brookes.

Luckasson, R., Coulter, D., Polloway, E., Reiss, S., Schalock, R., Snell, M., Spitalnik, D., Y Stark, J. (1992). *Mental retardation-definition, classification, and systems of support.* Washington, DC: American Association on Mental Retardation.

Mace, F.C. (1994). The significance and future of functional analysis methodologies. Special issue: Functional analysis approaches to behavioral assessment and treatment. *Journal of Applied Behavior Analysis, 27*, 385–392.

Malette, P., Mirenda, P., Kandborg, T., Jones, P., Bunz, T., & Rogow, S. (1992). Application of a lifestyle development process for person with severe intellectual disabilities: A case study report. *Journal of the Association for Persons with Severe Handicaps, 17*, 179–191.

McDonnell, J., & Laughlin, B. (1989). A comparison of backward and concurrent chaining strategies in teaching community skills. *Education and Training in Mental Retardation, 24*, 230–238.

McLaren, J., & Bryson, S.E. (1987). Review of recent epidemiological studies of mental retardation: Prevalence, associated disorders, and etiology. *American Journal of Mental Retardation, 92*, 243–254.

Mittler, P., Brouillette, R., & Harris, D. (1993). *World yearbook of education: Special needs education.* London: Kogan Page.

Mittler, P., & Serpell, R. (1985). Services: an international perspective. In A.M. Clarke, & J.M. Berg (Eds.), *Mental deficiency: The changing outlook* (4th Edition, pp. 715–787). London: Methuen.

Moon, M.S. (1994). *Making school and community recreation fun for everyone: Places and ways to integrate.* Baltimore, MD: Paul H. Brookes.

Moon, M.S., Inge, K.J., Wehman, P., Brooke, V., & Barcus, J.M. (1990). *Helping person with severe mental retardation get and keep employment.* Baltimore, Md: Paul H. Brookes.

Neel, R.S., & Billingsley,F.F. (1989). *Impact: A functional curriculum handbook for students with moderate to severe disabilities.* Baltimore, Md: Paul H. Brookes.

Newton, J.S., Ard, W.R., & Horner, R.H. (1993). Validating predicted activity preferences of individuals with severe disabilities. *Journal of Applied Behavior Analysis, 26,* 239–245.

Newton, J.S., Horner, R.H., & Lund, L. (1991). Honoring activity preferences in individualized plan development: A descriptive analysis. *Journal of the Association for Persons with Severe Handicaps, 16,* 207–212.

Nisbet, J. (1992). *Natural supports in school, at work, and in the community for people with severe disabilities.* Baltimore, Md: Paul H. Brookes.

O'Brien, J. (1987). *A guide to life-style planning: Using The Activities Catalog to integrate services and natural support systems.* Baltimore, Md: Paul H. Brookes.

O'Brien, J., & Lyle, C. (1987). *Framework for accomplishment.* Decatur, GA: Responsive Systems Associates.

O'Brien, J., & Mount, B. (1991). Telling new stories: The search for capacity among people with severe handicaps. In L.H. Meyer, C.A. Peck, & L. Brown (Eds.), *Critical issues in the lives of people with severe disabilities* (p. 89–92). Baltimore, Md: Paul H. Brookes.

O'Neill, R.E., Horner, R.H., Albin, R.W., Storey, K., & Sprague, J.R. (1991). *Functional analysis of problem behavior: A practical assessment guide.* Sycamore, IL: Sycamore Press.

Parmenter, T.R. (1993). International perspective of vocational options for people with mental retardation: The promise and the reality. *Mental Retardation, 31,* 359–367.

Parsons, M.B., McCarn, J.E., & Reid, D.H. (1993). Evaluating and increasing meal-related choice throughout a service setting for people with severe disabilities. *Journal of the Association for Persons with Severe Handicaps, 18,* 253–260.

Plata, M., & Chinn, P.C. (1989). Students with handicaps who have cultural and language differences. In R. Gaylord-Ross (Ed.), *Integration strategies for students with handicaps* (p. 149–176). Baltimore, Md: Paul H. Brookes.

Rowitz, L. (1990). International issues: An emerging trend. *Mental Retardation, 28,* iii–iv.

Rueda, R., & Martinez, I. (1992). Fiesta Educative: One community approach to parent training in developmental disabilities for Latino families. *Journal of the Association for Persons with Severe Handicaps, 17,* 95–103.

Rynders, J.E., Schleien, S., & Meyer, L. (1993). Improving integrated outcomes for children with and without severe disabilities through cooperatively structured recreation activities: A synthesis of research. *Journal of Special Education, 26,* 386–407.

Sailor, W., Wilson, W.C., & Gerry, M. (1991). Policy implications of emergent full inclusion models for the education of students with severe disabilities. In M.C. Wang, M.C. Reynolds, & H.J. Walberg (Eds.), *Handbook of special education: Research and practice* (Volume 4) (p. 175–196). Oxford: Pergamon Press.

Schleien, S.J., & Ray, M.T. (1988). *Community recreation and persons with disabilities.* Baltimore, Md: Paul H. Brookes.

Serpell, R. (1988). Assessment criteria for severe intellectual disability in various cultural settings. *International Journal of Behavioral Development, 1,* 117–144.

Skrtic, T. (1986). The crisis in special education knowledge: A perspective on perspective. *Focus on Exceptional Children, 18,* 1–16.

Snell, M. E. (1983). Implementing and monitoring the IEP: intervention strategies. In M.E. Snell (Ed.), *Systematic instruction of the moderately and severely handicapped* (2nd Edition) (p. 113–146). Columbus, Ohio: Charles E. Merrill.

Snell, M.E., & Brown, F. (1993). Instructional planing and implementation. In M.E. Snell (Ed.), *Instruction of individuals with severe disabilities.* New York: Macmillan.

Stainback, S., & Stainback, W. (1992). *Curriculum considerations in inclusive classrooms: Facilitating learning for all students.* Baltimore: Paul Brookes.

Stainback, S., Stainback, W., & Forest, M. (Eds.). (1989). *Educating all students in the mainstream of regular education.* Baltimore, MD: Paul H. Brookes.

Staub, D., Schwartz, I., Gallucci, C., & Peck, C. (1994). Four portraits of friendship at an inclusive school. *Journal of the Association for Persons with Severe Handicaps, 19,* 314–325.

UNESCO (1990). *Special needs in the classroom.* Paris: UNESCO.

UNICEF (1994). State of the world's children.New York: UNICEF.

U.S. Department of Education (1994). *To assure the free appropriate public education of all children with disabilities.* Sixteenth annual report to Congress on the implementation of the Individuals with Disabilities Act. Washington, DC: U. S. Department of Education.

Vandercook, T., York, J., & Forest, M. (1989). The McGill Action Planning System (MAPS): A strategy for building the vision. *Journal of the Association for Persons with Severe Handicaps, 14,* 205–215.

Vincent, L.J., Salisbury, C., Walter, G., Brown, P., Gruenewald, L.J., & Powers, M. (1980). Program evaluation and curriculum development in early childhood/special education: criteria of the next environment. In W. Sailor, B. Wilcox, & L. Brown (Eds.), *Methods of instruction for severely handicapped students* (pp. 303–328). Baltimore: Paul H. Brookes.

Warren, S.F., & Reichle, J. (Eds.). (1992). *Causes and effects in communication and language intervention.* Baltimore, MD: Paul H. Brookes.

Whitehurst, G.J. & Crone, D. A. (1994). Social constructivism, positivism, and facilitated communication. *Journal of the Association for Persons with Severe Handicaps, 19,* 191–195.

Wilcox, B. & Bellamy, G.T. (1987). *The activities catalog: An alternative curriculum for youth and adults with severe disabilities.* Baltimore, Md: Paul H. Brookes

Wolfe, P. (1994). Judgment of social validity of instructional strategies used in community-based instructional sites. *Journal of the Association for Persons with Severe Handicaps, 19,* 43–51.

Wolery, M., Ault, M.J., & Doyle, P.M. (1992). *Teaching students with moderate and severe disabilities: Use of response prompting strategies* (p. 261–297). White Plains, NY: Longman.

Wolery, M., & Haring, T.G. (1994). Moderate, severe, and profound disabilities. In N.G. Haring, L. McCormick, & T.G. Haring (Eds.), *Exceptional children and youth.* New York: Macmillan.

Ysseldyke, J.E., Algozzine, B., & Thurlow, M.L. (1992). *Critical issues in special education* (2nd Ed.). Boston: Houghton-Mifflin.

11

Learners With Language Impairments

KATHARINE G. BUTLER

Ph.D., Research Professor, Syracuse University, Syracuse, NY, USA

Learners with Language Impairments provides an international perspective on infants, toddlers, children, and adolescents with language impairment, their incidence and prevalence—which is sadly understated in most countries around the world—and their identification, diagnosis (referred to here as assessment), and classification (frequently subsumed under some other primary label). The role of the speech-language pathologist and other language specialists is discussed as it is reflected in the individualized planning process, particularly in the United States and Western Europe, with sidebar comments on other nations and their status in provision of services to those with language impairments. Treatment, e.g., intervention, is described in some detail. The role of families across the preschool and school-age years is described, as is the role of language specialists and classroom teachers; particularly, the role of the language of instruction (classroom discourse) as related to its impact on children's comprehension and production of language. Technology as it is used in assessment and in intervention for the severely impaired is briefly reviewed, as are transition services and follow-up studies. The efficacy issues in assessment, intervention, and delivery of services is a strand that runs throughout this chapter.

Introduction

As Wang, Reynolds, and Walberg have so cogently noted in the introduction to this section on *Distinct Disabilities*, there are some children with disabilities who require specialized services, services that may be provided in regular or special education classrooms, services that vary considerably depending on the individual's age, cognitive, social, emotional, cultural, and linguistic status as well as the context in which intervention or instruction is provided. Such specialized services may be significantly different across North

America, depending on state or provincial rules and regulations. However, services provided elsewhere around the world differ even more dramatically, varying from no services in some countries to very advanced services in others. Recent international developments have begun to close the wide gap between developed and developing countries. Throughout this chapter, attention to increasing similarities as well as important differences will be provided where appropriate.

This chapter addresses the most recent issues in *educational* placement of language-impaired children, adolescents, and young adults, taking into account the gradual melding of the health and education systems in the United States, particularly as it effects speech-language-impaired children. In addition, there are striking changes occurring at the federal level as political party changes at the Congressional level in 1994 have resulted in new players at the helm. Reauthorization of important federal legislation is hanging in the balance. Under currently existing legislation, health and education systems at the federal, state, and local levels have been undergoing rapid change.

Current Status of Services for the Language Impaired: Co-mingling of Health and Education Around the World

In the recent past, school systems in the United States have provided services to children labeled language or learning disabled, not only through public education dollars, but also through accessing private insurance company funding via parental health plans, and in some cases through public health dollars. As school districts found this to be a successful financial strategy, they have tended to establish contractual arrangements with private practitioners rather than adding to their speech therapy staff employees. While this results in a merging of education and health dollars, it does not result in a merging of staff. There may be unforeseen and unwanted consequences as "managed care" looms ever higher on the health horizon.

Following an aborted attempt by the federal government to overhaul health care in 1993 and 1994, for-profit insurance companies have established mechanisms for controlling the cost of health care, which has been spiraling out of control. "Managed care" has become a byword for doing more with less. For example, mental health services for children and adults were among the first to be effected (Hymowitz & Pollock, 1995), followed by other health care services, including speech-language pathology, particularly in hospital and rehabilitation settings.

Within the past year or so, momentous changes have been wrought by managed care experts. This presumably lower cost and more effective system for the provision of services has shaken the very foundations of health care in the United States. At the same time, there are massive attempts to restructure the schools and to go beyond mainstreaming to inclusive classrooms. This alters the setting for delivery of speech-language services to students previously serviced in special education settings. Since both speech-language and learning disabilities are high-incidence disabilities, the impact on school-based services is already occurring in this country and in other countries where budget deficits are the concern of the moment and of the future (Pickelle & Ramos, 1995). These authors point out that there are "common worldwide pressures of constricting financial resources, growing elderly populations, and increasing demands on the part of payors to be given outcomes data in the provision of quality health care. . . [which calls for breaking] down the turf issues between subspecialties. . . . Similar scenes are being played out in. . . Europe, Asia and South America" (Pickelle & Ramos, 1995, p. 11).

It may seem a bit strange to begin a chapter on educational approaches to quality education for all students by referencing the current status of health services, but, as we shall see, a merged delivery of service is likely to continue as schools also continue to struggle to give appropriate services to infants, toddlers, children, and youth. Readers, therefore, will see such terms from health care as treatment efficacy of patients and outcomes measures emerging in educational literature. "Efficacy" can be thought of as an ongoing, unbroken stream of activity seeking to prove the effectiveness of clinical procedures while "outcome measures" relate to observed clinical changes, i.e., patient/client characteristics, treatment method, provider characteristics, frequency, duration, cost, etc.) of the intervention provided (ASHA Task Force on Treatment Outcome and Cost Effectiveness, 1994). On the other hand, educational systems speak of remedial or compensatory education, learning rates, a Free and Appropriate Education (FAPE), Least Restrictive Environment (LRE), alternative approaches, assessment, diagnostics, classification, placement, and Individualized Educational Programs (IEPs) as terms of consequence. It should be noted that while schools speak of student progress, health and rehabilitation settings speak of outcomes evaluation, i.e., a systematic procedure for monitoring the effectiveness and efficiency with which results are achieved as well as customer satisfaction following termination of services (Commission on Accreditation of Rehabilitation Facilities, 1995). Given the wording of the CARF statement, it is easy to see that health is primarily oriented toward the adult, while education is oriented toward the child and the parents. Customer satisfaction in the education setting frequently refers to parental satisfaction and, over the past decade, to greater student involvesment.

The Professionals who Provide Language Services: What's in a Name?

Learners with language impairments are frequently served by an array of special educators both in and out of the classroom. Not only do regular and special educators "teach" language in the classroom, many other language specialists are part of the educational team. In early intervention programs for infants and toddlers (ages zero to three), speech-language pathologists (SLPs) frequently provide language intervention, as they do in early childhood programs (variously identified as serving three to five year olds or three to eight year olds, depending on the regulations of various states (Bowe, 1995; Rice & Wilcox, 1995). Many SLPs provide services in the schools at the elementary level, with additional numbers working with older children and adolescents throughout middle and high schools in conjunction with learning disabilities specialists, remedial reading teachers, and resource room specialists (RRS). SLPs are also found providing services in "whole language" classrooms (Norris & Hoffman, 1993) where educators stress that "all aspects of language form an indivisible and integrated system" (p. 15). In addition, SLPs provide services in special education classrooms to almost all special education students, including the mentally retarded, the physically disabled, the emotionally disturbed, the autistic, those with Attention Deficit Disorder (ADD), and in the few remaining Schools for the Deaf and Schools for the Blind, as well as private schools and settings that deal with Pervasive Developmental Disorders (PDD), in their role as a special educator or a member of the related personnel staff.

As in so many other professions, terminology varies over time and space. In the United States, for example, the profession grew out of the academic disciplines of university departments of Speech or Psychology. In the early years of the 20th century, the field was known as "speech correction" and practitioners were "speech correctionists," a somewhat unseemly term from a 1990s perspective. From the 1930s to the 1950s, speech correctionists dealt with children and adults whose speech was labeled "defective." The definition of who was "speech defective" was tripartite, including individuals whose speech (a) deviated so far from the speech of other people that it called attention to itself, (b) interfered with communication, or (c) rendered its possessor maladjusted (Van Riper, 1947). This definition lingered in the literature through the half century mark and longer. More recent viewpoints on the definition of language acquisition and its subsystems, and the factors that may result in language impairment, will be addressed in subsequent sections of this paper. As implied earlier, viewpoints regarding language and its impairments and those who provide language services vary mightily around the world.

Those who work with language impaired children are called speech therapists or logopedists in a fair number of countries. Variations on the theme are Fonoaudiologa in Argentina and Columbia, Audiologoped in Denmark, Orthophoniste in France and Lebanon, speech pathologist in Australia, speech clinician in Indonesia, communication clinician in Israel, Sprachheillehehrer in Austria, special education teacher in the People's Republic of China, and Defectologist in Russia (Stewart, 1991). As the appellation "audiologa" or "audiologoped" indicates, in some countries the professional is considered to be an individual who also works with hearing impaired children or adults. In other parts of the world, linguists or psycholinguists are engaged in working with children or adults who are language impaired (James, van Steebrugge, & Chiveralls, 1994; Slama-Cazacu, 1994). Neuropsychologists, in this country and elsewhere, may be found providing diagnoses and treatment in childhood as well as adult language disorders (Costa & Rourke, 1990).

In summary, whether one is called a SLPs, speech therapist, or logopedist, or any of the variations noted above, and whether the title means a few courses in language and its disorders, or the equivalence of a bachelor's, master's or doctoral degree, most practitioners address students' needs for assessment and intervention although their procedures and techniques for dealing with language disorders may be very different.

A Paradigm Shift: The Language Continuum

While for a number of years SLP or speech therapists dealt largely with disorders of oral language, its comprehension and its production, there has been a paradigm shift in the last five to ten years (Butler & Wallach, 1994a). That paradigm shift has focused on viewing children's language acquisition and its disorders at the discourse level and in multiple contexts. For a number of years, language specialists tended to treat children's language disorders at the level of the sound, the syllable, the word, and the sentence. This treatment was provided largely in the oral domain only (i.e., assessing and intervening in receptive and expressive language). Typically, treatment was provided in clinical settings with one-on-one or small-group therapy, the so-called pull-out programs in the schools. However, with increased knowledge of the changes that occur when school entrance is at hand, speech clinicians began looking at how teacher talk, text talk, and child talk are difficult for children with language impairments. SLPs also recognized that it was important to measure and treat language at the conversational level (to assist in social activities) and at communication above the level of the sentence as well as across discourse and text genres (Scott, 1994).

Language: At Home and at School

With the growing interest in looking at language as it effected the child's performance in the context of their daily lives, both the home and the school contexts have become the object of study. Since language itself is an arbitrary conventional code system, and there "is no systematic relationship between words, and the objects, events and relationships they encode," it is clear that "any particular sound sequence that constitutes any given word does not have any perceptual or otherwise meaning" (van Kleeck, 1994, p. 86). In other words, words mean only what the cultural and linguistic community says they mean.

Another aspect of the paradigm shift has been an increasing regard for linking spoken language with reading and writing, the new view of a language continuum that honors emerging concepts related to the primacy of oracy to literacy, and the reciprocal nature of spoken and written language (Wallach & Butler, 1994). This is the converse of the earlier picture of spoken language and written language as "two separate and distinctly different skills. . . in different modalities" (Kavanagh, 1991, p. vii). As Kavanagh indicates, the language continuum reflects that the skills of reading and writing, i.e., literacy, are built on and are extensions of the oral language acquired early in life, in fact, during infancy.

Practitioners Amid the Paradigm Shift

Research has taught those of us who provide language assessment and intervention that "the relationship between spoken and written language is reciprocal and dynamic," and that this recognition "is inconsistent with a view of service delivery that targets only spoken language deficiencies" (Kamhi, 1995, in press; Wallach & Butler, 1994). A considerable number of child language disorders specialists are now trained to assess and intervene along the "language continuum" as children move from infancy and prelinguistic communication to reading and writing and the acquisition of higher-level forms of literacy.

Parenthetically, preservice or inservice training varies from college to college, and from one country to another. In the United States, SLPs are trained in Schools of Education, Medicine, Allied Health, or Arts and Sciences at the master's and doctoral level. As the reader can well imagine, the placement of the university training program deeply affects the content of the program. While there is more similarity than difference in the United States, Canada, and the United Kingdom, where the profession has long been evolving and where national standards exist, the body of information imparted in training programs in other parts of the world may differ substantially. In addition, World Wars I and II and the subsequent Cold War period restricted research and the consequent knowledge base in the Eastern European countries and the then-U.S.S.R. This situation has now been reversed; there is evidence of language researchers and practitioners from Eastern European, African, and Mid- and Far Eastern countries contributing manuscripts to international journals and increasing attendance in international Congresses and multinational meetings. Large databases at university computer centers in several places around the world have made cross-cultural and cross-linguistic language research findings available to researchers and practitioners in almost all countries.

That is not to say that all practitioners are learning the same body of knowledge in the same way across nations. Far from it. There are tremendous differences related to cultural and native language(s) structural linguistic differences, level of government support for health and education, the degree of autonomy in the practice of one's profession, the views held regarding the need or desire to intervene at an early age, or, indeed, at any age. The opportunity to attain an education and to develop literacy skills is limited in many places around the globe. Education for girl children and their mothers is not easily available in a number of developing countries. The importance of a mother's level of education has been repeatedly been found to effect the literacy skills of her children.

For example, in Iceland, where there is a 200-year history of a high level of literacy, a recent study reports that both urban and rural families have long followed a tradition of families reading together, storytelling, and verse making (Taylor, 1995). In addition, educational requirements state that children learn Danish, English, and German in primary school. Taylor's ethnographic study found significant relationships exist between shared literacy experiences and higher educational levels of mothers. He cites the old Icelandic adage, "better shoeless than bookless," as still alive and well (p. 198). This study highlighted the strength of the Icelander families' cultural and linguistic interest in the learning of multiple languages, while retaining an interest in "we who are Icelandic."

The First Practitioner and His Student

Historically, the true study of children's disordered language was thought to begin with the work of Itard, a medical doctor who attempted to teach the "Wild boy of Aveyron" to speak in the late 1700s (Lebrun, 1980). The boy was found roaming nude in a French forest at the age of 12 and taken to a school for the deaf in Paris. There, Itard named him Victor and began long years of effort to teach him language. It is reported that Victor was able to understand "every-day language" but that he died at about 40 years of age in 1828, never having acquired expressive language.

A Modern-day Genie

Almost 200 years later, in the late 1970s, a modern-day Victor was discovered in California. A girl, given the name of Genie, reportedly spent her first 13 years in a closet, where she was held by her parents (Curtiss, 1977). She too learned to understand the language of here and now, her immediate environment. But she too failed to develop more formal language skills.

Child language experts now know how important very early caregiver-child social and communicative interaction is to the development of spoken language (Butler, 1986, 1994a). As noted by Snow (1984) more than a decade ago, a child's experience with "effective communication is the source of knowledge about the form and content [of language]. A child who could not already interact could never learn to communicate; a child who cannot communicate would never learn language" (p. 5).

Even though human communication is one of the most important aspects of living, it is too frequently ignored by countries that must first meet the needs of their people for adequate nutrition, housing, medical treatment, building primary schools, reducing the mortality rate of young children, and decreasing the average number of children per woman (UNICEF, 1995). Language impairments may be a low priority among emerging countries because it is a "hidden" disability—as, often, are reading and writing impairments. The problem is invisible to the unobserving eye. But to be without language is to be without the ability to interact with others—socially, emotionally, linguistically, and even cognitively.

The above sections have introduced the reader to the many faces of language impairment and those who provide language instruction and intervention. But, more generally, what is language?

A Definition of Language

Language has been described as "the very definition of what it means to be human and to be a developing member of the species" (Locke, 1992, p. 4). It should now be obvious that researchers have yet to fully understand how language is acquired. George Miller put it well when he said "we already know far more about language than we understand" (Miller, 1990, p. 7). It follows that we know far less about language impairment, since we have not been able to determine what the limits of variability are among normal children in the process of acquiring language. It is certain that we have much to discover about how children tread the language acquisition path, moving along the oracy to literacy continuum.

There is considerable disagreement among researchers as to how and why children are early or late "talkers" (Butler, 1991; Paul, 1991; Shatz, 1994; Whitehurst et al., 1991). Children growing up in literate societies are expected to understand and produce language through speaking, reading, and writing, and to do this at a level of increasing sophistication in the Information Age. A significant number of children in all countries have difficulty in learning to use language and using language to learn (Westby, 1985).

Children with language impairments encounter serious difficulties with processing and producing the five known aspects of language, which serve as the foundation of our language. These five systems of language are (1) phonology, (2) morphology, (3) syntax, (4) semantics, and (5) pragmatics. The next section identifies and discusses these five subsystems, which constitute our current understanding of language, as they affect children with impaired language.

Language Impairment

Language disorders or impairments were defined in 1982 by the American-Speech-Language Hearing Association in the following manner:
A Language Disorder is the impairment or deviant development of comprehension and/or use of a spoken, written, and/or other symbol system. The disorder may involve (1) the form of language (phonologic, morphologic, and syntactic systems, (2) the content of language (semantic system), and/or (3) the function of language in communication (pragmatic) system) in any combination.

1. Form of Language
   a. PHONOLOGY is the sound system of a language and the linguistic rules that govern the sound combinations.
   b. MORPHOLOGY is the linguistic rule system that governs the structure of words and the construction of word forms from the basic elements of meaning.
   c. SYNTAX is the linguistic rule governing the order and combination of words to form sentences, and the relationships among the elements within the sentence.
2. Content of Language
   a. SEMANTICS is the psycholinguistic system that patterns the content of an utterance, intent, and meanings of words and sentences.
3. Function of Language
   a. PRAGMATICS is the sociolinguistic system that patterns the use of language in communication, which may be expressed motorically, vocally, or verbally. (ASHA, pp. 949–950)

As can be seen in the above definition, "language contains a finite set of elements. The elements include words, which are themselves composed of elements—the sounds of a particular language. These elements are combined in predictable ways by phonological, syntactic, morphologic, and semantic rules to yield a potentially infinite number of sentences" (van Kleeck, 1994, p. 75).

Normally developing young children gradually become aware that language consists of elements that can be combined in systematic ways. However, language-impaired children have failed to make some of the necessary "connections" required to understand and use language with the necessary competence to be successful in childhood's most important tasks: learning to use language and then using language to learn.

An underlying assumption within and outside the educational system is that children come to school "ready to learn." Unfortunately, for a significant number of children (reported to be between 10% and 25%) that may not be true. Goals 2000, a federal initiative for improving education, has recognized this fact and has made "readiness to learn" a priority (Nelson, 1993). However, what is not well understood is that the language of instruction as presented by teachers in the classroom may be incomprehensible to children with language impairment, with limited proficiency in English, or children from cultural and linguistic contexts that do no enable these populations to be successful in responding to literacy demands. As immigration accelerates apace, it is not uncommon to find districts with 100 or more different languages among their students. The challenge for all school personnel is to support the learning of all children through language, spoken and written.

The above definition of language impairment may seem too complex. The examples below may help clarify the terms used to describe the three overarching aspects of language and its disorders: (1) *form*, (2) *content* and (3) *use*.

Difficulties with the form of language are phonology, morphology, and syntax. Each will be presented in turn although, in reality, children with language impairments show difficulties in all three areas for reasons that will become clear. For now, it is sufficient to realize that normal infants, toddlers, and preschoolers who have not yet mastered their native language may display speech sound distortions, omissions, or additions. If these deviations are not severe, the children may not be viewed as impaired during the first few years of life. Among normal children, "phonological production is largely complete by the age of four, although slight departures from standard pronunciation may continue until eight and one-half years" (Dickinson & McCabe, 1991, pp. 2–3). As Stoel-Gammon (1990) points out:

> Normal infants begin to babble as early as 6 months or as late as 9 months, first words appear anywhere from 10 to 15 months. As children grow older the amount of individual variation tends to decrease in areas such as vocabulary size, and phonology, making it easier to identify atypical development. (p. 21)

Jerry, a Student with a Phonological Impairment

An example from the author's clinical experience is of 7-year-old Jerry who told me, "Oh, my Dod, the kool nur cai to my hou wi no dood readon." (Oh, my God, the school nurse came to my house with no good reason.) In this particular case the school nurse had been dispatched to the boy's home by the principal when his first-grade teacher indicated that neighbors suspected child abuse—rather than to inquire about his evident phonological difficulties, e.g., omission of final consonants. The first-grade teacher had other concerns as well. Jerry was in serious academic difficulty in reading, writing, and spelling, even though he had been retained a year in kindergarten because his spoken language was frequently unintelligible. His teacher reported that Jerry was unable to match sounds to letters. As with other poor readers, he appeared to lack a sensitivity to or an appreciation of the sounds in words and could not produce phonological sequences (e.g., sh—i—p). There is a large body of evidence that indicates that language processing problems may underlie reading disabilities (Catts, in press; Fletcher et al., 1994; Stanovich, 1988). In Jerry's case, another factor may have been contributing to his difficulty, since co-morbidity is not uncommon among children with communication problems. Very recently, psychologists and SLPs working with physically or sexually abused children have found linkages between maltreatment and language disorders (see Knutson & Sullivan, 1993 and Snyder & Saywitz, 1993 for specifics). Whether or not Jerry is a victim of abuse is before a juvenile court, and he may be asked to give testimony. Of course, it is uncertain whether or not the judge will believe that Jerry can give credible testimony.

Ann: A Student with Morphological Difficulties

Young children learning English go through a telegraphic stage. In particular, children with language impairment are very late in mastering grammatical morphemes (Leonard, 1992). Grammatical morphemes include articles, plurals, third person singular, regular past verbs, irregular past verbs, etc. These are all frequently produced in errored form by language-impaired children, as are function words such as "is" and "the." Ann, an eight-year-old second-grader, has been diagnosed as a child with Specific Language Impairment (SLI). Mastering grammatical morphemes is especially difficult for children with SLI, although motor development, hearing, nonverbal language ability, and emotional functioning is within the normal range (Leonard, 1992). During a language arts session, her teacher explores Ann's verbal responses:

Teacher:	Use the word *evening* in a sentence, Ann.
Ann:	At the evening it was dark.
Teacher:	Mmmm. (Pause). What is a word to go with Venus or Mars?
Ann:	They is planets or places.
Teacher:	OK. Now, tell me about the story we read this morning *Where the Wild Things Are.*
Ann:	I have the worstest remembering mind (shaking her head no).
Teacher:	Well, let's have you tell us about this picture instead. (Holds up a picture of a small girl and a baby sitting together on the porch steps of an old wooden shanty.)

Ann: A little baby and a little girl. They was sitted on the step, holdin' on to each other, and they, and they, and they . . . Their mama doing laundry. And they, um, they wented inside and they plays, End.

While Ann's response to the picture is essentially correct, and she creatively inferenced that a mother is doing laundry and that the two children will go into the house to play, the teacher is unimpressed, viewing the disordered morphology as a critical flaw. Although Ann has provided an interpretation of what the picture means to her, she has failed to match form to content. It is a costly mistake in the ears and eyes of many beholders.

Charles: A Student whose Syntax is "Out of Order"

Syntactical disorders frequently combine with other aspects of language impairment, such as phonology, morphology, and semantic (word meaning) difficulties. However, Charles' case has been selected because he appears to have primary difficulty in ordering and combining the words, and seems unable to comprehend the relationships between words. Charles is 11 years old and in the fourth grade. He is being seen by an SLP and a Learning Disabilities Resource room teacher. Charles speaks little in the classroom, but is trying hard to meet the reading and writing requirements of his teacher. Below is a written story which he revised five times, each revision taking several hours to produce.

> Onece a puts a time. There were a big exbloseing on a plaent called Erth. Erth hand airplanes and ships and traines. It haves Relieding and cars, biceks, Stonys, shops, Airports. Weme there were exsiiblosings thease Space Sips came to Erth. The space ships blowed up Erth. Hafter the space pepol waet back to there Planit. Some yone saide we made a mastak. So the space pepel weant bake to Erth. The space pepel said wear sarie. A man said wo is going to clean up this mase? The space peple side we will clean up this meas. So the did. And the peple on Erth siade we will hepe. So they cean the planit and the planit looked good. the Space peple siade good buy and the weant back to there painit a they meaner dotder then a gane. The End.

Readers who are knowledgeable about emergent or invented spelling might classify Charles' written work as varying between letter-name and transitional spelling. Letter-name spelling is the ability to breaking a word into its phonemes and representing phonemes with letters of the alphabet (Temple, Nathan, Temple, & Burris, 1993). Temple et al. identify transitional spelling as similar to standard spelling; for example, it includes the use of silent letters, scribal rules, and so forth, but the child is uncertain of the proper spelling. Indeed, Charles is an uncertain speller, writer, and reader.

His language skills are far below those of the other students in his regular education classroom, but are similar to the skills of children seen by his resource room teacher. With the example of Charles' out-of-order syntax, his difficulty with phonological awareness and morphology, we now turn our attention to the second of the three areas noted previously, the content of language, known as semantics, the system that patterns the content of an utterance, the meanings of words and sentences in discourse. In the context of the schools, semantics is frequently subsumed under the study of vocabulary, also referred to by language specialists as the lexicon or the words stored in memory (Juel, 1990).

A Student with Semantic Problems

Edgar, at 17, already the father of two children, is receiving vocational training in an alternative high school, and is receiving SLP and LD services. His job coach notes that various attempts to place Edgar have foundered. Potential employers complain that he will not or cannot follow directions. During the language evaluation, the SLP asked:

Examiner:	How would you change a tire?
Edgar:	(10″ pause) Well . . . um . . . I change a wheel Saturday night coz when I went outside, the wheel was flat. An' me an' my frien', we hadda change it.
Examiner:	Good . . . tell me exactly what you did.
Edgar:	Well . . . umm . . . first my frien' he got the . . . thing that you use . . . um the thing for puttin' under the . . . um . . . (20″ pause)
Examiner:	(Prompting Edgar) The jack?
Edgar:	Yeah, that's it. I knowed what it was but I just couldn't think of it.

Thus, we leave Edgar, able to change a tire but unable to find the appropriate lexical terms to describe his procedural knowledge. Word-finding difficulties often accompany semantic impairments. The next and final system of language is language use or function, a relative newcomer to the assessment and treatment of language impaired children.

Leonard: A student with a Pragmatic Impairment

Lenny, 13, is referred for a language evaluation. There has been an ongoing constellation of complaints from peers, teachers, and school administrators throughout his childhood years. Medical records indicate normal birth, but early and chronic otitis media (ear infections). Lenny was slow to develop language and also evidenced some level of dysfluency. His grandmother reported that he was late in developing fine and gross motor skills as well as oral language. In preschool he had difficulty "focusing" and later was thought to be "hyperactive." By second grade, he had been enrolled in four schools. Private tutors were engaged to work with him on reading and writing. Later, activities were undertaken to develop his motor and social skills. Even later, psychological evaluation revealed a significant difference between performance and verbal skills. His academic performance dwindled and his classmates evidenced dislike for his inappropriate social behavior.

As with so many children whose language and behavioral profiles match Lenny's, a careful language evaluation reveals difficulty with the use of language. Some examples from the diagnostic evaluation are cited below:

1. Lenny's responses to the Multiple Contexts Subtest of the *Test of Word Knowledge* included both successes and failures. "Blunt" was defined as "Somewhat stupid": "retreat" was defined as "to fall back, to escape. . . like a vacation, I mean": "Landslide" was defined as "oversaturated."

2. Lenny's responses to an informal evaluation of idioms, i.e., nonliteral uses of language that were well understood by average sixth graders, yielded the following:
   E: What does "getting carried away" mean?
   L: To drift with the conversation, like. . .
   E.   What does "lend me a hand" mean?
   L: First I think of the hand, a prosthesis. Then I think of "help me", then working on my chores, "give me a hand".
   E: What does "let the cat out of the bag" mean?
   L: Make someone real mad.

3. Metaphor comprehension is important to measure, since it provides information on how the student comes to know words, his linguistic ability, knowledge base, and mental capacity for language. Lenny's responses included the following response to the metaphor, A wave in the ocean is a curl of hair.
   L: (12″ of silence)
   E: What does that make you think of?
   L: (Moves hands to designate wave patterns. (45″ silence)
   L: (Smiles) Little ones are a bad perm. (50″ silence) Now I've got it. . . let's draw it. . . the waves are the head and the earth may be the face. (Takes pencil and sketches an animal figure that he eventually labels as a turtle)

Pragmatic impairments are difficult to modify since most individuals' use of pragmatic rules are carried out below the level of consciousness. Meta-pragmatics, on the other hand, is the consciousness of such rules. Children who make statements to their peers reflecting this conscious awareness say things like: "Don't use your loud voice in the classroom" or "You're not supposed to interrupt" (van Kleeck, 1994, p. 56). Milder pragmatic problems are often thought by listeners to reflect rudeness, poor manners, or lack of parental training. More serious pragmatic problems are thought by observers to reflect a child's loss of self-control, irrationality, or oppositional behavior.

Incidence and Prevalence

The incidence and prevalence of language impairment is obscured by poor data collection. Little currently accurate national data exists in most countries. The data that has been collected tends to be specific to the agency or association that collects it. Health and education statistics differ, as do professional areas of specialization (medical, including psychiatric, hospital based versus rehabilitation based, education, including or excluding special education, etc.). While the terms *incidence* and *prevalence* are often used interchangeably, this is technically incorrect. Incidence refers to the number of new cases in a particular period of time while prevalence is the number of all new and old cases of a disease or occurrence of an event during a particular period of time (Anderson, 1994).

For example, children are identified at local, state, and national levels as having a primary disorder such as retardation, autism, learning disabilities, attention deficit disorder, traumatic brain injury, cerebral palsied, cleft palate, deafness or visual impairment, pervasive developmental disabilities, emotionally disturbed, and so forth. All, or almost all, also have a speech or language disorder, but this is viewed as "secondary" to the primary condition, and this information (i.e., co-morbidity figures) is not provided in federal reports. Experience has shown that children who are served in special education settings also have speech or language difficulties as can be identified in their listening, speaking, reading, and writing performance in the classroom.

Bello (1994) indicates that in 1991 approximately 42 million people in the United States were affected by hearing loss or other communication disorders. Of these, 28 million individuals had a hearing loss and 14 million had a speech, voice, or language disorder (National Deafness and Other Communication Disorders Advisory Board, 1991). It has been estimated that hearing, speech, and language disorders cost an estimated $30 billion each year through education and medical services and lost productivity (National Institute of Neurological Disorders and Stroke, 1992).

Further, it has been estimated that 6 to 8 million individuals have some type of language disorder. Since language disorders affect children and adults differently, statistics are kept separately by the National Institutes for Health. Another difficulty arises when attempts are made to determine the incidence of childhood language problems. Because of our early propensity to judge only expressive language problems, without considering disorders of comprehension, figures in the area of children's language frequently reflect spoken language disorders and do not address the overlapping co-morbidity of children whose language difficulties span speaking, reading, and writing.

For many years, disorders of "listening" have been referred to by teachers as auditory processing disabilities, by which they seem to mean that the child has difficulty following oral directions and fails to attend to classroom discourse. Audiologists and some SLPs refer to the same behavior as a Central Auditory Processing Disorder (CAPD). As Friel-Patti (1994) points out, auditory processing deficits "generally refer to difficulties processing the speech signal in the absence of any permanent peripheral hearing loss, although conductive hearing losses are sometimes thought to be a contributing factor" (p. 371). She also notes that "evidence given over the years has generated confusion about the role of auditory processing factors in language development and disorders. . . .[T]he absence of a coherent theory addressing the contribution of auditory processing to language acquisition as well as accounting for difficulties experienced by some language impaired children clearly impairs advancement in this area" (p. 371).

Dyslexia, a developmental language disorder, is also a frequently misunderstood disability. The word dyslexia comes from two Greek roots—*dys*, meaning poor or inadequate, and *lexia*, meaning verbal language. The Orton Dyslexia Society (ODS) is an international nonprofit organization that has been helpful in promoting an understanding of this language impairment. ODS estimates that as many as one in ten people may be considered dyslexic. They describe dyslexia as a Specific Language Disability that is associated with a mind that learns differently. This difference is reflected in the individual's difficulty learning by conventional means. A recent definition by ODS (1995) indicates that

> dyslexia is a neurologically based, often familial disorder which interferes with the acquisition and processing of language. It is manifested by difficulties in recerotive and expressive language, including phonological processing, in reading, writing, spelling, handwriting, and sometimes in arithmetic. . . Although dyslexia is lifelong, individuals with dyslexia frequently respond successfully to timely and appropriate intervention. (p. 3)

Thus, dyslexia is considered one type of a developmental language disorder, but is rarely reported in the general literature on language acquisition and disabilities.

In summary, while incidence and prevalence figures are variable, language disorders in children, as reported by federal health agencies, have indicated that about 2% to 3% of preschool children and about 1% of school-age children suffer from language disorders in Grades 1–12. Two thirds of language or speech disorders occur in boys (National Institute of Neurological Disorders and Strokes, 1988).

Other nations also experience difficulty in determining the incidence and prevalence of speech, language, and hearing impairments. Egypt, for example, has one communication disorders educational program at Ain Shams University. A pilot study, carried out in Greater Cairo on a sample of 600 school children (6 to 12 years of age) revealed an incidence of communicative disorders of 16.5% (Massouad, Kotby, & Rashid, 1979). Ain Shams Center for Communicative Disorders has analyzed their case load for the last 16 years and reports that 63% of those cases are language disorders (including hearing impairment), 22% are speech disorders, and 15% exhibit voice disorders. Finally, special education services are provided to 4,300 students with mental retardation and 5,900 children with hearing impairment (ages 6–18), in a total of 1,145 special education classrooms (Kotby & El-Shobary, 1992).

It is doubtful whether incidence and prevalence figures can be assumed to be stable across nations and across languages. A myriad of factors contribute to language disorders, not the least of which is hearing loss resulting from chronic middle ear disease. Studies have documented racial/ethnic differences in otitis media, and racial heritage is now an accepted increased-risk factor for the disease (Kavanagh, 1986; Stewart, Anae, & Gipe, 1989). Recent work in Hawaii and the islands of the South Pacific (e.g., Samoa, Papua New Guinea, Guam, and Micronesia) reveal the previously unrecognized importance of middle ear disease as a contributor to speech-language difficulties (Chin-Chance, 1988; Heath, Plett, Tibbetts, & Medeiros, 1987). A number of point-prevalence studies have found an extremely high prevalence of chronic middle ear disease among young children, and reduced language competence in children as young as one year of age. Otits media ranges from an incidence of 9% to 42%, and incidence of perforation was found to range from 3% to 13% (Stewart et al., 1989). It is also reported that Hawaiian, part Hawaiian, and Samoan children failed the annual hearing screening at a much higher rate than did children of Japanese, Caucasian, and Filipino ancestry. Twice as many Hawaiian children failed the screening and almost three times as many Samoan children failed. Part of the problem is thought to be the limited health care resources in the region. For example, in Micronesia— which covers 3 million square miles and 2,000 islands—as of 1989 there were no audiologic services; no ear, nose, and throat physicians; no indigenous speech-language pathologists; and very limited pediatric care (Stewart et al., 1989). In 1994, one SLP was identified as providing services through the New Tribes Mission, Papau, New Guinea (Spahr, 1993–1994).

Etiologies have been cited as accounting for the racial/ethnic differences in the prevalence of otitis media and the subsequent growing evidence of a correlation between middle ear disease, hearing impairment, and language impairments (Hasenstab, 1987). Speech, language, and hearing are inextricably linked in the development of normal language. A deficit in one area of communication competence may result in significant difficulty in the others.

Identification

Identification of language disorders is but one of the tasks of SLPs. Most frequently referred to as assessment or evaluation, identification frequently is a two- or three-step process, with parents, grandparents, or teachers expressing concerns for the child's delay in language acquisition. Less frequently, concerns are expressed by pediatricians or primary physicians. In either case, a screening of the child's speech, language, and hearing is often the first step, followed by a lengthier assessment if the initial screening indicates difficulties in cognition and/or language. There may even be a third interaction initiated, as the language specialist explores in depth the child's phonology, morphology, syntax, semantics, and pragmatics.

School systems have increased their efforts to identify children in need of special education services. Identification of communication disorders in the preschool years (ages three to five) has taken place with increasing frequency. In the late 1980s and the 1990s, the advent of early intervention services has led to the development of interdisciplinary teams in hospitals, agencies, and preschools. If these teams include SLPs, children with speech, language, and hearing disabilities may be identified during infancy and toddlerhood when seen by knowledgeable professionals. For example, low-birthweight infants are increasingly seen by early intervention teams, and their subsequent growth and development followed as agencies have undertaken collaborative efforts (50% of infants weighing less than 750 grams are now surviving [Bauchner, Brown, & Peskin, 1988]). However, seamless service across a child's first five years has not come easily (Apter, 1994), even though the 1986 Public Law 99–457 (Part H)—an amendment to the Individuals with Disabilities Education Act—has mandated services to children with developmental disabilities from birth through two years of age (Laadt, Burno, Lilley, & Westby, 1994).

Appropriate services for infants born at developmental risk and their families requires the efforts of a multi- or transdisciplinary team. No one professional is capable of providing for all the exigencies that occur to families of at-risk or disabled infants and toddlers. This is particularly true since the assessing of communication behaviors in prelinguistic infants is not an easy task, even for those with specialized knowledge. Language, by its very nature, is difficult to quantify. Intimate knowledge of the patterns of vocalizations that occur in the first 12 months of life is an important clue to adults attempting to identify neonates and infants with speech-language problems. Parents can provide important information on language acquisition and are now involved in the screening and assessment process to a much greater extent than formerly. Research appears to indicate that parental involvement in developmental screening provides important information. For example, when a parent questionnaire was combined with an early screening inventory, the combined measure improved the specificity from .83 to .94 and decreased the rate of false positive from 70% to 50% (Henderson & Meisels, 1994). These authors report that although 50% remains a high figure, it does represent a significant decrease in overreferrals. The United States' "zero through two" initiative has been preceded by a number of other countries' efforts to provide screening and follow-up in the earliest years of life. For example, the Netherlands has excellent services for children in the early years of life and their parents (Goorhuis-Brouwer, 1994, personal communication).

Recent research shows there are very early indicators of infants' comprehension of language (Foster, 1994) prior to the development of first words. Infants must acquire world knowledge before acquiring word knowledge. Words become meaningful as they become attached to concepts which are somewhat larger. This means that infants and toddlers first gain understanding of events. For example, understanding that the sound of the microwave door closing or the noise of boiling water in a pot signals that a warm bottle is imminent comes before the spoken word "bo-lle" (bottle) emerges; that Mother's saying "Time to go bye-bye" precedes traveling to the babysitter; or that being fed and bathed signals Daddy's arrival home. There is much language learning that can be observed prior to an infant's expression of his/her first words. In fact, parent-infant interactions may "say" volumes to an astute observer of linguistic and nonlinguistic signs of communication (Kenworthy, 1994). Identification, then, begins with observation, followed by a screening that, in turn, may precede an in-depth assessment of the comprehension and production of language, and associated communicative behaviors. Particularly complex cases may require multiple appointments to reach a valid diagnosis.

Diagnosis and Classification

As noted earlier, an assessment or evaluation may render a diagnosis, which may be multifaceted, since language impairments or delays are frequently a visible sign of an otherwise invisible speech-language or hearing impairment. Despite rapid advances in the identification and diagnosis of hearing loss, children are frequently not identified as hearing impaired within the first two years of life. Despite rapid advances in psycholinguistic research and our understanding of normal and disordered language acquisition, parents and health and educational professionals remain uninformed and tend to delay seeking a definitive diagnosis of any number of conditions associated with language disorders. Hearing loss is just one of them.

The general public is relatively unaware that speech-language pathologists not only analyze spoken language comprehension and production, but also are permitted to view reading and writing comprehension and production within a holistic framework. School-based clinicians have recognized for some time that the unrecognized core of the curriculum assumes that children come to school with intact language skills. The unspoken assumption is that the primary purpose of schooling is to continue to develop children's skill in *using language to learn*, once they have mastered *learning to use language.*

Literacy and academic success are not only contingent on using language in speaking, reading, and writing via the sound, syllable, word, and sentence level, they also require that children from the earliest years of their lives "manage" their spoken and written discourse within the context of communication interactions in multiple settings. Conversation may seem effortless to adults, but is a complex series of events for young children. As Weiss (1995) notes, "holding a successful conversation requires that all of the form and content features of the message have to be present and accounted for, and, in addition, the rules of conversation must be applied" (p. 19). As Crystal (1987) concludes, the components and levels of speech-language competence interact, and when there is difficulty in one part of the language system—for example, in syntactical structure or morphology—there will be fewer resources to handle other components, such as phonology or fluency. Weiss (1995) proposes that as the demands on young children to enter into more adult-like conversations increase (as do demands to initiate and maintain topics, to assume conversational roles, and to repair verbal messages gone awry), infants and toddlers must also learn turn management, i.e., the ability to maintain discourse messages across turns. In addition, dyadic communication with the primary caregiver soon gives way to responding to multiple conversational partners. Now the child must attend to a number of other speakers concurrently, reducing the child's capacity to incorporate other aspects of language comprehension and production. The child may encounter similar difficulties as the school years approach, when reading and writing place further demands on the child's developing linguistic processing skills. It is the fortunate child who comes to school ready to learn.

School-age children encounter the language of instruction when they enter the educational system. This may be a time of difficulty since school demands for literacy increase rapidly. As Silliman and Wilkinson (1994) stress, "To a great extent within classrooms, the discourse created by teachers and students determines both what is learned and how learning takes place. Scaffolding. . . support[s] and/or provides opportunities for students to develop and refine literacy skills" (p. 27). Scaffolding refers to the guidance provided by an adult or peer through verbal communication, a communication that assists the student in doing something he/she could not do without assistance (Cazden, 1988). However, scaffolding, unless appropriate to the child's comprehension level, may be negative rather than positive. All learners have had the experience of failing to understand teacher directions or the explanation provided in written texts. Children with language disorders frequently exhibit failure to meet instructional discourse demands. For example: "Put down your pencils, pick up your copy of the worksheet on the back table, complete the first nine addition problems and be sure to answer the story problems on the back of the sheet. When you have finished, put the paper on my desk and choose an activity in the book corner for the time remaining before recess." Such complex directions are relatively common and may not represent appropriate scaffolding for many 6 or 7 year olds. Language assessment, therefore, needs to move beyond the "piecemeal" measurement of smaller units of discourse, as measured by many current standardized language tests. Language specialists now focus on evaluating larger units of discourse, such as storytelling, reading, and writing, the use of expository texts, and the child's ability to recall prior experiences and link them with new knowledge being presented during classroom instruction. Peterson & McCabe (1994) put it well when they say that "if children are able to construct decontextualized narratives when they get to school, they will have an easier time acquiring literacy. . .in the decontextualized style of language valued at present in formal North American schooling. . ." (p. 947).

Since many children with limited English proficiency or language differences (but not disorders), and children from culturally diverse backgrounds may "look" as if they are language learning disabled, speech-language pathologists, school psychologists, and teachers need to evaluate language along the continuum from oracy to literacy and across spoken and written domains. Although standardized tests are frequently required by school districts and state education agencies, they must be supplemented with curriculum-based and dynamic assessment procedures, viewing the child's language competence across multiple settings and under various scaffolding conditions (Nelson, 1994; Palinscar, Brown, & Campione, 1994; Silliman & Wilkinsonson, 1994b; Wallach & Butler, 1994b).

Dynamic assessment has been of particular interest to language specialists as they attempt to grapple with the difficulties in using standardized language tests to effectively measure language impairments and learning disabilities. Disentangling language and learning has proved to be fraught with false negatives and positives. As Lidz (1991) has pointed out, "the focus of dynamic assessment is on the assessor's ability to discover the means of facilitating the learning of the child, not on the child's demonstration of ability to the assessor" (p. 9). In addition, she states that "dynamic assessment is typically contrasted with static assessment. This reflects the fact that dynamic assessment focuses on learning *processes*, in contrast to the traditional assessment focus on already learned *products*. When product is the outcome there is no information regarding the reason for failure or the learner's ability to achieve. There are also no guidelines or implications for intervention to connect the assessment with the intervention" (pp. 3–4). Lidz asserts that although curriculum-based assessment permits examiner awareness of the next educational step, it does not provide information on how the student learned or failed to learn the responses provided (p. 4). Dynamic assessment may well be the wave of the future, but its value in classification is as yet unknown.

Classification Systems in America

For the past 20 years, students have been placed in a number of educational categories for the purpose of assignment to a continuum of service settings. Once established, the classification system and the method of collecting data has remained stable. High-incidence disabilities have included Learning Disabilities (which, by definition, included children with listening and speaking disabilities), speech impairment (which included children with phonological and language impairments), and mental retardation (which also includes a large number of individuals with language and cognitive impairments). Serious emotional disturbance has been an increasingly important classification (sometimes known as behavioral disorders, and often encompassing children with spoken and written language impairments). Sensory impairments have largely been known as low-incidence impairments, including the blind and the deaf (both groups are subject to a variety of language impairments, particularly the deaf and other hearing impaired). Orthopedic impairments and other health related impairments are relatively low incidence; but, with the addition of children with Attention Deficit Disorders over the past few years to the category of "Other Health Impaired," this category may increase substantially (Wagner, 1991).

Another "new" category is that of Traumatic Brain Injury (TBI). Head trauma, of course, is not new; however, its inclusion as a category has assisted parents of children with head trauma to receive special education services, including assessment and intervention for the cognitive-communicative disabilities that are "typically secondary to more general cognitive and self-regulatory impairment associated with damage to the frontal lobes" (Szekeres & Meserve, 1994, p. 22). Sohlberg and Mateer (1989) emphasize that language specialists assessing TBI patients must systematically review the patients' functional systems according to current neuropsychological theory, using a process-oriented approach to rehabilitation planning. They stress that the use of formal tests is simply to enhance clinical observations of the patient's performance on everyday tasks. Treatment costs of TBI at the current incidence rate of 69 per 100,000 is projected to cost $8 billion a year for new cases of those hospitalized with TBI in any given year (Brooks, Lindstrom, McCray, & Whiteneck, 1995). Recent research on TBI students returning to school following physical recovery indicates that the student likely will have difficulty with language, "resulting in disorientation, disorganization of verbal activities, stimulus-bound responses, reduced capacity for learning, and reduced ability to process incoming information" (Blosser & DePompei, 1989, pp. 67–68). Regular educators, and even special educators who have had experience with other handicaps, will find themselves nonplussed by the student with TBI as he/she reenters school "since the combination of deficits found in head-injured students cannot be as easily categorized and defined" (Blosser & DePompei, 1989, pp. 68–69). Teachers as well as the family of students with TBI require counseling and support (DePompei & Williams, 1994).

Another category that has not made its way into federal or state classification systems if that of the language and learning needs of individuals with Severe Speech and Physical Impairments (SSPI), many of whom are cerebral palsied or have sensory or cognitive impairments. Assessment and intervention for SSPI make use of alternative and adaptive communication (AAC) systems (Steelman, Pierce, & Koppenhaver, 1994). SLPs who work with children with SSPI are considered specialists in a subcategory that requires not only an excellent understanding of the development of child language, but also of AAC systems, synthetic speech output, specialized computer adaptations, specialized knowledge of software selection, rehabilitation engineering, and physical access refinements (Higginbotham, Scally, Lundy, & Kowarksy, 1995).

A final area that also is not delineated in national categories, but that has been a subject of intense research in the United States, Scandinavia, and the United Kingdom, is Specific Language Impairment (SLI). SLI is an unresolved puzzle, according to Rice, Wexler, and Cleve (1995), who note that children with SLI "seem to have the necessary prerequisite competencies needed for language acquisition, yet their language milestones are delayed in emergence and a protracted period of time is needed for acquisition of some fundamental linguistic features" (p. 850). Other researchers, utilizing a Language Delay (LD) approach, note that SLI youngsters demonstrate a delay in all aspects of language (Lahey, Liebergott, Chesnick, Menyuk, & Adams, 1992). Still others postulate a language processing breakdown (Leonard, 1989), while recent investigations indicate there may be a genetic base for SLI (Lahey & Edwards, 1994).

Children with SLI are thought to have difficulty learning language despite apparently normal emotional, social, cognitive, and motor development. During their school years they are likely to be labeled Specifically Learning Disabled (SLD) (Bashir & Scavuzzo, 1992). A recent study in Sweden indicated that morphological as well as syntactic differences were noted among SLI children speaking Swedish compared to studies of English-speaking SLI children. Cross-linguistic studies are needed, since it would appear that the word order of a language may affect the symptomatology noted in North American studies (Hansson & Nettelbladt, 1995).

Another classification system used in the United States by health professionals is the *Diagnostic and Statistical Manual of Mental Disorders* (DSM-IV), published by the American Psychiatric Association and used by professionals in the mental health fields. The DSM-IV, which is used by SLPs who provide diagnoses and classification within medical settings, offers five axes on which to record the biopsychological assessment of a client (Morrison, 1995). The five axes include: Axis I, Mental Disorders; Axis II, Personality Disorders and Mental Retardation; Axis III, Physical Conditions and Disorders; Axis IV, Psychosocial and Environmental Problems (including education problems and problems related to the social environment); and Axis V, Global Assessment of Functioning (GAF). Since the DSM-IV follows the medical model of illness, it frequently is not considered appropriate for educational settings. Included among the DSM-IV disorders are Attention Deficit/Hyperactivity Disorder, Autistic disorders (including Asperger's disorder), Pervasive Developmental Disorders, Learning Disorders, and Communication Disorders (which includes Expressive Language Disorders, Mixed Receptive-Expressive Language Disorders, Phonological Disorders, Stuttering, and Communication Disorders Not Otherwise Specified). The codes provided are extremely limited when it comes to the area of communicative disorders, and many known conditions are neither cited nor coded for reimbursement.

International Classification (World Health Organization)

Although it is claimed that the DSM-IV sets the worldwide standard, many countries use the International Classification of Impairments, Disabilities, and Handicaps (ICIDH) (World Health Organization, 1990). A recent study conducted by the Netherlands Institute of Primary Health Care in Utrect used the ICIDH as a method of classifying and registering speech therapy assessment (Raaijmakers, Dekker, Dejonckere, & van der Zee, 1995). The authors report that the ICIDH was useful in establishing reliable diagnoses of impairments. However, a review of the report indicates that most of the clients identified were adults, and the authors indicate shortcomings in classification of some of the diagnostic assessments. It would seem, therefore, that whatever the classification procedure used, whether in the United States or elsewhere, there is room for improvement based on ongoing research. In the case of the U.S. venture into managed care, such research is underway in various clinical trials and outcome studies.

Individualized Planning

Individualized planning by an SLP or a language assessment team begins prior to meeting the child and his parents, his teacher(s), and perhaps his siblings and peers. Planning the evaluation of an individual child is done within the framework of the clinician's theoretical understanding of how language is acquired, how it can be measured, the tools one uses for such measurement, and the interpretation given to the data collected. Although the planning ought to take place after a review of a case history—which would provide medical and developmental data, reports of the family and the school, and, if possible, an observational period in the contexts of the child's everyday environment—the reality is that much data will be missing and many opportunities lost. As Neisworth and Bagnato (1988) point out, the "passage of Public Laws 94–142 and 99–457 requires. . . multidimensional interdisciplinary evaluation procedures" (p.23), and they note that "it is naive and often misleading to employ a singular criterion in clinical child assessment. . ." (p. 24). However, if professionals are to "do the job," they will need to use multiple measures, obtain data from multiple sources, and evaluate multiple domains. Evaluating the language of the young child requires that one look at more than "language" as heretofore defined. Assessment must include socioemotional, gross and fine motor, and self-care domains at the minimum. As Bailey and Wolery (1989) indicate, effective assessment covers developmental and behavioral domains, involves the parents as assessment partners, and is ecologically valid, among other factors. Assessment should be undertaken in as naturalistic a setting as possible, and observation of children's routines in playing, eating, and other daily activities is important, as is exploration of academic tasks during the school years.

The importance of socioeconomic status remains a constant factor related to language development and growth. Recently reported longitudinal research with very young children within their family setting confirms this. Hart and Risley (1995) report additional evidence that middle class families provide a greater number of experiences to their children, and it is the frequency of language experiences, rather than race/ethnicity or gender, that contributes to the rapid trajectory of children's acquisition of words. "Children born into homes with fewer economic resources have fewer experiences. . . and the consequence is that they learn fewer words and acquire a vocabulary of words more slowly" (Bloom, 1995, p. xi). Hart and Risley (1995) not only provide important information that stresses the critical nature of early language learning, but also propose interventions to equalize early experience. They also provide data suggesting that there is a strong link between children's language both at age 3 and at age 9, and data at the high school level that suggests students who have a limited vocabulary (not necessarily language disordered) are thus unable to deal with advanced textbooks (Hart & Risley, 1995). These findings support the idea that the oracy-to-literacy continuum is a valid concept that needs to be considered when assessing children from birth to adulthood.

It should be apparent from the above discussion that a single SLP will have great difficulty completing a comprehensive language evaluation on which to construct a treatment plan (typically an Individualized Educational Plan). Nevertheless, there are many situations in which there are no other options. Recognizing the difficulties thereof, other language specialists often undertake the task, most frequently by using standardized tests (sometimes of doubtful reliability and validity). Today there are literally hundreds of tests, some good, some bad, with a sizable number somewhere in between. These are "static tests" that measure what is currently known, not what the child is capable of learning. For that reason, dynamic assessment has come to the fore. As noted earlier, dynamic assessment approaches attempt to measure the student's potential for change, i.e., how well he/she performs given varying degrees of assistance (Palincsar, Brown, & Campione, 1994). In language assessments, audio- and videotapes are helpful in capturing ongoing interactive language performance as well as documenting progress during the intervention process. Running records can be particularly helpful as an example of a narrative tool, permitting clinicians to accurately obtain a chronological record and possibly leading to further measurements of domains that require exploration (Wilkinson & Silliman, in press). Joint viewing by SLPs and special and regular education teachers can lead to new insights and to modifications in intervention planning stages. In addition, little else is as telling as videotapes to document changes in narrative and expository skills, which can be clearly delineated on the screen and in transcripts of the child's emerging language skills.

Individualized Planning

Planning for intervention and the selection of diagnostic methods and materials pays large dividends, as does planning for long-term goals on the IEP. Each requires a level of expertise and a knowledge of this particular child's language profile and the contexts in which learning most readily occurs. Beyond the IEP is its execution; too many IEPs are stuffed into drawers and consulted rarely, if at all. The concept of IEP planning is brilliant, but its execution tends to be less so. Those who implement the plans often fall back on preprepared materials—constructed by themselves or colleagues—or commercially produced programs. This type of activity is not individualized programming. Interventions planned for groups with varying language abilities may be less than satisfactory for the child under consideration. If individuals are not served on an individual basis, then their assets and needs must be considered within the context of small-group instruction. Matching unique profiles may be difficult for a language specialist and almost impossible for a teacher with a large group to consider. Burnout of learning disabilities specialists and SLPs, not to mention the stress on regular education teachers engaged in teaching in inclusive settings, is rarely fully appreciated by advocates of full inclusion. There are times in children's lives when they need the attention of, and scaffolding by a concerned adult. As Kaufman (1995) points out, "the best available research belies the claim that the most effective instruction is provided in heterogeneous groups" (p. 23).

Traditionally, SLPs have been trained as "clinicians" who, like physicians, see patients one by one, conduct diagnostic tests, and design individualized treatment. In most cases, the assessment of children with speech and language disorders continues to be individually conducted. More recently, clinicians have been encouraged or required to place their assessments "in context," although there was a movement as early as the 1950s and 1960s to provide speech and language services in the classroom (Van Riper & Butler, 1955). "In context" refers to viewing children's language within their environment— their home and family, the school and classroom, and the community. Family-focused assessment and intervention is now the expected norm in many infant, toddler, and preschool settings, with parents providing much more than their child's health and academic history and perhaps an interview. Early interventionists are now considering parents part of the assessment team as well as the intervention team, and these shared activities have been incorporated into many health and education settings (Bennett, Nelson, & Lingerfelt, 1992).

SLPs are also being urged by administrators to move their interventions, if not their assessments, into full inclusion settings, where services are provided only in the classroom. Assessment largely remains confined to settings outside the relatively noisy classroom. Research evidence indicates that speech, language, and hearing are best evaluated in quiet, wherein the student can hear without unnecessary acoustic and visual distractions. However, classroom observation and data collection are highly recommended, if for no other reason than that the classroom is the site of academic instruction and of interaction with teachers and peers. Obviously, good planning requires that the child's comprehension and production of spoken, written, or read language should be measured in a variety of settings, as should the student's ability to interact with adults and peers. Pragmatic and social skills can and do play an important role in all aspects of a student's functioning, placement, and programming.

Individualized planning which results in an individualized educational program and, it should be added, individualized instruction and services, is at the heart of special and remedial education. It is also the parents' fervent hope that such attention to individual needs will exist in regular education settings, a matter sometimes forgotten in planning for "the class."

Treatment and Intervention

What is identified as treatment or rehabilitation by other health or related service settings is typically identified as intervention by school-based personnel in speech-language pathology. It is not just a matter of semantic choice, but of identifying through collaborative consultation (Prelock, Miller, & Reed, 1995) with teachers and other education personnel the specific needs of the children with communication disorders and those at risk for language and learning problems. School teams succeed to the extent that they are enabled to become involved in role extension, role enrichment, and role expansion. Relatively few teams approach the ideal of role release, very possibly because few have enough time to carry out the collaborative process to its fullest extent. It is apparent, however, that school-based SLPs (as well as their colleagues in other settings) must become much more familiar with the curriculum of the classroom as the language-disordered child is faced with new academic tasks. It is also important that the teacher's awareness of communication breakdowns leads to modifications of instructional strategies across academic areas (Prelock, Miller, & Reed, 1992). One of the recently recognized difficulties of school districts contracting out services to private practitioners is that there is little, if any, time for team building and sharing of curriculum goals and the unique profile of the language-impaired child. (For an exhaustive review of the origins and outcomes of collaborative consultation, see Coufal, 1993).

It is difficult to present an overview of intervention, although there are a plethora of "programs" that promise successful results for children with language impairment. SLPs either individualize intervention based on the five specified areas (phonology, morphology, syntax, semantics, and pragmatics), or select a program that appears to meet the needs of a particular child or small group. Some choose to provide intervention procedures that are compatible with the classroom context, for example, whole language or decoding. Others may choose a philosophically compatible approach, one that mirrors their accepted theory of language acquisition and instruction. The closer the program selected is to the child's speech-language and literacy needs and the greater the clinician's ability to modify the program to meet the individual child's language and learning needs, the more likely that the intervention will be successful.

It is important to note that "there is no agreed-upon, conventional account of children's language acquisition. Instead there are two opposing points of view, with associated differences in scientific methods" (Rice & Wilcox, 1995, p. 16). The developmental view holds that language stems from cognition and therefore is placed within the cognitive domain. Its hallmarks include an emphasis on meanings and uses, with children viewed as self-directed learners, and is acquired in sociocultural contexts—but not "taught." Therefore, it is presumed that children learn the use of language in a variety of social contexts. Language is embedded within the context of family-child interactions and child-child interactions as they occur across settings, including preschool (Bricker, 1993). The implications of this theoretical position are clear when viewing certain early childhood curricular materials. In addition, ways of talking with children and the influence of parent and teacher language on children's language acquisition also have implications for teachers and early interventionists. Current evidence suggests that directive adult talk (whether by parent or teacher) may have unintended consequences. The developmental perspective views the acquisition of language itself as being the reinforcing agent, with no need for explicit praise and reinforcement for early linguistic attempts (Rice & Wilcox, 1995). Indeed, there is some evidence that adults who give many commands and drills, whether they are parents or teachers, may actually discourage language development (Cross, 1991).

The opposing point of view is referred to as the linguistic approach, which emphasizes the more formal aspects of language acquisition (Pinker, 1994). This view holds that language is innate and is not the same as the social and cognitive domains. Children are thought to have an innate grammar which, although universal, must be "tuned" to their native language (Rice & Wilcox, 1995). Much of what is viewed as important therefore is identified through a series of linguistic analyses. Rice and Wilcox note that they do not see a convergence of these two perspectives (developmental vs. linguistic) now or in the near future, but they maintain that both have something to offer those who must assess and intervene with children at risk for or already exhibiting language impairments. (Readers are referred to Rice and Wilcox's 1995 text for excellent suggestions for building a language-focused curriculum for the preschool classroom). It should also be noted, as Sergeant (1995) states, that speech-language impaired children—even those who show great progress in the preschool—should be followed up in the elementary grades and that long-term advocacy is necessary if the child is to have continued and appropriate services as needed.

Wallach and Butler (1994) emphasize that language intervention must go well beyond the sole consideration of oral language intervention. Although oral language is as much a part of literacy learning and acquisition as written language, the primacy of printed language overrides the primacy of oral language at certain times and in certain situations. Most language specialists are aware that the need for language to be explicit is reduced when speakers and listeners are face to face, but that the need increases when only the written word is present. The literate repertoire that exists within the printed pages of books is far wider than that which is part of much of spoken language. School texts and school teachers for the most part express themselves in a literate/syntactic mode. Wallach and Butler (1994) stress that "reading and writing encourage children to become linguists. As long as English remains an alphabetic system and as long as print maintains its conventions, young readers must do at least two related things: (1) they must come to terms with speech-to-print differences; and (2) they must bring their language knowledge to the surface by talking about language and analyzing it" (p. 8). Therefore, there is a role for all language specialists to play in children's years in elementary and secondary education: to provide both prevention and intervention services to children and adolescents who are at risk or who have language impairments. Methods and materials are abundant. The necessity of fitting the methods and materials to the child, and not the reverse, creates the greatest challenge.

Programmatic Arrangements

Much of what has been said in the previous section addresses the viability of programmatic accommodations to be made for infants, toddlers, children, and adolescents with language impairments. As reported throughout this chapter, linkages with families are a high priority, particularly during the early years of life and into the school-age years. However, a few caveats may be in order. As Bowe (1995) points out, "early intervention and preschool special education must, by statute, be family-focused. . . . So much stress has been placed on valuing and empowering the family that now there is a risk of denigrating the role of the ECSE worker" (pp. 499–500). Bowe further states that the federal provisions in IDEA provide families very special roles in program implementation. Although the statute envisions a partnership between families and ECSE programs (Goodman, 1994), the professional must not abdicate all responsibility. In short, the point of the exercise is joint decision making.

Bowe (1995) also discusses programmatic issues, such as questions regarding the appropriateness of some of medical interventions and the mislabeling of infants and young children—as well as the as yet unmet need to "prove" that early childhood services and programs are effective. It is typically and intuitively understood that the earlier the intervention, the more successful. Unfortunately, the best-controlled research has yet to conclusively prove this statement. Qualitative studies have added new information, but due to their uncontrolled nature they have not provided the necessary documentation regarding the efficacy of early intervention.

There are, of course, many small studies that appear to demonstrate efficacy of various programmatic approaches to language impairment. Among them is one that demonstrates the effectiveness of a collaborative consultant approach to basic concept instruction over an 8-week period in a kindergarten classroom (Ellis, Schlaudecker, & Regimbal, 1995). This study followed up the promising work of Russell and Kaderavek (1993) and Seifert and Schwarz (1991) with Head Start children. As pointed out by Nelson (1993), school-age children who have experienced language problems in the preschool "begin to fail in the development of written language abilities and are identified as learning disabled or as exhibiting poor readiness abilities for reading and writing. They are either referred back to the SLP, or more often, to a "reading specialist" (p. ix). Nelson indicates that if an SLP and a reading specialist collaborate, the SLP will attend to subtle metalinguistic abilities (e.g., phonemic awareness, metaphoric language, ambiguous language, etc.) while the reading specialist provides intervention in discrete reading abilities (e.g., letter-sound correspondences and rhyming). Meanwhile, the teacher may approach the child's reading difficulties from yet another perspective. Not surprisingly, Nelson recommends an integrated curricular approach to intervention—such as that found in Norris and Hoffman's (1993) text focusing on a whole-language approach. Another suggested program in language instruction proposed for teachers is that of using instructional discourse to manage the vagaries of children with language disorders in a mainstream classroom. Suggestions for teacher-student dialogue are provided in a lesson plan format, preceded by a discussion of the importance of teacher use of language in moving children from understanding "the here and now" to the higher levels of "then and there" and the sophisticated text strategies needed for instruction in science, history, mathematics, etc. (Blank, Marquist, & Klimovitch, in press).

Programs for working with older children and adolescents which have enjoyed some success are represented by (1) the work of German (1993), who has developed a curriculum for intervention with children and adolescents with word-finding difficulties, (2) the matching of the curriculum content with a "therapeutic Language Arts" component at the middle school and secondary level (Ehren, 1994), and (3) instructional approaches to a variety of problem-solving strategies appropriate for those students with spoken and written language impairments (Blachowicz, 1994). In addition, Scott (1994) notes that adolescents cannot become experienced with types of language that they do not experience, and must do so in order to survive within the school discourse contexts. She also comments that British schools, when compared to American schools, are viewed as "more cognizant of the need for literacy training in the spoken mode (their term for oracy training), hence their interest in cross-mode and cross-genre assessment at the discourse level" (p. 246). Westby (1994) points out that there are many cultural differences within and between nations in the comprehension and production of story text structure by older students, adolescents, and adults. For example, Native American narratives are very different from most Western narratives, since they do not involve sequence, causality, and succession. When the written work of Australians who speak English, Arabic, and Vietnamese was compared, it also revealed great differences in the level of plot and the use of detailed descriptions (Soter, 1988). Westby's (1994) comments on the effects of culture on genre, structure, and style of both spoken and written text are useful to those undertaking language interventions with culturally diverse groups.

Uses of Technology

Except for a devoted group of researchers and clinicians in the area of Augmentative and Alternative Communication who work with severely language impaired and nonspeaking individuals, the use of computer-based assessment and intervention techniques has been sparse. This is slowly changing. In 1992, Miller, Frieberg, Rolland, and Reeves published an article indicating the results of their decade-long series of projects aimed at implementing computerized language sample analysis (LSA) in public schools. Earlier, they developed the Systematic Analysis of Language Transcripts (Miller & Chapman, 1982) which has greatly assisted clinicians in evaluating lexical, morphological, syntactic, semantic, and pragmatic performance and interpreting conversation and narration. This is important work since linguistic analyses are time consuming, and their use has therefore been restricted in the public education sector.

Dollaghan (1992) also has presented a variety of recent advances in measuring spontaneous language, while Schwartz (1991) draws together a number of current procedures for integrating microcomputers into language assessment and intervention. The role of computers in promoting literacy in children with severe speech and physical impairments has been well documented (Yoder & Koppenhaver, 1993), as have the expectations of teachers and parents (Light & McNaughton, 1993). Nevertheless, there is much to be accomplished before technology becomes a mainstay of language programming and services.

Transition and Follow-up Studies

Transition and follow-up studies of language-impaired students are limited. As noted throughout this chapter, as children grow older, their label often changes from language impaired to learning disabled, and less often to mental retardation—or they become known as disabled readers. Therefore, the reader may find documentation of transition services to individuals labeled with the so-called "primary label" in other sections of this text. Although language impairments remain an indisputable problem, and reflect that fact that language impairments do not disappear as students grow older, such students are rarely provided assistance under the provisions of IDEA, Section 504 of the Rehabilitation Act of 1973 or the Carl D. Perkins Vocational and Applied Technology Act of 1990 (Stromski, in press). Clinical observation and case history data appear to suggest that difficulties in comprehending and expressing language—particularly in written form—remain as a life-long impairment. Earlier sections of this paper have reflected the mixed follow-up studies from early childhood studies. Rigorous scientific study thus far has failed to undergird the past efforts to provide positive efficacy data for many aspects of early intervention. A happy ending to the story of children with language impairment and its sequelae still eludes us.

References

A Committee of Orton Dyslexia Society Members (Winter, 1995). Definition of Dyslexia, *Perspectives*, p. 3.

American-Speech-Language-Hearing Association. (1982). Definitions: Communicative disorders and variations. *Asha*, 24:11, pp. 949–950.

Anderson, K. N. (Rev. Ed.) (1994). *Mosby's Medical, Nursing and Allied Health Dictionary, Fourth Edition*. St. Louis, MO: Mosby Year Book, Inc.

Apter, D. S. (1994). From dream to reality: A participant's view of the implementation of Part H of P. L. 99-457. *Journal of Early Intervention*, 18.2, pp. 131–140.

ASHA Task Force on Treatment Outcome and Cost Effectiveness. (1994). Collecting Outcome Data: Existing Tools, Preliminary Data and Future Directions (pp. Unnumbered). Rockville, MD: American Speech-Language Hearing Association.

Bashir, A. S. & Scavnuzzo, A. (1992). Children with language disorders: Natural history and academic success, *Journal of Learning Disabilities*, 25, pp. 53–65.

Bauchner, H., Brown, E. (1986). Premature graduates of the newborn intensive care unit: A guide to follow-up, *Pediatric Clinics of North America*, 35, 1.207–1.226.

Bailey, D. B., Jr. & Wolery, M. (1989). *Assessing infants and preschoolers with handicaps*. New York: Merrill Publishing.

Bello, J. (1994). Prevalence of Speech, Voice, and Language Disorders in the United States, *Communication Facts, 1994 Edition*, p. 2. Rockville, MD: American Speech-Language-Hearing Association.

Bennett, T, Nelson, D. E., & Lingerfelt, B. V. (1992). *Facilitating Family-Centered Training in Early Intervention*. Tucson, AZ: Communication Skill Builders.

Blachowicz, L. Z. (1994). Problem-solving strategies for academic success. In G. P. Wallach & K. G. Butler (Eds). *Language Learning Disabilities in School-Age Children and Adolescents: Some Principles and Applications.*(pp. 304–322). Needham Hghts: MA: Allyn & Bacon.

Blank, M, Marquist, A. & Klimovitch, M. (In press). *Directing School Discourse*. Tucson, AZ: Communication Skill Builders.

Bloom, Lois. (1995). Foreword, *Meaningful Differences in the Everyday Experience of Young American Children* (pp. ix–xiii). Baltimore: Paul H. Brookes Publishing Co.

Blosser, J. L. & DePompei, R. (1989). The head-injured student returns to school: Recognizing and treating deficits, *Topics in Language Disorders*, 9:2, pp. 67–77.

Bowe, F. G. (1995). *Birth to Five: Early Childhood Special Education*. New York City: NY: Delmar Publishers.

Brooks, C. A., Lindstrom, J, McCray, J. & Whiteneck, G. G. (1995). Cost of medical care for a population-based sample of persons surviving traumatic brain injury, *Journal of Head Trauma Rehabilitation*, 10:4, pp. 1–13.

Butler, K. G. (1991). (Ed.). Late bloomers: Language development and delay in toddlers, *Topics in Language Disorders*, 11:4.

Butler, K. G. (1984). Language research and practice: A major contribution to special education (pp. 272–302. In R. J. Morris & B. Blatt (Eds.) *Special Education: Research and Trends*. New York City, NY: Pergamon Press.

Butler, K. G. (1986). *Language Disorders in Children*. Austin, TX: Pro-Ed.

Butler, K. G. (1994a). (Ed.) *Early Intervention II: Working with Parents and Families*. Gaithersburg, MD: Aspen Publishers, Inc.

Butler, K. G. & Wallach G. P. (1994b). Keeping on track to the twenty-first century (pp. 418–428). In G. P. Wallach & K. G. Butler (Eds.), *Language Learning Disabilities in School Age Children and Adolescents*. Needham Heights, MA: Allyn & Bacon.

Catts, H. W. (In press). Defining dyslexia as a developmental language disorder: An expanded View, *Topics in Language Disorders*, 16:2.

Cazden, C. B. (1988). *Classroom discourse: The language of teaching and learning*. Portsmouth, NH: Heinemann.

Chin-Chance, S. (1988). The performance of selected ethnic groups in Hawaii on achievement tests, *Pacific Education Research Journal*, 4:1, pp. 53–58.

Commission on Accreditation of Rehabilitation Facilities (CARF). (1995). *1995 Standards for Medical Rehabilitation*. Tucson, AZ.

Committee of ODS Members' Definition, *Perspectives*, Winter, 1995, p. 3.

Costa, L & Rourke, B. P. (Eds.) (1990). Editorial Policy III, *Journal of Clinical and Experimental Neuropsychology*, 12:2, p. 181.

Coufal, K. L. (1993). Collaborative consultation: A problem-solving process, *Topics in Language Disorders*, 14:1, pp. iii–100.

Cross, T. G. (1984). Habilitating the language-impaired child: Ideas from studies of parent-child interaction, *Topics in Language Disorders*, 4, pp. 1–14.

Crystal, D. (1987). Towards a "bucket" theory of language disability: taking account of inter-action between language levels, *Clinical Linguistics & Phonetics*, 1:1, 7–2.

Curtiss, S. (1977). *Genie*. New York: Academic Press.

DePompei, R. & Williams, J. (1994). Working with families after TBI: A family-centered approach, *Topics in Language Disorders*, 15:1, pp. 68–81.

Dickinson, D. & McCabe, A. (1991). The acquisition and development of language: A social interactionist account of language and literacy development (pp. 1–40). In J. F. Kavanagh (Ed.), *The Language Continuum: From infancy to literacy*. Parkton, MD: York Press.

Dollaghan, C.A. (1992). Analyzing spontaneous language: New methods, measures and meanings, *Topics in Language Disorders*, 12:2.pp. ix–92.

Ehren, B. J. (1994). New directions for meeting the academic needs of adolescents with lan-guage learning disabilities. In G. P. Wallach & K. G. Butler (Eds.), *Langauge Learning disabilities in School-Age Children and Adolescents: Some Principles and Applications* (pp. 393–417). Needham Hghts., MA: Allyn & Bacon.

Fletcher, J., Shaywitz, S., Shankweiler, D., Katz, L., Liberman, I., Stuebing, K., Francis, D., Fowler, A., & Shaywitz, B. (1994). Cognitive profiles of reading disability: Comparisons of discrepancy and low achievement definitions (pp. 6–23). *Journal of Educational Psychology*, 86.

Foster, S. (1994). The development of discourse topic skills by infants and young children (p. 18–32). In K. G. Butler (Ed.), *Early intervention II: Working with parents and families*. Gaithersburg, MD: Aspen Publishers.

Friel-Patti, S. (1994). Auditory linguistic processing and language learning. In G. Wallach and K. Butler (Eds.), *Language learning disabilities in school-age children and adolescents: Some principles and applications* (pp. 373–392). Needham Heights, MA: Allyn & Bacon.

German, D.J. (1994). Word finding difficulties in children and adolescents. In G. P. Wallach & K. G. Butler (Eds.), *Language learning disabilities in school-age children and adolescents: Some principles and applications* (pp. 323–347). Needham Heights, MA; Allyn & Bacon.

Goodman, J. (1994). "Empowerment" versus "best interests": Client-professional relationships. *Infants and young children*. 6:4, pp. vi–x.

Goorhuis-Brouwer, S. M. (August 28, 1992). Personal communication, Child Language Committee, University Hospital, Gronigen, The Netherlands.

Hart, B. & Risley, T. R. (1995). *Meaningful Differences in the Everyday Experience of Young American Children*. Baltimore, MD: Paul H. Brookes Publishing Co.

Hansson, K. & Nettelbladt. (1995). Grammatical characteristics of Swedish Children with SLI, *Journal of Speech and Hearing Research*, 38, pp. 589–598.

Hasenstab, M. S. (1987). *Language, learning and otitis media*. Boston: Little Brown.

Heath, R. W., Plett, J. D., Tibbetts, K. A. & Medeiros, P. H. (1987). Hearing dysfunction in Hawaiian preschoolers: Its relation to educational achievement and family characteristics. Unpublished Report.

Henderson, L. W. & Meisels, S. J. (1994). Parental involvement in the developmental screening of their young children: A multiple-source perspective, *Journal of Early Intervention*, 18:2, pp. 141–154.

Higginbotham, D. J., Scally, C. A., Lundy, D. C., & Kowarsky, K. (1995). Discourse comprehension of synthetic speech across three augmentative and alternative Communication (AAC) output methods. *Journal of Speech and Hearing Research*, 38:4, pp. 899–901.

Hymowitz,C. & Pollock, E. J. (1995). Psychobattle: Cost-cutting firms monitor couch times as therapists fret, *The Wall Street Journal*, CXXXIII, No. 8, pp. A1, A9.

James, D., van Steenbrugge, W., & Chiveralls, K. (1994) Underlying deficits in language-disordered children with central auditory processing difficulties, *Applied Psycholinguistics*, 15:3, pp. 311–328.

Juel, C. (1990). The role of decoding in early literacy *instruction on* and assessment. In L. M. Morrow & J. K Smith (Eds.), *Assessment for instruction in early literacy* (pp. 135–154). Englewood Cliffs, NJ: Prentice-Hall.

Kamhi, A. (In press) What role should SLPs play in the assessment and intervention of written language? In G. P. Wallach & Butler, K. G.'s Chapter in Learning disabilities: Moving in from the Edge, *Topics in Language Disorders*, 16:1.

Kaufmann, J. M. (In press). Why we must celebrate a diversity of restrictive environments. In *Learning Disabilities Research and Practice*.

Kavanagh, J. F. (1986). (Ed.). *Otitis media and child development*. Parkton, MD: York Press.

Kavanagh, J. F. (1991). (Ed.) *The language continuum: From infancy to literacy* (p.vii). Parkton, MD: York Press.

Kenworthy, O. T. (1994). Caregiver-child interaction and language acquisition of hearing-Impaired children. In K.G. Butler (Ed.) *Early Intervention II: Working with Parents and Families*, pp. 123–144.

Knutson, J. G. T. & Sullivan, P. M. (1992). Communicative disorders as a risk factor in abuse, *Topics in Language Disorders*, 13:4, pp. 1–15.

Kotby, M. N. & El-Shobary, A. (1992). The Egyptian state of the art in communication pathology, *Working Papers in Logopedics & Phoniatrics, No. 8*, pp. 45–52.

Laadt-Bruno, G, Lilley, P. K. & Westby, C. E. (1994). A collaborative approach to Developmental care continuity for infants at risk and their families (p. 3–16). In K. G. Butler (Ed.), Early Intervention: Working with infants and toddlers. Gaithersburg, MD: Aspen Publishers.

Lahey, M. & Edwards, J. (1995). Specific language impairment: Preliminary investigation of factors associated with family history and with patterns of language performance, *Journal of Speech and Hearing Research*, 38, pp. 643–657.

Lahey, M., Liebergott, J., Chesnick., M. Menyuk, P., & Adams, J. (1992). Variability in children's use of grammatical morphemes, *Applied Psycholinguistics*, 13, pp. 373–398.

Lebrun, Y. (1980). Victor of Aveyron: A reappraisal in light of more recent cases of feral speech, *Language Sciences*, 2:1, 32–43.

Leonard, L. B. (1989). Language learnability and specific language impairment in children, *Applied Psycholinguistics*, 10, pp. 179–202.

Leonard, L. B. (1992). The use of morphology by children with specific language impairment: Evidence from three languages. In R. S. Chapman (Ed.) *Processes in Language Acquisition and Disorders*. St. Louis, MO: Mosby Yearbook.

Lidz, C. S. (1991). *Practitioner's Guide to Dynamic Assessment*. New York: The Guilford Press.

Light, J. & McNaughton, D. (1993). Literacy and augmentative and alternative communication (AAC): The expectations and priorities of parents and teachers, *Topics in Language Disorders*, 13:2, pp. 33–46.

Locke, J. L. (1993). *The child's path to spoken language*. Cambridge, MA: Harvard University Press.

Mahaffey, B. (July 20, 1995). Personal communication.

Massouad, A., Kotby, M. N. & Rashid, A. M. (1979). Epidemiological study of communicative disorders among Egyptian school children. Presented before the 8th Congress of the Union of European Phoniatricians, Roszeg.

Miller, G. (1990). The place of language in scientific psychology, *Psychological Science*, 1, pp. 7–14.

Miller, J. & Chapman, R. (1982). *SALT: Systematic analysis of language transcripts-Harris computer version* [Computer program]. Madison, WI: University of Wisconsin-Madison, Waisman Center, Language Analysis Laboratory.

Miller, J. F., Freiberg, C., Rolland, M-B., & Reeves, M. A. (1992). Implementing computerized language sample analysis in the public school, *Topics in Language Disorders*, 12:2, pp. 69–82.

Morrison, J. (1995). *DSM-IV made easy: The Clinician's Guide to Diagnosis*. New York: The Guilford Press.

National Deafness and Other Communication Disorders Advisory Board. (1991). *Research in human communication (NIH Publication No. 92–3317)*. Bethesda, MD: National Institute on Deafness and other Communication Disorders.

National Institute of Neurological Disorders and Stroke. (1992). *Profile*. Bethesda, MD: Author.

Nelson, E. B. (1993). *The national education goals report. Volume One: The national report, 1993: Building a nation of learners* (pp. x–xi). Washington, D.C.: U. S. Government Printing Office.

Neisworth, J. T. & Bagnato, S. J. (1988). Assessment in early childhood special education: A typology of dependent measures (pp. 23–50). In S. L. Odom and M. B. Karnes (Eds.) *Early Intervention for Infants & Children with Handicaps: An Empirical Base*. Baltimore: Paul H. Brookes Publishing Co.

Nelson, N. W. (1994). Curriculum-based language assessment and intervention across the grades. In G. P. Wallach & K. G. Butler (Eds.) *Language learning disabilities in school-aged children and adolescents: Some Principles and Applications* (pp. 104–131). Needham Heights, MA: Allyn & Bacon.

Nelson, N. W. (1993). Foreword (pp. vii–ix) in J. Norris and P. H. Hoffman, *Whole language Intervention for school-age children*. San Diego, CA: Singfular Publishing Group., Inc.

Norris, J. & Hoffman, P. H. (1993). *Whole language intervention for school-age children*. San Diego, CA: Singular Publishing Group, Inc.

Palincsar, A. S., Brown, A. L., & Campione, J. C. (1994). Models and practices of dynamic Assessment. In G. P. Wallach & K. G. Butler (Eds.) *Language learning disabilities in school-aged children and adolescents: Some principles and applications*, (pp. 132–144), Needham Heights, MA: Allyn & Bacon.

Paul, R. (1991). Profiles of toddlers with slow expressive language development. *Topics in Language Disorders*, 11:4, pp. 1–12.

Peterson, C., & McCabe, A. (1994). Assessment of preschool narrative skills, *American Journal of Speech-Language Pathology*, 3, pp. 44–56.

Pickelle, C. & Ramos, T. (1995). Rehab across the pond, *Rehab Management*, 8:4, p. 11.

Pinker, S. (1994). *The language instinct: how the mind creates language*. New York: William Morrow and Company.

Prelock, P. A., Miller, B. L., & Reed, N. L. (1995). Collaborative partnerships in a Language in the Classroom program, *Language, Speech, and Hearing Services in Schools*, Vol, 26, pp. 286–291.

Raaijmakers, M. F., Dekker, J., Dejonckere, P. H. & van der Zee, J. (1995). Reliability of the assessment of impairments, disabilities and handicaps in survey research on speech therapy, *Follia Phoniatrica et Logopaedica*, 47:4, pp. 199–109.

Rice, M. L., Wexler, K. & Cleave, P. L. (1995). Specific language impairment as a period of extended optional infinitive, *Journal of Speech and Hearing Research*, 38, pp. 850–863.

Rice, M. L. & Wilcox, K. A. (1995). *Building a language-focused curriculum for the preschool classroom, Volume 1, A foundation for lifelong communication*. Baltimore, MD: Paul H. Brookes Publishing Company.

Russell, S. & Kaderavek, J. (1993). Alternative models for collaboration, *Language, Speech and Hearing Services in Schools*, 24, pp. 76–78.

Schwartz, A. H. (1991). Integrating microcomputer applications into clinical practice, *Topics in Language Disorders*, 11:2, ix–96.

Scott, C. (1994). A discourse continuum for school age students: Impact of modality and genre (pp. 219–252). In G. Weallach and K. Butler (Eds.) Language learning disabilities in school-aged children and adolescents. Needham Heights, MA: Allyn & Bacon.

Seifert, H. & Schwarz, I. (1991). Treatment effectiveness of large group basic concept instruction with Head Start students, *Language, Speech, and Hearing Services in Schools*, 22, pp. 60–64.

Sergent, F. (1995). Life after LAP (pp. 181–198). In M. L. Rice & K. A. Wilcox (Eds.), *Building a Language-Focused Curriculum for the Preschool Classroom, Volume I, A Foundation for Lifelong Communication*. Baltimore: MD: Paul H. Brookes Publishing Co.

Shatz, M. (1994). *A toddler's life: Becoming a person.* New York, NY: Oxford University Press.

Silliman, E. R. & Wilkinson, L. C. (1994a). Discourse scaffolds for classroom intervention (pp. 27–52). In G.P. Wallach and K. G. Butler, (Eds.) *Language learning disabilities in school-age children and adolescents: Some principles and applications.* Needham Heights, MA: Allyn & Bacon.

Silliman, E. R. & Wilkinson, L. C. (1994b). Observation is more than looking (pp. 145–173). In G. P. Wallach and K. G. Butler (Eds.) *Language learning disabilities in school-age children and adolescents: Some principles and applications.* Needham Heights, MA: Allyn & Bacon.

Slama-Cazacu, T. (1994). Quo vadis, psycholinguistics? *International Journal of Psycholinguistics.* 10:2 [28]. pp. 203–216.

Snow, C. E. (1983). Foreword: Language development and disorders in the social context, *Topics in Language Disorders,* 4:4, p. v.

Snyder, L. S & Saywitz, K. J. (1993). Child abuse: Cognitive, linguistic, and developmental considerations, *Topics in Language Disorders*, 13:4, iii–88.

Sohlberg, McK. M. & Mateer, C. A. (1989). The assessment of cognitive-communicative functions in head injury, *Topics in Language Disorders*, 9:2, pp. 15–33.

Soter, A. O. (1988). The second language learner and cultural transfer in narration. In A. C. Purves (Ed.), *Writing across languages and cultures.* Newbury Park, CA: Sage.

Spahr, F. (1993–1994). (Ed.) *American Speech-Language-Hearing Association Membership Directory, 1993–1994.* Rockville, MD: American Speech-Language-Hearing Association.

Stanovich, K. (1988). The right and wrong places to look for the locus of reading disability, *Annals of Dyslexia*, 38, pp. 154–177.15:1, pp. 21–36.

Steelman, J. D., Pierce, P. L., & Koppenhaver, D. A. (1994). The role of computers in promoting literacy in children with severe speech and physical impairments (SPPI). In K. G. Butler (Ed.) *Severe communication disorders: Intervention strategies* (pp. 200–212). Gaithersburg, MD: Aspen Publications.

Stewart, B. A. (1991). *International Directory of Education for Speech-Language Pathologists,* First Edition. Rockville, MD: American Speech-Language Hearing Association.

Stewart, J. L, Anae, A. P. & Gipe, P. N. (1989). Pacific Islander children: Prevalence of hearing loss and middle ear disease, *Topics in Language Disorders*, 9:3, pp. 76–83.

Stoel-Gammon, C. (1991). Normal and disorders phonology in two-year-olds, *Topics in Language Disorders 11:4*, pp. 21–32.

Szekeres, S. F. & Meserve, N. F. (1994). Collaborative intervention in schools after traumatic brain injury. *Topics in Language Disorders*, 15:1, pp. 21–36.

Taylor, R. L. (1995). Functional uses of reading and shared literacy activities in Icelandic homes: A monograph in family literacy, *Reading Research Quarterly*, 30:2, pp. 194–219.

Temple, C., Nathan, R., Temple, F. & Burris, N. A. (1993). *The Beginnings of Writing*, 3rd Ed, pp. 101–120. Needham Heights, MA: Allyn and Bacon.

United Nations Children's Fund. (1995). *The State of the World's Children.* Oxford, England: Oxford University Press.

van Kleeck, A. (1994). Metalinguistic development. In G. P. Wallach & K. G. Butler (Eds.) *Language Learning disabilities in School-Age Children and Adolescents: Some Principles and Applications.* Needham Heights, MA: Allyn and Bacon.

van Riper, C. (1947). *Speech correction: Principles and Methods, 2nd Ed.* New York, NY: Prentice-Hall.

van Riper, C. & Butler, K. G. (1955). *Speech in the Elementary Classroom* (pp. 1–276). NY: Harper and Row.

Wagner, M. (1991, September). *Drop outs with disabilities. What do we know? What can we do?* Menlo Park, CAS: SRI International.

Wallach, G. P. & Butler, K. G. (1994a). Creating communication literacy and academic success (pp.2–26). In Wallach, G. P. & Butler (Eds.) *Language Learning Disabilities in School-Age Children and Adolescents.* Needham Heights, MA: Allyn & Bacon.

Wallach, G. P. & Butler, K. G. (1994b). Reflections: On being observant and dynamic in school-based contexts and elsewhere (pp. 174–178). In G. P. Wallach & K. G. Butler (Eds.) *Language learning disabilities in school-aged children and adolescents: Some principles and applications.* Needham Heights, MA: Allyn & Bacon.

Weiss, A. L. (1995). Conversational demands and their effects on fluency and stuttering, *Topics in Language Disorders,* 15:3, pp. 18–31.

Westby, C. E. (1985). Learning to talk—talking to learn: Oral-literate language differences. In C. S. Simon (Ed.) *Communication skills and classroom success: Therapy Methologies for Language learning disabled students,* pp. 181–218. San Diego, CA: College-Hill Press.

Westby, C. E. (1994). The effects of culture on genre, structure, and style of oral and written tests. In G.. P. Wallach and K. G. Butler (Eds.) *Language Learning Disabilities in School-Age Children and Adolescents: Some Principles and Applications* (pp. 180–218). Needham Heights, MA; Allyn & Bacon.

Whitehurst, G. J., Fischel, J. E., Lonigan, C. J., Valdez-Menchaca, M. C., Arnold, D. C., & Smith, M. (1991). Treatment of early expressive language delay: If, when and how, *Topics In Language Disorders,* 11:41, pp. 55–68.

World Health Organization. (1990). *International Classification of Diseases, 10th Revision.* Geneva, Switzerland: Author.

Yoder, D. E. & Koppenhaver, D. A. (1993). (Eds.) Literacy learning and persons with severe speech impairments, *Topics in Language Disorders,* 13:2, iv–95.

Section 3

Support Systems

The focus of Section 3 is on the important supports required by the fields of special and remedial education. As in any other field of human services, the operation of quality programs depends on many resources and supports that may not be directly apparent in the base programs. Each of the chapters of this section presents major challenges to policymakers and to all stakeholders who are concerned with quality programs. Each rests upon a knowledge base in which research and inventive practices are important. The future of education for students with special needs depends very much on the adequacy of the knowledge base in these areas and upon the extent to which practices in legislative halls, universities, public agencies, and schools are up to state-of-the-art standards.

The first chapter of Section 3 provides a general review of costs and funding systems for all aspects of special and remedial education. Account is taken of the history of specialized school programs and of needed changes in support systems as programs develop over time. A note of urgency is expressed in this chapter about the high costs of many of the special school programs and the possibility, perhaps even the likelihood, that cutbacks in funding will be required in the near future. A proposed set of criteria for an adequate funding system is presented.

Perhaps more than in any other aspect of education, the field of special education has led the way in collaboration with parents. This has not always been accomplished in smooth and thoughtful ways. Indeed, there has been a worrisome turning to the courts to solve some difficult issues between educators and parents. Nevertheless, much has been learned and much progress has been made in linkage of teachers and parents. Parents have joined forces and become strong advocates for programs and for their improvement—a story told in this chapter.

Nothing works well in special and remedial education except as there are well-prepared teachers who conduct programs in good school situations. The third chapter in this section provides an extended discussion of trends and issues in specialized teacher preparation and, indeed, in the preparation of all educators. There are developments and issues of enormous importance in this domain, all given attention in this exceptionally important chapter.

The fourth and final chapter of this section and of the entire volume covers developments of collaborative arrangements between schools, parents, and various community agencies—as in mental health, welfare, general health, corrections, churches, etc. These developments give every indication of becoming significant and permanent aspects of services to children and their families. Implications of this movement are very important for all stakeholders: teachers, parents, teacher-preparation centers, social agencies, health workers, and more. Clearly, this is one of the major moving elements in today's programs for exceptional and at-risk students.

12

Funding

MAYNARD C. REYNOLDS
Professor Emeritus of Educational Psychology, University of Minnesota
and Senior Research Associate, National Center on Education in the Inner Cities, Temple
University, USA

This chapter focuses on the costs and expenditures of categorical education programs, particularly special education. Some consideration is also given to the recently revised Chapter 1 program for economically disadvantaged students and to the provisions for minority-language children. Although we have a sizeable literature on these topics, it is not a research literature in the main; there is very little opportunity for experimental research in the domains considered here. Nevertheless, it is possible to examine funding practices and how they correlate with program policies and developments. This territory is much influenced by economic trends, changing expectations for schools, and sociopolitical factors. The history of program developments also has clear implications for the design of funding systems. While categorical funding is intended to lead programmatic changes, the needed changes themselves change over time. The chapter is heavily oriented to observations in the United States and, perhaps, neglects activities in other parts of the world. Yet the relations between resources and programs appear to be similar in all places (e.g., Ainscow, 1991). A proposed integrated view of funding strategies concludes this paper, taking into account history, management of incentives, and purposes of categorical funding in today's schools.

Introduction

The way we pay for categorical education programs is as important as the amount of money provided. Programs tend to follow dollars—not totally, but to an important degree—which means that if discrepancies exist between policy (intentions) and money flows, the outcomes of the programs can be best predicted by tracing the dollar flow (Parrish, 1993; Wood, Sheehan, & Adams, 1984). The result can be most unfortunate programmatically. Sad and costly misjudgments have been made in setting up funding systems in the past.

For example, funds have been established (in one state) for the construction of new regional schools for disabled students just when the movement toward inclusion was reaching its height; school systems have moved ahead on "mainstreaming" policies when only set-aside special-class placements were receiving special education funds; or schools have moved very slowly into early education and preventive programs for at-risk students, although the advantages for children were clearly defined.

The number of categorical funding streams is enormous and growing. In 1991, 77 federal programs for children and families were funded with $100 million or more per year (Gardner, 1994). The sixteenth report to Congress on the implementation of special education laws cited a survey by the Carolina Policy Studies Program which reported "as many as 44 sources of funding" for early intervention programs and 25 laws and programs addressing the same target population (U.S. Department of Education, 1994, p. 43). In California in 1992–93, "at least 57 separate categorical programs" provided $5.1 billion to the schools (California State Legislative Analyst's Office, 1993, p. 1). The reports required by federal authorities list 13 special education categories. Categorical systems undoubtedly contribute to the fragmentation of schools as organizations.

If programs follow dollars, it is important that funding systems be designed to serve as tools for leadership. That is the main theme of this chapter. Part of the problem is that so few people pay attention to the design of funding systems; program designs, consequently, tend to be considered superficially and to be simply carried over from one funding period to the next by tradition or worse. For example, the establishment of "medicaid waiver" policies (in the 1980s) seems to have caused shifts in millions of dollars, correlated with the deinstitutionalization of thousands of severely disabled children who were delivered to local schools for educational services.[1] School leaders initially were hardly aware of the changes and were not involved in designing them. Copeland and Iversen (1981) observed that "such changes, as more and more of them come along, tend to define a new policy for the states which grows as if guided by an invisible hand" (p. 68); "... a process ... [tending] to produce a system that becomes immobilized in its own contradictions" (p. 69).

Special education funding tends to be narrowly categorical, a situation that has its critics but which also produces advocates—parents and others—who often "shaped the evolution of these programs and their current condition as well" (Casanova & Chavez, 1992, p. 68).

---

[1]States were authorized to apply to the federal government for waivers authorizing them to shift money which had gone to residential institutions to support of community services. This resulted in shifts of many "patients" from institutions to communities where they began living in group homes or other facilities. The shift of money was to welfare or human service agencies to pay for community services, but not to schools.

General school administrators often feel driven to increase dollar inflow to their districts; at the same time they oppose categorical funding systems as if they were almost venial sins. They would prefer to have all money for schools "left on a stump in the dark of the night" so that they can pick it up and do their own categorizing without constraints. In fact, of course, all funds are categorized ultimately: should one buy more pencils or raise teachers' salaries? Thus the argument is not over categories per se but over who should do the categorizing.

Another facet of doubt has been created by professionals who argue that much of the categorical approach is invalid programmatically. That is, most special education programs involve classification and placement procedures for children which have little if any clear relation to educational needs (Shepard & Smith, 1981; Ysseldyke, Algozzine, Richey, & Graden, 1982).

In an important study of child classification and placement practices by a distinguished panel (created by the National Academy of Science) one conclusion was that "it is the responsibility of the placement team that labels and places a child in a special program to demonstrate that any differential label used is related to a distinctive prescription for educational practices...that lead to improved outcomes" (Heller, Holtzman, & Messick, 1982, pp. 101–102). The panel went on to say that there was "no justification for categorical labeling that discriminates mildly retarded children from other children with academic difficulties" (p. 87). The panel's findings raise doubts about at least 75% of the special education organizational arrangements and about the distinction of major elements of special education from the Chapter 1 program.

Another rising concern these days is with the fact that categorical funding operates on the input side. That is, money flows in when a student is identified by certain characteristics as qualified and placed in a special education program. Whether the program will be or is demonstrated to be beneficial for the student seems not to matter for funding purposes (National Association of State Boards of Education, 1992). A related and troubling matter is that "educators know very little about how well [categorical] programs work. Many evaluations are not evaluations as such, but operational reviews" (California State Legislative Analyst's Report, 1993, p. 6). Hence many persons want funding decisions shifted to variables that reflect outcomes of programs, and they want improved evaluations of programs. Why use and pay for categorical programs unless and until they show that they work?

What Are the Costs and Expenditures?

Costs and expenditures in categorical programs are coming under close and threatening public scrutiny these days. Special education, in particular, has become a page one topic for major newspapers. The *New York Times* in 1991

(Berger, 1991) reported that 25% of New York City's total expenditure for public schools was linked to special education and that "...in the current fiscal crisis when savings of $505 million must be found, cutbacks are being confined almost entirely to regular classrooms. Special education is protected against reductions by a web of state mandates and court rulings" (p. 1).

The *Washington Post* reported that Montgomery County (Maryland) special education programs were "inefficient" and that, in 1994—a year in which 275 parents in the county appealed their children's school placements—costs "skyrocketed to nearly $1.4 million" just for attorney fees and court settlements. Nearly 800 of the Montgomery County special education students in that year were in private schools, supported by public funds (Beyers, 1995, pp. A1, A16).

A series of four front-page articles in the *Minneapolis Star Tribune*, beginning December 4, 1994, carried the startling headline, "Average Kids are Losing" (Hotakainen & Smetanka, 1994, p. 1). The articles, which summarized special education expenditures for the State of Minnesota, portrayed trends over the decade 1982–83 to 1992–93 as follows:

- Total expenditures in special education increased from $227 million to $602 million.
- Totals increased by 92.5% for general K-12 education; 165% for special education. In that same period, inflation (CPI) was 45%.
- Enrollment increases in regular and special education were about equal (10.3% and 10.1%).
- Federal supports increased by 86% but at peak represented less than 6% of total costs.
- State supports reached about $200 million in 1992–93, a 94% increase over the decade.
- Local expenditures were $370 million in 1992–93, a 247% increase in a decade.

For the nation as a whole, the average per-pupil cost of K-12 education in 1990–91 was $5,266 (Supplement, 1994). For students with disabilities an estimated additional average of $6,845.80 was spent. In that same year, 4.8 million students were enrolled in special education, yielding a total estimated cost of $58 billion in the entire nation for the education of disabled students. About $6 billion of federal funds have been spent on Chapter 1 programs in each recent year and still more funds have gone to other categorical programs.

It is difficult to specify the true costs of special education and other categorical programs because, in part, different elements are included in the various studies. Questions arise about (a) how to subtract the costs for students who are served part time in regular education programs? (b) identification and diagnostic procedures? (c) special transportation? (d) buildings and facilities? (e) adaptations to buildings to ensure program accessibility? (f) attorneys and litigation? etc. However, a common inference is that *on average* public schools spend about 2.3 times as much per pupil in special education as for students in regular education (Chaikind, Danielson, & Brown, 1993). This estimate is based mainly on the most comprehensive and reliable of special education cost studies by Moore, Strang, Schwartz, and Braddock (1988).

Some of the specific findings of the Moore study were as follows:

- Average per-student expenditures for transportation in special education was seven times higher than for a student in regular education ($1,583 v. $234), probably because of fewer students per vehicle, use of special ride-along aides, etc.
- Federal funds constitute about 5% of total expenditures for special education.
- The district with the highest per-student expenditures outspent the district with the lowest outlays by five to one.
- About 5% of the total student population received special educational and psychological assessments annually, at an estimated average cost of $1,230.
- Average expenditures for special education varied widely according to the type of administrative arrangements. In resource programs average per-pupil expenditure just for special education was $1,325; average expenditure in self-contained programs (i.e., special classes) was $4,233; average cost per student in residential programs was $28,324.

A Bit of History

Special education programs developed rapidly in the various states under the changing provisions of state laws throughout the 1950s and 1960s, decades that were characterized by a rapid expansion in the national economy. This was a period in which finding exceptional children and bringing them into the schools was a primary concern. Funding systems of that period quite appropriately emphasized per-pupil payoff units. A further surge in program development occurred following the enactment by the federal government of Public Law 94-142 in 1975.

This unprecedented entrance of the federal government into the education of disabled children involved promises of major financial supports; yet actual federal appropriations have consistently fallen far short of "authorizations." Nevertheless, a full load of federal regulations was delivered, turning state and local administrators—as some people believe—into surrogates of federal authorities. Special education is cited as one of the extreme cases of "proceduralism," meaning that norms for mere compliance procedures often have surpassed those for truly substantive matters in the field (Reynolds & Birch, 1988). One major study of administrative behavior concluded that "the job of categorical program managers is geared to keeping the funding agencies satisfied rather than managing the education programs" (Hannaway, 1985, p. 64).

The economic climate in the period of entry by the federal government (1975 and immediately following years) was mixed, which made it difficult to expand education in general, but it was balanced by a decline in the number of children entering the schools. The baby boomers of the late 1940s and early 1950s had completed their mandatory schooling and fewer children were entering the lower grades. In 1975, about 51 million children were enrolled in the nation's schools, a figure that dropped to 44 million a decade later. The proportion of households including children dropped even more sharply. In about the same decade (1974 through 1983) the number of teachers and other staff members employed in special education programs in Minnesota nearly doubled (95%) while regular secondary and elementary education teachers declined by 25% and 9.5%, respectively (Newsletter, 1984). The story was similar in most states: general education was shrinking, while special education was expanding.

In this period (1975 et seq.) many regular education teachers who had received "pink slips" turned to special education for employment and many school psychologists, social workers, and other school specialists were placed in the special education funding streams. The rate of enrollment of severely disabled students increased in public schools due, in part, to the medicaid waiver policies. The sharply advancing costs of special education since then undoubtedly reflect this increased enrollment of severely disabled students. Data provided by Moore et al. (1988) indicate that per-pupil expenditures in special education were advancing more than twice as rapidly as those of regular education (in constant dollars, 10% vs. 4%, 1977 to 1985).

Another change during this period was the rapid takeover by special education of one of the programs historically provided, or at least initiated, in regular education: the so-called "learning disability (LD)" program. This enormous development represents a cost shift for remedial programs from regular to special education (Hartman, 1981). Some resentment seems to be building up among regular educators toward the present efforts to return many LD students—but not the categorical funds—to mainstream educators.

All through recent decades there has been a tendency to enlarge the definition or inclusiveness of virtually all disability categories. For example, Siegel (1987), writing about the fields of speech and hearing, commented as follows:

> In recent years it has been argued that the province of speech-language pathologists and audiologists should include reading problems, tongue thrust, swallowing, feeding abnormalities and clients so severely handicapped that *communication* is equated with simple sensory stimulation. From a period when the profession was narrowly exclusive, we have become almost recklessly inclusive in the willingness to work in so many areas (p. 308).

Over the past several decades in which special education has grown to its present large and doubtful dimensions, some quite unhealthy trends have occurred in the lives of children. While divorce rates and the numbers of single-parent households have soared, teachers have found themselves meeting increasing numbers of children who are experiencing grief and limited care. In many places, but especially in inner-city situations, economic disinvestments, joblessness, and poverty have increasingly correlated with crime, child abuse, and drug addiction. Children in such environments often lead frightful lives of little hope, and the teachers who serve these children lose credibility as the children lose hope. Families that show resilience despite adversity tend to move out of decaying communities, often to the suburbs, leaving only "remnant populations and communities trapped by their economic irrelevance" in diminished labor markets (Bartelt, 1994, p. 2). Teachers prefer not to teach in such situations and growing numbers of students become reluctant to attend school. School suspension and dropout rates increase. Correlations of problems with racial differences complicate the issues enormously.

These difficult life situations are often interpreted as clinical problems of individuals; as a result the professionals, including school psychologists, have moved in with labels, separations, and sometimes doubtful prescriptions. Here is where we see the increasing numbers of emotionally or behaviorally disturbed students and elevated burnout rates for teachers. The knowledge base for the education of such children is meager; just maintaining an acceptable level of order and decency in classrooms is demanding and expensive. Teaching tends to be simplified to minimize the problems, and program success may be measured simply by reduced referrals to the principal's office. Too often, education falls short of aggressive teaching in forward-looking ways.

When analyzed fully these situations can be identified as complex reflections of deep and broad social changes in, for example, the economy (e.g., globalization, downturns in manufacturing jobs, etc.). The problems of the children in these environments transcend the ability of any one institution or agency to solve them. Yet schools have been stretched to try to be helpful–to provide breakfast as well as lunch for students; to offer after-school care for those who otherwise would return to empty apartments or homes; and to offer counseling to students grieving over problems of family discord and neglect—all at high expense to school budgets. But these efforts are not enough; so we see schools struggling to link their functions to those of social and health agencies. The demands for extra-familial services have grown rapidly in virtually all human service agencies, but there is much disjointedness and inefficiency in their operations.

Some signs in the context of school-community linkages are promising. For example, the revisions of 1989 in Title V of the Social Security Act dramatically alter the Maternal and Child Health Services; "States are now required to spend 30% of the funds...on children with special health care needs and to improve the service system for these children" (Ireys & Nelson, 1992, p. 321). Specific expectations have been created concerning the coordination of maternal and child health programs with school programs. In a number of states, schools have found ways of using Medicaid funds to pay for the health-related assessment and treatment functions associated with special education (Lewin/ICF, & Health Policy Consultants, 1991). These funds may decline as revisions in the Medicaid program—much in prospect currently—are made.

Another example is provided in Kentucky where broadly coordinated Family Resource and Youth Service Centers are being established, giving priority to school areas in which many students are eligible for free or reduced-cost lunches. In California, a set of Healthy Start Programs is being linked with Head Start programs. In many places, "grassroot" movements are in evidence to support broad coalitions of schools and other agencies, and state and federal policies are moving to support these developments. Private organizations, such as the Casey Foundation, are helping to explore possibilities in such ventures. The promise is for more cohesive "one-stop" services to children and families. The story is only beginning and most likely will reveal many difficulties. But the efforts may help to constrain the expansion of direct school-supported services by bringing other agencies and their resources into the cooperative endeavors with the schools (Behrman, 1992).

Hence, some of the growth and high expense of special education must be seen in the broad context of societal, even global, changes. These changes are widely regarded as the remote root causes of many of the problems now evident in the schools; but they are also the start of very broad public efforts for reform, such as those occurring in the current Empowerment Zones and Enterprise Communities (U.S. Departments of Agriculture and Housing and Urban Development, 1993). These are territories for easy cynicism about the likelihood of success, but we have hardly any choice but to try—and to try very hard for progress. Educators are being called into service in these quite new and difficult situations.

When we consider changing the funding streams for school programs, we must be aware that some school districts, especially those in the large cities, face more than an average share of problems and challenges. "Cities experiencing the greatest population losses and, by extension, a more diluted tax base, must simultaneously carry an increasingly costly educational system" (Bartelt, 1994, p. 3). The large cities and deprived rural areas will need more than average amounts of special funding to meet the needs of the children and families they serve.

All of these elements of recent history bear upon the needs for specialized school programs. As changes occur in school programs, design considerations for the funding systems that support them must keep pace.

The Purposes of Categorical Funding

It may be helpful at this point to consider the purposes of categorical funding. Why should some parts of education receive special or categorical funding? Why not "leave all the money on a stump" for local education agencies (LEAs) to apportion and simply establish procedures, some mandatory and specialized, to insure adequate services for all students? Are the purposes of funding different in the 1990s than they were a decade or two ago? The answers to these questions encompass the general purposes of categorical funding. At the time of this writing (the mid-1990s), as we look to the new millennium, these purposes of categorical funding are suggested:

1. To achieve the development of programs in high priority but neglected areas; for example, to encourage schools to develop programs for disabled students. This purpose suggests that categorical funding systems should be judged by the developments they engender.
2. To support the maintenance of essential but high-cost programs; for example, to encourage school districts to continue highly structured, intensively staffed, high-cost programs for multidisabled and severely disabled students.
3. To bring decisions on child placement into accord with the least restrictive environment (LRE) principle.[2] For example, funding might be so structured that it reduces the incentives to send a child to a zero-cost state institution when a community-based program would be more appropriate and in accord with the LRE principle; or, to favor placement in a regular school in the child's neighborhood over one in a remote place.

Design Features

In considering design features for categorical funding, something must be said about degree of categorization. Funding systems should not encourage multiple narrow categorical streams. Rather, each stream should be as broad

---

[2]The concept of excess cost frequently is brought up in this context. That idea is that state and/or federal categorical aids to LEAs ought to be sufficient to make their costs of educating disabled students equal to those for other pupils, thus permitting placement decisions to be made strictly on professional grounds. It should be noted that even if funding is at the "excess cost" level, the terms of the funding may be varied and have great importance to the quality of services provided. Any method of special education funding discussed in this paper could be set at an estimated "excess cost" level. What is suggested here is that funding systems should not totally neutralize placement decisions; instead, they should introduce bias toward placement in the least restrictive environment (LRE).

as possible and still lead toward basic goals. B. F. Skinner faced an analogous problem in defining a *response* (see Skinner, 1938). He could have used a definition that detailed the most minute aspects of "muscle-twitching," a narrow definition, or he could have focused on molar aspects of behavior. His decision was to deal with neither extreme but, rather, to focus on the level of response that would produce a smooth learning curve. Thus he chose to measure the extent of lever-pushing among his animal subjects in the Skinner box, but not which muscle group or paw the animal used to push the lever. Similarly, categorical funding should be designed to produce a smooth curve of development toward the system's goals. Unfortunately, the field of special education has gotten itself into a fair amount of wasteful "muscle-twitching" in the narrow categories it has adopted and in the ways it apportions its funds.

A distinction of importance can be made between "earmark" and "set-aside" approaches to categorical funding. An example of an "earmark" is the requirement that some minimum percentage of vocational education funds be spent on programs for disabled and disadvantaged students. The "earmarked" funds are attached to more general funds and prompt the general school leaders to plan and conduct the necessary special programs. Other examples include the historic 15% "earmark" for disability-related features of ESEA Title III programs (for innovative projects); or the requirement that disabled pupils be included in Headstart programs. These "earmarks" have the effect of causing special programs for disabled students to be weaved into the main fabric of education without establishing separate funds or separate operating systems.

By contrast, most categorical funds flow to local schools from state and federal levels in a "set-aside" fashion. The money comes from a special fund that is separately appropriated and disbursed. Special channels and standards are specified. Separate bureaucrats at all levels administer the funds and the programs. Even if there is no inherent requirement that the *programs* be "set aside" just because the funds are, the tendency is for programmatic separation to creep in whenever funding systems have a high degree of separateness. Categorical programs, under these conditions form a "second system" of education. Today, as we attempt to implement the *least restrictive environment* principle, *earmarks* appear to have advantages over *set asides* as a funding strategy.

The Relation of Cost Studies and Funding Systems

Studies in the costs of categorical programs have become quite common, but the methodology used in such studies has itself become a problem. Early studies in the field of special education, such as those by Rossmiller and Moran (1973), suggested that findings on the variance in costs across school districts were more interesting than those on central tendencies. They found very large differences among districts in the costs of special education programs. Their study included only communities that had been named as having good special education programs. The investigators compared the costs for educating disabled children in each of several categories with the costs of educating nondisabled children and came up with data such as that shown in Table 12.1.

Table 12.1 A Comparison of Per Pupil Costs of Regular and Special Education in High Cost and Low Cost School Districts

	Cost Ratios: Special Ed/Regular Ed		
Disability Category	Lowest Cost District		Highest Cost District
Visually Disabled	1.05	to	11.45
Emotionally Disturbed	1.58	to	11.64
Hearing Handicapped	1.05	to	5.88

The data show, for example, that in some communities the education of visually disabled students cost only 5% more (1.05) than the education of their nondisabled peers; in other communities the cost was more than 11 times higher (11.45). Clearly, reporting "average" per-pupil costs for education of pupils in the several categories in the face of such variations in cost would be largely meaningless.

Rossmiller's calculations were based on per-pupil costs. Had the data been framed in terms of other units (e.g., per-hour costs of instruction in braille), it is likely that the costs would have been less divergent across communities. Great variability in per-pupil costs probably can be accounted for by the differences in functions performed by the special education staff, such as amount of direct instruction provided, time spent in screening and preventive programs, time spent in consultation with parents, and so on.

Costs and funding systems, however, can be related without using the same variables. For example, in computing the costs for any given program one might list every detail of expenditure. But it would be foolish to force all schools into such detailed accounting procedures. At one time a major accounting firm was associated with a special education system that used EESEUs (Ernst and Ernst Student Educational Units). The idea was to observe the arrangements and resources used for each special education student in each 10-minute segment or unit of the school day, to record the costs for each unit, and then to aggregate the units for each child and category. For example, a disabled student might be enrolled in a regular class, along with some 24 other students, for 14 units per day. For each of those units the costs attributed to the teaching of the special education student would be 1/25th of the teacher cost. Aggregates of EESEUs for students assigned to special education would represent the costs of special education. The whole procedure would have had the absurd result of forcing the employment of one or more EESEU Clerks in every school building.

Funding systems, however, can be based on a relatively simple unit, for example, the costs of teachers' salaries. The level of reimbursement of salaries could be set to approximate the "excess cost" (or any other level of cost) of special education. The important point is that cost studies and expenditure programs are not necessarily based on the same variables.

Criteria for Categorical Special Education Funding Systems

Given the purposes for categorical funding, it is essential to specify the criteria that should be considered in choosing a particular system of funding. Over the years, a number of proposals have been made for such criteria (Alexander & McQuain, 1984; Hartman, 1981; 1992; Jones & Wilkerson, 1974). Some such criteria are discussed below. Clearly the criteria listed here correlate with one another to an extent, but they have some independent use in thinking about a total system.

Simplicity and Efficiency

Other things being equal, a simple system should be preferred over a complex one. This means that the record-keeping requirements should be brief, simple, and unambiguous. Dollars and time should be saved whenever possible. The present operating systems are not simple; they probably produce more information and less accountability than are desired.

Increases in federal rules and regulations have added complexity over recent decades. According to an Office of Education (OE) administrator, in a period of less than five years there was "more than a tenfold increase in the number of documents (OE) published annually in the *Federal Register* from 32 in fiscal year 1972 to an estimated 368 in fiscal year 1977" (Commission on Federal Paperwork, 1977). The increases in federal rules and monitoring procedures correlate with the increases in official publications and with complexity.

Fiduciary and Programmatic Trace

A good funding system obviously should provide records on the fiduciary side, to ensure that funds are managed honestly. But a good system also should provide programmatic traces; that is, it should demonstrate that money allocated for special programs actually reach the intended targets. Unfortunately, many systems fail to leave "programmatic traces"; for example, categorical money may go into the general funds of a school district and never find the way to the programs intended to be supported.

Developmental Tool

A good funding system should create effective tools for educators who are responsible for program development in the special field of concern. For example, when a special education administrator meets with the school board to convince the members that teams of teachers and support staff are needed in every school, he or she is in a more persuasive position if most of the costs would be covered by state and federal funds. This criterion suggests that a categorical funding system should create leverage at exactly the point where essential decisions are necessary to accomplish the purposes of the categorical system.

Cross-District Equity

A good system should help to equalize opportunities for the education of students, irrespective of the wealth or location of the school district. Categorical funding should be joined closely with approaches made in general school funding to achieve equity across districts in provisions for schooling. An "urban multiplier" feature may be needed to support the extraordinary level of intensive instruction required in inner-city situations (McCarthy & Sage, 1982).

Early, Preventive Programs

A serious deficiency of many special education programs is that they serve only children who are "full-blown casualties"; for example, children who were seriously emotionally disturbed or are learning disabled. In the case of the so-called learning-disabled pupil, it is a common part of the identification procedure to require a large discrepancy between "expectations" for the child's learning rate and the actual achievement rate. The unhappy fact is that such discrepancies increase only through neglect during early years of schooling. It is, indeed, unforgivable that educators must "wait around" for the growth of such discrepancies before intervening. Earlier efforts, oriented to prevention of "full-blown" casualties, are needed.

Flexibility

A good funding system has flexibility so that it can support a high degree of flexibility in local school systems' developments of specialized programs. There are many ways of organizing such programs. For example, in some communities the special teacher may not have a self-contained class, preferring to work in a support role (as a second teacher) in regular classes. Or, a special teacher might work in an itinerant role to serve selected students who are mainstreamed in many various programs of a vocational school. Other communities may not even have a vocational school. Flexibility is needed so that variations of many kinds can be implemented at the discretion of local educators and parents.

Avoids Labeling

A good funding system avoids requirements that children be labeled. A special teacher should be enabled to work with students who, for whatever reason, are not responding well to existing instructional programs, or who may be inattentive and troublesome. Usually there is little profit for anyone in categorizing such students as *mentally retarded, learning disabled,* or *emotionally disturbed.* The idea of labeling children as the first step in providing special education is bankrupt! The teacher's work must proceed on the basis of information about instructional problems and needs, not gross and often demeaning classifications. Many labels cause deep hurts in and resentments by children and their families. Yet, ironically, many of our funding and accountability systems still require classifications and labels. The development of new systems that do not depend on labeling will require efforts to persuade policymakers and advocates that such new systems will work.

Focus Leadership on Quality

Special education programs are managed by administrators. The funding system can waste their attention and energies on trivia or draw their concerns to topics that are vitally important to the quality of programs. For example, if a system requires fine distinctions in categorizing children and forces adherence to impermeable boundaries for each classification, an administrator inevitably is drawn into regulating the boundaries and developing related information systems. This focus can be quite useless and demoralizing. It may influence the quality of people who can be recruited to work as special education administrators.

In contrast, a system that centers attention on broad community plans and the deployment of well-qualified staff members who are given wide discretionary authority in developing services for children, may enhance the likelihood of securing high-quality administrator-leaders. Recently, a high rate of turnover has been discerned among state directors of special education programs. One can hypothesize that excessively burdensome procedural functions are one cause of the rapid burnout of such administrators.

Avoids "Bounty Hunt" Mentality

A good funding system does not create incentives for finding and labeling more and more children to be served in special education. The system should encourage the identification of all children who need intensive or special forms of education but, at the same time, it should also encourage and support preventive programs without financial disincentives for such programs. We need new approaches in funding that help school systems to develop and sustain programs that produce positive outcomes for children and thus reduce rosters of students considered to be exceptional.

In Accord with the Least Restrictive Environment Principle

Some funding systems "pay off" only when students are displaced from mainstream school programs for full- or part-time placement in separate educational programs. This practice is inconsistent with both legal requirements and the strongly held values of many parents, students, and educators. It is extremely important that categorical funds be available to help develop and sustain programs of intensive and special kinds when they are provided in regular schools and classes. Recent legal revisions in relation to Chapter 1 programs (for children of the poor) have opened opportunities for the support of schoolwide remedial programs, even those that would be joined to special education programs. In the views advanced here, this is a desirable trend. Even in the case of immigrant students and those who have language-related problems, it is argued increasingly that "the best way to help...is to strengthen the school system that serves them, not to create new categorical programs" (McDonnell & Hill, 1993, p. xiii).

Link Special and General School Funding

A good categorical funding system should link up easily with funding for regular education as well as with the full set of other categorical funding systems. Increasingly, the trend is to coordinate categorical programs with the development and improvement of school systems as a whole. This movement requires funds to be shared, just as other functions are. Such linkages also facilitate capping or limitations in special education funding, a matter of much concern in many places.

Present Funding Systems

Several kinds of funding systems are now in operation in the United States. Each can be examined against the criteria specified in the preceding section. No funding system is adequate if it is underfunded, of course. Federal funding has fallen far behind the promises first made in PL 94-142. At the time of this writing (March-April, 1995) there are many uncertainties about federal funding for special education and many related programs, such as those supported under Medicaid. We hear talk of the devolution of federal programs, suggesting that even more funding burdens may fall upon state and local agencies.

Five forms of special education funding are examined here (see Table 12.2). Each has possible variations. Each is distinguished on the basis of the *unit* that triggers dollar flow to local educational agencies.

1. *Per pupil* unit. Each disabled child is so identified and may be placed in a specialized program. Such placement qualifies the school district for categorical funds. Where such a system operates there is a massive, child-by-child micro-level triggering of dollar flow, depending on the number of children labeled and served. It rewards "child find" activities and special placements.

Table 12.2 Ratings of Five Funding Systems on Selected Criteria

Criteria for Categorized Funding Systems	Funding Systems				
	Per Pupil	Per Pupil with Index	Programmatic Units	Personnel in Approved LEA Plan	Flat Grant
Simplicity and efficiency	−	−	+	+ +	+ +
Provides both fiduciary and program traces	−	−	+	+	0
Centers leadership on quality	−	−	+	+ +	0
Serves as developmental tool	−	−	+	+	−
Cross-district equity	−	−	+	+	0
Supports early, preventive program	−	−	0	+	0
Flexibility	+	−	−	+	+ +
Avoids labeling	−	−	0	+	+ +
Avoids "bounty hunt" mentality	−	−	0	+	+ +
Is in accord with LRE principle	−	−	0	0	+
Links conveniently with funding for regular education	−	−	0	0	+

2. *Per pupil with index.* This system operates like the per pupil system except that the pupils consigned to the different categories qualify their districts for differing amounts of money. For example, an educable mentally retarded pupil might qualify his district for 2.2 times the number of dollars provided for nondisabled students whereas more severely retarded children may have an index of 2.51. (These figures were used in one state.) Usually the index is based on *average* per-pupil costs of instruction in each category, as revealed in cost studies. In view of research findings on the variance across districts on costs of narrowly defined categorical units, this approach to funding may be considered flawed and unlikely to prosper.

3. *Programmatic unit.* Here an element of the program becomes the unit of funding. For example, at one time the so-called "Master Plan" for California schools set a fixed dollar amount of state support for each resource room. The unit amount was sufficient to pay a teacher and to purchase necessary supplies (Brinegar, 1975). Sometimes this approach is one of those labeled "resource-based."

Another possibility might be to pay some dollar amount for each hour of services by, for example, psychologists, braille teachers, or physical therapists. Paying off on such programmatic units was one of the possible approaches recommended by Nicholas Hobbs (1975), following his in-depth study of special education programs.

4. *Personnel in approved plan unit.* The major and most stable cost element in all categorical school programs is the salaries for personnel. Usually, staff salaries make up about 80% of the direct costs of special school programs. Some states pay a proportion of the costs of personnel (e.g., Minnesota). This system is also described in some literature as "resource-based."

5. *Flat grants.* This system appears to be growing in justification and use. Pennsylvania recently adopted such a system. It is sometimes called "population-based" funding (Center for Special Education Finance, 1994). For example, a state legislature may decide that each school district should receive an "earmarked" amount equal to 15-20% of the general education supports. A requirement might be that the earmarked funds be used to pay for categorical programs.

Flat grants may be an intermediate step toward the purging of categorical funding and use of procedural requirements to ensure services of quality to *all* students. If flat grants are awarded on the basis of the number of students enrolled in special education, it could have a "bounty hunt" effect and also cause the rapid growth of relatively low-cost programs (e.g., speech correction, for which one specialist may be expected to serve as many as 50 students). It is, perhaps, better to base flat grants on total school populations and then, possibly, to make adjustments to achieve cross-district equity. A study by the Inspector General of the U.S. Department of Education (1994) recommends that federal supports for special education be changed to flat grants based on total school population, but with a modification to enhance funding for districts with many students from impoverished families.

Beginning the Evaluation of Funding Systems

A remaining challenge is to look at each of the five funding systems in relation to the criteria specified earlier. Table 12.2 summarizes the author's ratings of each system on each criterion. Ratings are made on a five-point scale ranging from double-negative (–) to negative (-), to neutral (0), to positive (+), and, finally, double positive (++). Table 12.2 is obviously complex to read; some of the ratings are undoubtedly arguable.

On virtually every criterion the *per pupil* and *per pupil with index* systems, as shown in Table 12.2, are rated negatively. The U.S. Department of Education uses a count of disabled students as the determiner of special education funding levels, a procedure that may have been justified in past years when "finding" children who needed special help was of high concern. But no longer. Because "the goal of universal access to education has met with such great success that it now lacks power as a national ideal" (Bakalis, 1987, p. 10), it is time to move on to other forms of categorical funding. We can expect increasing difficulties if we persist in using per-pupil systems that involve so many "negatives."

Resource-based systems have the advantage of giving leaders a strong developmental tool that can be used, for example, to foster preventive programs and to avoid "bounty hunt" procedures. In the case of flat grant procedures, if money is distributed to schools, each can then be required to pay for its categorical programs within the limits of the dollars awarded. A deduction from a particular school's allotment can be made if a student is sent to a center of any kind outside the home school. In this way, a strong incentive for providing programs within the home school can be created.

Some strong voices are advocating a different approach—one that would simply give vouchers to parents permitting them to choose the schools in which to enroll their children. It seems likely, however, that what is designed as parent choice could quickly be turned into administrative choice if schools are permitted to exclude children who present problems. A voucher system that supposedly encourages parent choices yet permits rejection decisions by schools seems likely to produce a kind of tribalized total school system. Like-minded parents could join together to create schools that serve only their children. Students with special needs would be likely to suffer in such a system. Obviously, advocates for disabled and at-risk children should be exceedingly cautious about placement schemes that could result in exclusionary decisions affecting the schooling of their children. Certainly the possible effects of vouchers should be carefully analyzed and debated when revisions in funding schemes are considered.

Reducing Costs

An assumption made here is that while the level of funding for special education and most other categorical programs may remain stable, it is quite possible that they will be reduced, at least in the near term. Special education funding is "on the table" for very close inspection and reform at all levels. The current political and economic climate is not favorable to increases in public funding for anything except, possibly, prisons. The national debt is enormous and the likely devolution of federal functions to states and local agencies is ominous. On the other hand, opportunities for revising funding systems at all levels are more open than usual; the implications for policy are very significant.

On the basis of the analysis offered in earlier sections of this chapter, some of the means by which efficiencies could be achieved in special education, in related support systems, and in other categorical programs are as follows:

1. *Press forward aggressively on early and preventive education programs.* Studies by Schweinhart and Weikart (1980) and a summary of research by Lazar and Darlington (1982) indicate, for example, that special education placements could be cut in half if strong preschool education were provided to at-risk children. Fortunately, progress is being made on this front. Similarly, there is good evidence that early, highly intensive work can help to prevent major problems in learning (Slavin, Madden, Karweit, Livermore, & Dolan, 1990; Wasik & Slavin, 1993). Widespread development of early education for all children is needed; the general provision of early education will make it possible to provide less segregated programs for disabled and at-risk children.

2. *Drop child labeling processes, especially for those considered to have mild disabilities.* The reference here is especially to students labeled as "educable mentally retarded" or "learning disabled" and to those who exhibit mild behavior problems. Programs for such students can be combined across current categories and can also encompass students now served in Chapter 1 programs (Brouillet, undated); all of these children can be absorbed in enriched regular education programs. The combination will involve considerable work to strengthen regular education programs in such matters as team teaching and cooperative grouping. Reorganization of large schools into minischools also looks very promising. Perhaps most important will be to provide intensive programs in the very earliest days of school enrollment, as noted above. A procedure that takes account of pupil learning rates, but without labels, has been described by Reynolds, Zetlin, and Wang (1993).

3. *Close special day schools which serve only disabled pupils and provide for the accommodation of disabled pupils in regular schools.* This move undoubtedly will require some clustering within selected regular schools of, for example, children who need braille instruction. This is not a proposal for

so-called "full inclusion," the enrollment of all students full-time in regular classes. A "full inclusion" policy is not justified for all severely and profoundly disabled pupils and represents self-defeating planning rigidity. The elimination of special day schools also would result in savings, mainly in transportation costs. At last count the Los Angeles public school district operated 18 separate schools for disabled students and the costs for transporting students to these schools are undoubtedly enormous; yet at the same time deep retrenchments have been made in that district's general school budget.

4. *Transform special programs (e.g., for students considered to have "mild disabilities") into intensive time-limited arrangements* that can be tested for effects and then continued or discontinued on evidence of progress. There has been a tendency for some special placements to be longstanding and for the modes of operation to focus on problem-minimizing operation, rather than on aggressive forms of instruction. Initial placements in special programs might be limited to a single semester for purposes of providing intensive help in specific academic or social skills, after which the pupils would be returned to regular programs. Placement in the special program could be continued if the students' learning progress is evident and promising but not sufficient to justify full return to regular education. The essential idea here is to remake most special education into extremely intensive and time-limited operations.

5. *Transform funding systems to include supports for successful programs.* This proposal assumes that evidence can be produced to show how programs prevent or overcome learning problems. The idea is to provide incentives rather than disincentives for the operation of successful special programs. Slavin and Madden (1991) have provided a procedure that is applicable to Chapter 1 programs. Heistad and Reynolds (in press) have suggested an approach that is applicable to a meld of several categorical programs. Fry (1990) has suggested a scheme for rewarding schools that operate successful "pre-referral" interventions; that is, programs that solve problems by the careful use of regular school personnel and procedures that result in fewer referrals to high-cost special programs.

6. *Work aggressively for broad coalitions of agencies to serve children and families.* Schools cannot hope to meet the complex needs of all children and families with only their own resources. They need outside expertise to deal with such problems as parental counseling, drug-addiction treatment, child neglect and abuse, juvenile delinquency, and much more. It would be a saving for schools and a gain for families if coalitions of agencies were formed to provide broadly coherent, effective and efficient services, both public and private, to families with special needs. Beginning efforts to this end are evident in many communities.

7. *Encourage the use of "waiver for performance" procedures* at both federal and state levels to permit programmatic experiments in services to special

populations in the schools. Such waivers should authorize the commingling of funds now allotted to the several categorical programs. We have ample reason now to combine LD and Chapter 1 programs and to incorporate them into very intensive regular education programs at the earliest levels; but the combination will require waivers of existing rules and regulations. Historically, the U.S. Department of Education has had less authority to grant waivers than any other cabinet department. This situation started to change with the legislation of 1994, however.

8. *In the case of "flat grants," allocate funds to individual schools and require each school to pay for the special education it uses.* This means, for example, that when educators send a child to an outside facility for education, the referring school would be responsible for covering the costs out of its flat grant funds. The purpose of this procedure would not be to neutralize cost factors in placement decisions, but to introduce a bias in favor of placements in regular schools. A modified version of this policy might be needed to provide for more central or districtwide funding of extremely special programs such as severely emotionally disturbed students.

9. *Revise IEPs (Individualized Educational Plans)* to emphasize mutual commitments by teachers and parents to improving the lives and learning of children. Current procedures overemphasize the commitments of the schools and underemphasize commitments by parents. When parents and educators do not agree on IEPs they should not take immediate adversarial stances. Each wants what is best for each child. Appeal should be to an ombudsman or a team of mediators, people who can help to make decisions on useful educational alterations in the child's situation. In any case, the resort to formal judicial procedures and the courts should be an absolute last resort; justices are experts in law, not education. These suggestions are directed to serving children better, maintaining as large a measure of trust among all parties as possible, and saving the high expense of formal due process.

An Integrated View

If programs follow dollars, then funding systems should be regarded as tools for leadership. In the early days of special education, funds were needed to aid in the identification of unserved and poorly served children and to support the creation of programs for them. In that context, using a pupil unit to trigger special education funds was appropriate.

A different kind of need for leadership became apparent in most schools in the 1980s. Virtually all children were then in school. The new situation called for a funding system focused on programs; thus, resource-based funding systems emerged. Increasingly, however, it was apparent that too much separation and the disjointedness of burgeoning categorical programs were becoming a major problem. Also, a movement for the general reform of schools and the possibility of bypassing the categorical programs gained momentum, most notably in the very complex and large special education programs.

Now, in the mid-1990s, a flat grant program—"earmarked" as a part of general support for the schools—is indicated for special education and most other categorical programs. Local school districts should be given flexibility in the use of allotted funds with, perhaps, the following few constraints: literally *all* children must be given opportunities for appropriate education; parents must be guaranteed a partnership role with educators in planning programs; and definite evidence of program outcomes must be obtained and reported.

The day is here for linking categorical funding systems to general school supports and for purging the entire system of the excesses of narrow funding streams, disjointedness, and "proceduralism." The recommendations offered here differ only in details from a far-seeing plan advanced by Nicholas Hobbs (1975) two decades ago: "Funds for the support of public school programs for children in need of special assistance should be provided on the basis of a flat percentage of the total appropriation (federal, state, and local) for all children" (p. 265). What we need today, given the threats of retrenchments and much school reorganization, are decisions that will create strengthened, efficient, and cohesive school systems to serve all students.

Conclusion

Many people take very seriously the categories of education that are applied to children, as if the categories "carve nature at its joints." School busses still pass one another on dusty country roads and busy urban streets, at high expense, to deposit children at various categorical stations. The more children so involved, the more money school districts receive from the federal government and from many states. Nevertheless most of this activity is of doubtful merit!

Clearly, these are turbulent times for the funding of public services, including the schools. We are in a period when funding for special and regular education may be seen as competitive, but increased collaboration and systemic reforms are desired by many educators across all levels of education. The extraordinary concern with the costs of special education at this time may cause pell-mell cutbacks; what we need, however, are carefully considered changes. It should not be surprising if some special educators and parents' advocacy groups go into a defensive mode and try to protect all special programs and funds in their present forms. Many people are well paid for maintaining the status quo. Political battles, including moral claims and counter claims, can be predicted. Our most important need, however, is *openness* to change and close attention to program efficiencies, but always with the awareness that the first priority of concern must be the life and learning of children.

Programs follow dollars. Funding systems should be designed to support timely leadership.

References

Ainscow, M. (Ed.). (1991). *Effective schools for all.* London: David Fulton.

Alexander, M.D., & McQuain, S. (1984). *An analysis of state special education finance formulas.* Blacksburg VA: Virginia Polytechnic Institute and State University.

Bakalis, M.J. (1987). Power and purpose in American education. *Phi Delta Kappan, 65*(1), 7–13.

Bartelt, D.W. (1994). The macroecology of educational outcomes. *CEIC Review, 3*(1), 2–3. (Published by the Center for Research on Human Development and Education, Temple University, Philadelphia)

Behrman, R.E. (1992 Spring). School linked services. *The Future of Children, 2*(1). Entire issue, published by the Center for the Future of Children, The David and Lucille Packard Foundation, Los Altos, CA.

Berger, J. (1991 April 30). Costly special classes serve many with minimal needs. *New York Times,* pp. A1; A12.

Beyers, D. (1995 January 27). Montgomery's special education called inefficient, costly. *Washington Post,* pp. A1; A16.

Brinegar, L. (1975, September 25). *Comparison of current and future special education support systems.* Address in Bloomington, MN.

Brouillet, F.B. (undated). *Coordinating programs for students with mild learning problems.* Olympia WA: Office of the Superintendent of Public Instruction.

California State Legislative Analyst's Office. (1993). *Reform of categorical education programs: Principles and recommendations.* Sacramento: Author.

Casanova, V., & Chavez, S. (1992). Sociopolitical influences on federal government funding of gifted and talented and bilingual education programs. *Educational Foundations, 6*(4), 45–73.

Center for Special Education Finance (1994). *The CSEF resource, 2*(1). (See also *1*(2), Fall 1993.)

Chaikind, S., Danielson, L.C., & Braven, M.L. (1993). What do we know about the costs of special education: A selected review. *Journal of Special Education, 26*(4), 344–370.

Commission on Federal Paperwork. (1977, April 18). *Report on education.* Washington DC: Author.

Copeland, W.C., & Iversen, I.A. (1981, June 20). *Refinancing and reorganizing human services* (entire issue). Human services: Monograph series. Washington, DC: Department of Health and Human Services.

Fry, D. (1990). *Prereferral funding: A model for promoting system level change.* Paper presented at Conference of National Association of School Psychologists in San Francisco, April 1990. Available ERIC ED 337 471.

Gardner, S. (1994). *Reform options for the intergovernmental funding system: Decategorization policy issues.* Washington, DC: The Finance Project (1341 G Street NW, Suite 820, Washington, DC 20005).

Hannaway, J. (1985). Administrative costs and administrative behavior associated with categorical program. *Educational Evaluation and Policy Analysis, 7*(1), 57–64.

Hartman, W.T. (1981). Estimating the costs of educating handicapped children: A resource-cost model approach—Summary report. *Educational Evaluation and Policy Analysis, 3*(4), 33–47.

Hartman, W.T. (1992). State funding models for special education. *Remedial and Special Education, 13*(6), 47–58.

Heistad, D.H., & Reynolds, M.C. (in press). 20/20 analysis: A city-wide computer-based application. *Education and Urban Society.*

Heller, K.A., Holtzman, W.H., & Messick, S. (Eds.). (1982). *Placing children in special education: A strategy for equity.* Washington, DC: National Academy Press.

Hobbs, N. (1975). *The futures of children.* San Francisco: Jossey Bass.

Hotakainen, & Smetanka, (1994 December 4). Average kids are losing. *Minneapolis Star and Tribune,* p. 1.

Inspector General, U.S. Department of Education. (1994). *ED can allocate special education funds more equitably.* San Francisco: Office of the Regional Inspector General for Audit, Region IX. *Audit control No. A0928255)*

Ireys, H.T., & Nelson, R.P. (1992). New federal policies for children with special health care needs: Implications for pediatricians. *Pediatrics, 80*(3), 321–327.

Jones, P.R., & Wilkerson, W.R. (1974). *Options for funding special education.* Paper presented at the National School Finance Conference, March 17–19, 1974, at Disney World, Florida.

Lazar, I., & Darlington, R. (1982). Lasting effects of early education. *Monographs of the Society for Research in Child Development, 47*(2–3), Serial No. 195.

Lewin/ICF, & Fox Health Policy Consultants. (1991). *Medicaid coverage of health-related services for children receiving special education: An examination of federal policies.* Washington, DC: U.S. Department of Health and Human Services.

McCarthy, E.F., & Sage, D.D. (1982). State special education fiscal policy: The quest for equity. *Exceptional Children, 48*(5), 414–419.

McDonnell, L.M., & Hill, P.T. (1993). *Newcomers in American schools.* Santa Monica CA: Rand.

Moore, M.T., Strang, E.W., Schwartz, M., & Braddock, M. (1988). *Patterns in special education delivery and cost.* Washington, DC: Decision Resources Corporation.

NASBE (National Association of State Boards of Education). (1992). *Winners all: A call for inclusive schools.* Alexandria VA: Author.

Education Student Affairs Office, College of Education, University of Minnesota. (1984). *Newsletter.* Minneapolis MN: Author.

Parrish, T.B. (1993). *CSEF* (Center for Special Education Finance *Brief,* (2), 1–4.

Reynolds, M.C., & Birch, J.W. (1988). *Adaptive mainstreaming.* New York: Longman.

Reynolds, M.C., Zetlin, A., & Wang, M.C. (1993). 20/20 analysis. *Exceptional Children, 59*(4), 294–300.

Rossmiller, R.A., & Moran, T.H. (1973). *Cost differentials and cost indices: The assessment of variations in educational program costs.* Gainesville, FL: National Educational Finance Project and the Institute for Educational Finance.

Schweinhart, L., & Weikart, D. (1980). *Young children grow up: The effects of the Perry Preschool Program on youth through age 15.* Ypsilanti, MI: High/Scope Educational Research Foundation.

Shepard, L., & Smith, M. (1981). *Evaluation of the identification of perceptual-communication disorders in Colorado.* Boulder, CO: University of Colorado.

Siegel, G.M. (1987). The limits of science in communication disorders. *Journal of Speech and Hearing Disorders, 52,* 306–312.

Skinner, B.F. (1938). *The behavior of organisms: An experimental analysis.* New York: Appleton-Century-Crofts.

Slavin, R.E., & Madden, N.A. (1991). Modifying Chapter 1 program improvement guidelines to reward appropriate practices. *Educational Evaluation and Policy Analysis, 13*(4), 369–379.

Slavin, R.E., Madden, N.A., Karweit, N.L., Livermore, B.J., & Dolan, L. (1990). Success for all: First-year outcomes of a comprehensive plan for reforming urban education. *American Educational Research Journal, 27,* 255–278.

Supplement to *Teaching Exceptional Children*. (1994). Reston, VA: The Council for Exceptional Children, 26(3), 3.

U.S. Departments of Agriculture and Housing and Urban Development. (1993). *Building communities: Together*. Washington, DC: Authors.

U.S. Department of Education. (1994). *Sixteenth annual report to Congress on the implementation of the Individuals with Disabilities Education Act*. Washington, DC: Author.

Wasik, B.A., & Slavin, R.E. (1993). Preventing early reading failure with one-to-one tutoring: A review of five programs. *Advances in Education Research*, 1(1). (Published by the Office of Educational Research and Improvement, U.S. Department of Education, Washington, DC).

Wood, R.C., Sheehan, R.J., & Adams, J.M. (1984). State special education funding formulas: Their relationship to regular education placement. *Planning and Changing*, 15(3), 131–143.

Ysseldyke, J., Algozzine, B., Richey, L., & Graden, J. (1982, Winter). Declaring students eligible for learning disability services: Why bother with the data? *Learning Disability Quarterly*, 5, 37–44.

13

Parents and Advocacy Systems: A Family Systems Approach

ELENA C. PELL
Program Planning and Development Consultant, USA

ELENA P. COHEN
Deputy Director, Family Impact Seminar, USA

This chapter will examine the dimensions of the family system and explore an ecology that places the family within a larger environment of informal and formal systems of service and supports for families with children with special needs. The first part of this chapter presents a conceptual framework with which to approach families with children with disabilities in their multiple contexts. The framework proposed conceptualizes the family as an environment within an environment. The second part of the chapter describes the multiple roles that families play in the various systems that form the larger ecology of family life, and how those systems might better serve families through recognizing the value of family-centered approaches. Finally, we present a set a principles to help move towards a more family-centered approach to policies and programs.

Introduction

Clearly, the parental role is pivotal in supporting as well as advocating on behalf of a child with special needs. In recent years, there has been a growing recognition that parent involvment with schools and other systems needs to build on the strengths of family systems (U.S. Department of Education, 1994). This chapter will examine the dimensions of the family as a primary support system in the universe of services and supports for children with special needs. The family system will also be explored within the context of a larger environment of informal and formal systems of service and supports for families with children with special needs. The examination of family as "an environment within an environment" is a notion adapted from United Nations analysis of the form and function of family life throughout the world

(United Nations, 1993). We believe this approach to understanding the "ecology" of families provides a useful framework to better understand how systems of support can enter into more effective partnerships with families to ensure the best outcomes for children with special needs. Hence, we will examine both principles and tools for working systemically to provide appropriate supports to families, and how parents and families play a vital role in adjusting and challenging the system to better meet the needs of the primary support systems for children with special needs.

The second part of the chapter describes the multiple roles that families play in the various systems that form the larger ecology of family life, and how those systems might better serve families through maximizing the utilization of family-centered approaches. In the sense in which we are using it, the term *family centered* refers to "a combination of beliefs and practices that define particular ways of working with families that are consumer driven and competency enhancing" (Dunst, Johanson, Trivette, & Hamby, 1991, p. 115). Finally, we present a set a principles to help guide the move of practice and programs towards a more family-centered approach.

The concept of working comprehensively with whole families has received increased attention by practitioners, poverty experts, and policymakers (Schorr, 1989; Weiss & Halpern, 1990). The ecology of the family-centered approach understands the role of the family as an environment within an environment, the necessity for a continuum of interventions that emphasizes the vital role of family in the support of children with special needs, and how vital efforts to strengthen family stability and well-being and the family's capacity to fulfill its functions need to be grounded in a clear understanding of the role of family strengths and resources in meeting the needs of children. The work developing family approaches has resulted in a series of principles indicating that programs should be comprehensive and integrated to create a family-centered system of care (Staton, Ooms, & Owen, 1991).

In early intervention and special education programs for young children with special needs, the family-friendly beliefs that began to be espoused by practitioners in the late 1960s and early 1970s included respect for family members, recognition that the family is the major constant in a child's life, and an understanding that children's needs cannot be fully met unless the family joins with the program in meeting those needs. During the late 1980s, there was a major "reconceptualization" of the family's role in the delivery of services for children with disabilities. Spurring this change was the Reauthorization of the Education for All Handicapped Children Act (Public Law 99-457), the 1986 federal law that created the Part H program and greatly expanded the preschool Part B program. A new view of family involvement began to evolve, a view that emphasized family involvement not so much in assisting with service delivery as in determining priorities and in deciding what services would be delivered and how they would be provided.

Families moved, in this reconceptualization, from positions of subservience to professionals to positions of decision makers and full partners in planning. However, in 1995, the vast majority of states are still struggling with strengthening the capacity of local programs to provide family-centered intervention and support services, especially in rural and poorly accessible areas, and flexible delivery options that meet the multiple needs of children and families (Federal Interagency Council, 1995).

Over time, these and other efforts have succeeded in raising consciousness about the different roles of families in designing, implementing, evaluating, and advocating children's services. But there has not been a clearly defined framework to assess the future effects of public or private policies, programs, or components on the various contexts in which families are simultaneously included. (Policies and programs include decisions made and implemented at the federal, state, and local levels and in the executive, legislative, and judicial branches of government. Policy or program components are enabling legislation, regulations court interpretations, funding incentives and disincentives, administrative practices and procedures, staffing, and details of service delivery such as location and hours of operation [Ooms & Preister, 1988].)

Families with children with special needs are multidimensional and have distinct, interdependent domains of functioning that intertwine with each other. The domains in which families function are influenced by the family's structure, social and economic resources, and stability and the community resources that form the context within which every family lives.

Unfortunately, policies and programs that do not use a family-centered approach sometimes place differential and selective emphasis on one or more domains of family functioning. Often, programs address first—and with greatest intensity of effort—those needs that appear to policymakers, providers of services, and/or families to be the most pressing. The reasons that policies and programs emphasize one domain over another include the professional orientation and expertise of the individuals designing and implementing the program, the available resources, the philosophical orientation of a given program, and the perceptions regarding families' or children's immediate and pressing survival needs. Often, the implicit assumption, or hope, is that once the most urgent issues are addressed, the other less pressing but important aspects will be solved (Hanft, 1989). While taking one domain or part of the family ecology out of its systemic context may solve problems, a lack of understanding of the dynamic, changing reality can lead to unintended and unwelcome outcomes for the family, including a sense of powerlessness and lessened ability to identify and tap resources.

Part of the basis for our approach to working with families lies in the fundamental structure and logic of the overall family support system. "The United States lacks a normative system of supports for families and their children.... [Its system] is based on the development of specific reactive services to remediate problems after they occur, not on...prevention with remediation as a back-up" (Stephens, Leiderman, Wolf, & McCarthy, 1994, p. 7). The system of services and supports for families is, for the most part, a fragmented patchwork of policies and programs that disrupts as much as it shores up the family since families of children who are disabled, ill, or at risk spend extraordinary time and effort trying to negotiate multiple disconnected systems of services and support. There is a growing recognition of the need to define new ways to configure a system of services and supports for families.

Indeed, these systems change all the time. "Systems change is continual and inevitable—systems evolve in response to funding priorities and streams, demographic shifts, the emergence of new technologies, and political exigencies" (Stephens et al., 1994, p. 9). The question remaining is, Where are families actually placed within this change?

Viewing the family as a multidimensional ecology—an environment within an environment–also puts the notion of change in a very different context. "Families have become the system by which social change is internalized, adapted, and then activated in our society" (Graves & Gargiulo, 1993, p. 47). Viewing the family as an environment within an environment will enable us to better understand how this process does–and ideally should–take place.

The Family as an Environment

In its broadest sense, a family entails a range of psychological and social dimensions. Families are both psychological as well as socioeconomic units. Every individual has a family of origin, and most people live within a family context for their whole lives.

Thinking of the family as an environment helps in understanding the many complexities of the interplay among culture, form, developmental stage, and economics in families with children with disabilities. Simultaneously, considering the family as a social unit with definite functions and responsibilities within an environment helps to recognize that the family is embedded in a larger social, cultural, economic, and physical environment upon which it depends to meet the needs of the child with disabilities as well as those of the other family members.

Children with disabilities affect and are in turn affected by their families. From a family systems perspective, there is a reciprocal relationship between children with special needs and their families. While it is true that the severity of a child's disability can contribute to the experiences of the family, this fact is mediated by many other considerations including the characteristics and resources that the family has to meet its needs and the needs of the children with disabilities.

Family Resources

Family resources are the physical, economic, social, and spiritual assets that support families in their daily lives. In the case of families with children with disabilities, resources include everything that increases their ability to meet their needs and the needs of their children across all life areas. The most critical resource for any family relates to financial stability. The lack of economic resources impedes a family's ability to meet the needs of its members in basic life areas such as shelter, nutrition, physical and mental health, education, employment, and earned income.

Changes in the American economy and in American society have made it increasingly difficult for families to provide for the basic physical needs of their members. As a result of these changes, a growing number of families are unable to supply their children with the necessities, or are only able to do so with the help of public assistance (Zill & Nord, 1994). As of 1991, 14.3 million children under the age of 18—nearly 22% of all children—were living below the official poverty line. By contrast, 11% of working-age adults and 12% of senior citizens 65 years and over were living in poverty (Zill, 1993).

Family Structure and Family Forms

In recent years, great concern has been voiced about the American family's ability to perform its functions. This concern centers in large part on troubling changes in family behavior patterns in the United States, notably the increased prevalence of divorce, unmarried child rearing, declining marriage rates, the abandonment of offspring by many unmarried or remarried fathers, and the devastating effect that cocaine and alcohol abuse and violence have wreaked on family life in the inner cities (Zill, 1993).

Not all of the changes in family behavior have been negative ones. For example, today's parents are staying in school longer and limiting the number of children they produce. Thus, although more children are living with single mothers, those mothers are more likely to be high school graduates. And today's children have fewer siblings with which to compete for parental attention. Other family changes that people see as positive include the movement toward greater equality between husbands and wives in family decision making and the increased willingness of men to take on the burdens of child rearing, housework, and child care (Zill & Nord, 1994).

Because there are so many different types, it is difficult to precisely define the family. It has been defined as "any group of persons which cares for a child" and as "a group of persons...regarded as a family under the legislation and practice of a State" (United Nations Committee on Civil and Political Rights, 1991; Barber, Turnbull, Behr, & Kerns, 1989; United Nations, 1992). However, these definitions, like many similar ones, do not include all families. For example, some families never have children; others only rarely live together over their whole life cycle: children move out, spouses may separate, and so forth. A person can live in one household but have some of his/her family members living in another household (Tseng & Hsu, 1991). The following are some of the most common structures and forms of families in the United States.

Nuclear Families. Nuclear families are composed of the "basic" family members: parents and their (unmarried) children. This is the most frequently observed form of a household across the United States. Nuclear families may take two forms: one biological and the other social. The biological element consists of the two parents and a child. The two adults may or may not live together, but the child's biological nuclear family is defined by their union. The social nuclear family is not necessarily the same and is created when two people establish a relationship, by marriage or by cohabitation. Even if the children move out, the "empty nest" is still considered a social nuclear family. While a nuclear family offers a number of possible advantages in some societies, the limited number of family members makes it quite vulnerable to internal and external pressures such as those confronted by families with children with special needs (Barber et al., 1989; United Nations, 1992).

The biological nuclear family has additional variations. The most common variation is the one-parent family, which consists of a mother or a father and one or several children. Single-parent families are formed through the death of one spouse, divorce, migration separation, or a decision by the partners not to live together. Most one-parent families are female-headed, but the number of male-headed nuclear families appears to be rising (National Commission on Children, 1993; Family Impact Seminar, 1995).

Whether or not the single-parent household becomes a personal or social disaster can depend on the availability of sufficient material resources and supportive social networks, as well as on the culturally structured attitudes toward it. However, in general, the social conditions of genuine single-parent families are generally not as good as those of two-parent families. Single-parent families with children are six times more likely to be poor than married-couple families with children; single parents often bear the sole responsibility for child-rearing, and they usually experience more social isolation (Zill and Nord, 1994).

Extended Families. Extended family refers to a household composed of a nuclear family (parents and their unmarried children) plus other subgroups, for example, married siblings (and their children, if any, and the parent(s) of either the husband or wife). An extended family does not need to be large—a grandparent and grandchild living together is an extended family based arrangement. However, extended family usually refers to numerous people either living together or otherwise having frequent, intimate interaction.

One type of extended family is the kinship family, which usually consists of even larger units than a three-generation family. In addition to three generations, other relatives may belong to the same household and be regarded as members of the family. Closely related, tribal families are usually built on a social rather than a biological basis (United Nations, 1992).

The extended family is adaptive under certain kinds of economic and social conditions. The main economic advantages of this type of family are that the extended family can provide a larger number of workers than the nuclear family. In the case of children with special needs, there is also the value of companionship and support, since daily activities are carried out jointly by members of the family working together (Barber et al., 1989).

Reorganized Families. There are several forms of remarriage, or living together under common law, after one or several earlier relationships. Spouses may have a child or children who may or not live with them. Children can have several half-sisters and half-brothers, half-grandparents, and other nonbiological relatives.

Associated with the increase of divorce and remarriage, the number of stepfamilies is rising remarkably in our society (Visher & Visher, 1982). Stepfamilies differ structurally from biological families in several ways. Stepfamily experiences can include that of important losses, past family histories, parent-child bonds that predate the new couple relationship, the existence of a biological parent who is geographically removed, and children who are members of two households in which no legal relationship exists between stepparent and stepchild.

Family Functions

Because the family is the primary institution for reproducing the species and rearing the next generation of citizens, society relies on families to perform functions crucial to the survival and development of children, especially very young children. Family functions include the establishment of emotional bonds, sustaining caring relationships, providing economic support, socialization and education of children, protection of vulnerable members, and providing recreation and play. Everybody pays a price when families fail to perform or perform badly the functions society expects them to carry out. From an economic point of view, public funds then have to subsidize services that family members cannot or will not provide.

Family Interaction

What differentiates families from other social groups are the emotional, sociocultural, and legal relationships and responsibilities between the various members: spouses, parents, children, siblings, and relatives. In speaking of a family there is usually reference to the existence of one or several of these relationships, which have also been defined as subsystems of the family.

Marital Subsystem. In general couples find it problematic to maintain a healthy relationship while raising a child with special needs. "Stress affects the characteristics of family relationships. In general, the greater the stress, the more intense the relationships; the more intense the relationships, the stronger [is the] tendency toward emotional fusion and the more emotional influence among family members" (Beal & Chertkov, 1992). Marital conflicts are more prevalent in families having members with disabilities than in those whose members do not have a disability (Gaith, 1977). The burden of care and other responsibilities imposed by the child can rob one or both parents of the time and energy to maintain a healthy relationship.

Parental Subsystem. The parental subsystem (generally the mother/child dyad) has received the most attention from professionals concerned with early education. Less attention has been paid to father/child interactions, thus ignoring the father's role in the child's development, the effects of the child's disabilities on the father, and the ultimate impact these relationships will have on others within the family, although this seems to be changing (Family Impact Seminar, 1995). However, in general, mothers and fathers of children with disabilities may feel inferior in their parental roles.

Sibling Subsystem. As with parents, brothers and sisters must learn to make adjustments both initially and over the span of a lifetime because of the special needs of their sibling. They must learn to accept that a disproportionate amount of their parents' time and energies and of the total family resources must be deflected to meet the extra demands imposed by the child (Barber et al., 1989).

Extrafamilial Subsystem. Many families have to rely on their extended families for emotional and psychological support, information, or assistance with their children with special needs. Sometimes families pull away when they find that others are unable to understand their problems. Other times friends and relatives may avoid interaction with the family, due in part to their discomfort with the disabled person (Barber et al., 1989).

Family Life Cycle

The family life cycle can be conceptualized as a series of developmental stages representing periods in which the needs of the family are generally similar and the services and programs are both similar and stable. Several life-cycle models (Carter & McGoldrick, 1980; Terkelsen, 1980) and, recently, life-course models such as new family forms, human diversity, and environmental diversity have been proposed (Germain, 1994; McGoldrick & Carter, 1993; Tseng & Hsu, 1991; United Nations, 1994). Basically these models describe the family life cycle in the stages related to formation, extension, completed extension, contraction, completed contraction, and dissolution. Each of these stages has special implications for every member of the family when any one member has a disability.

Typically the stages are defined in terms of the ages of the children, particularly of the oldest child. In order to be useful, the family life-cycle model must be adapted and modified to reflect the variations both within the model and between different cultural and social groups. Depending on the population and the purposes for which the model is used, the number of stages can be either reduced or expanded. For example, additional stages, or substages, can be inserted. Although presumably families move through the life-cycle stages from beginning to end, various alterations can occur.

Table 13.1 is a general classification of nuclear family life-cycle stages in the last quarter of the twentieth century. It highlights the key transitional issues that are confronted during each stage. It is important to note that these are not universal breakdowns. Families can assume many varied configurations over their life spans, and it is increasingly difficult to determine which family life-cycle patterns are "normal," causing great stress for family members who have few consensually validated models to guide the passages they must negotiate (McGoldrick & Carter, 1993).

Table 13.1 The Stages of the Family Life Cycle[1]

Family Life Stage	Stressors and Emotional Tasks
Marriage establishment	Socially recognize rituals do not necessarily phase coincide with psychological needs of spouses and families involved. Tasks: (1) how to maintain and withdraw ties to families of origin in a culturally appropriate way.(2) Commitment to new system.
Childbearing phase	Stress is especially prevalent if there is strong cultural meaning attached to childbearing and if there are specific social expectations associated with bearing a child. TASK: Accepting new members into the system.
Child growing phase	Different philosophies about how to raise the child. Triangulation with the child. TASK: Proper and successful rearing and disciplining of the child.
Child leaving phase.	Different views on the issue of a child leaving the parents' home. Intense attachment of one of the parents to the children (or a particular child) so separation becomes troublesome. TASK: Launching children and moving on.
The empty nest phase	Spouses feel lost in their new life, stress from separation from work-related networks. TASKS: Establish a new, close relationship. Accepting multitude of exits and entries into the family system.
Widowing phase	Depression at the loss of lifelong partners, frustration as result of aging difficulties.

[1]Adapted from Carter & McGoldrick, 1993; Tseng and Hsu, 1991.

Both individual and family development have in common the notion of phases associated with developmental tasks. Phases are marked by the alternation of periods of *horizontal* (stable) and *vertical* (developmental transition) periods. During the horizontal periods, structures are developed and enriched based on the seeds planted in preceding transition periods. Transition periods are potentially the most vulnerable because previous individual and family structures and rules are reappraised in the face of new developmental tasks that may require discontinuous change rather than minor alterations.

The notion of horizontal and vertical periods is useful in linking the needs of the child with disabilities with those of development of the family. In general, the birth or onset of a disability exerts enormous pressure on the family system. Analogous to the addition of a new family member, the occurrence of disability sets in a restructuring process of socialization to the disability. Symptoms, loss, and the demands of shifting or acquiring new disability-related roles all require a family to refocus inward. This pull of the disability causes different normative strains on both the family and the family members' stages of development.

The tendency for a disability to pull a family increases with the level of incapacitation or risk of progression. The ongoing addition of new demands as a disability progresses keeps a family's energy focused inward, often impeding or stopping the natural life-cycle evolution of other members. After a certain stability has been reached, relapse periods may increase family stress. Also, the on-call state of preparedness dictated by many disabilities keeps part of the family focused on the disability, despite asymptomatic periods, hindering the natural flow between phases of the family life cycle.

When the onset of the disability coincides with a transitional period in family life (e.g., early child rearing), it can foster a prolongation of this period. Transition periods are characterized by upheaval, rethinking of prior commitments, and anxiety about change. As a result, those times hold a greater risk for the disability to become unnecessarily embedded or inappropriately ignored in planning of the next developmental phase. For instance, a disabled infant can be a serious roadblock to parents' preconceived ideas about mastery of child rearing. Also, family members frequently do not adapt equally to a disability. Each member's ability to adapt and the rate at which he or she does so is related to the individual's own developmental stage and role in the family.

Ethnicity and the Family

Most discussions of culture and families focus on ethnicity as an important determinant of family behavior. Ethnicity, race, and religion strongly influence family beliefs (McGoldrick, Pearce, & Giordano, 1982). Significant ethnic differences regarding health beliefs typically emerge at the time of a major health crisis, including a disability. However, Falicov (1988) offers a more comprehensive definition of culture as a "set of shared world views and adaptive behaviors derived from simultaneous membership in a variety of contexts, such as an ecological setting (rural, urban, suburban), religious background, nationality and ethnicity, social class, gender-related experiences, minority status, occupation, political learnings, migratory patterns, stage of acculturation, or values derived from belonging to the same generation, partaking of a single historical moment, or particular ideologies" (p. 336). Cultural differences result from simultaneous contextual inclusion (participation and identification) in different types of groups. Since families partake of, and combine, features of the many contexts to which they belong, there is a need to differentiate among universal, transcultural, culture-specific, and idiosyncratic family behaviors.

A family's belief about what is normal or abnormal and the importance members place on conformity and excellence in relation to the average family have far-reaching implications in the context of an adversity such as a child with disabilities. Family values that allow for "problems" without self-denigration have a distinct advantage in utilizing outside help and maintaining a positive identity in the face of a disability or chronic illness. If seeking help is defined as weak and shameful, the process undercuts this kind of resilience. In the case of disabilities where the use of professionals and outside resources is necessary, a belief that pathologizes this normative process adds insult to injury.

Families with strong beliefs in high achievement and perfectionism are prone to apply standards in a situation of disability where the control they are accustomed to is impossible. In addition, there are additional pressures to keep up with normative, socially expected developmental milestones of age-peers or other young couples. The fact that life-cycle goals may take longer or need revision requires a flexible belief about what is normal and healthy. To effectively sustain hope, particularly in situations of long-term disability, demands an ability to embrace a flexible definition of normality.

Successful coping and adaptation are enhanced when a family believes in a biopsychosocial frame for disability. This highlights the importance of the initial "framing event" and whether professionals actively normalize the biopsychosocial interplay, thereby helping to undercut pathologizing family and cultural beliefs. Rather than a shameful liability, a family can approach a biopsychosocial interplay as an opportunity to make a difference and increase its sense of control.

A family's beliefs about the cause of a disability need to be assessed separately from its beliefs about what can affect the outcome. The context within which disability occurs is a very powerful organizer and mirror of a family's belief system. If every family member is asked for his/her explanation about what caused the disability, responses will reflect a combination of medical information and family mythology. Negative myths may include punishment for prior misdeeds, blaming a particular family member, a sense of injustice, genetics, negligence by parents, or simply bad luck. Usually causal attributions that evoke blame, shame, or guilt are particularly important to uncover. Such beliefs make it extremely difficult for a family to establish functional coping and adaptation to an illness. Decisions about treatment then become compounded and filled with tension.

Family Sense of Mastery over Disability

It is critical to determine how a family defines mastery or control in general and how they transpose that belief to situations of disability. Mastery is similar to the concept of health locus of control, which can be defined as the belief about influence over the course or outcome of a disability. An internal locus of control orientation entails a belief that an individual/family can affect the outcome of a situation. An external orientation entails a belief that outcomes are not contingent upon the individual's or family's behavior. Families that view disability in terms of chance believe that when illness occurs it's a matter of luck and that fate determines recovery. Those who see health control as in the hands of powerful others view health professionals, God, or sometimes "powerful" family members as exerting control over their bodies and disability course.

A family's belief about mastery strongly affects the nature of its relationship to a disability or illness and to the service system. Beliefs about control predict certain health behaviors, particularly treatment compliance, and suggest the family's preferences about participation in its family member's treatment and healing process. Families that view a disability's course or outcome as a matter of chance tend to establish marginal relationships with providers largely because their belief system minimizes the importance of their own or the professional's impact on the process.

Many factors influence the extent to which families define a mastery of the larger systems in which they interact. Among the factors that influence a family's sense of mastery are time since immigration, the impact of generational acculturation conflict, place of residence, socioeconomic status, education, upward mobility, political and religious ties to an ethnic group, and the languages spoken by family members.

The Family Within an Environment

A Multidimensional Ecology of Families

Families provide a place for membership and identity; economic support and welfare; socialization and learning; mental and physical health and well-being; protection, caring, and affection; spirituality; recreation; cultural socialization; and a place where social responsibility, norms, rules, and appropriate behavior are learned and reinforced. Yet the family is embedded in multiple systems of support that are vital to a child's socialization, health, and well-being.

It is well documented that extrafamily social support and resources form an important source of aid for meeting families' needs (Dunst, Trivette, & Deal, 1988. Also see Cohen & Syme, 1985; Fisher, Nadler, & Whitcher-Alagna, 1983 cited in Dunst et al 1988). A multidimensional view of family support places the family in the ecological context of systems that make up "daily life." All children and families grow up in the midst of a "web of institutions" that includes family, neighborhood, school, social and health agencies that serve children, local government, and private employers (Davies, 1988). Both formal and informal support systems play vital roles.

The extended family, consisting of grandparents, aunts and uncles, cousins, and others who participate in caring and resource roles, is a participant in and is reinforced by the neighborhood, social setting, church, and various other institutions that make up the primary systems of informal support with which families have the greatest direct contact. These primary supports are surrounded by a larger system of formal community support services, including schools and institutions devoted to health, mental health, safety, and civic life, as well as local government and other entities that exist within communities. These systems are embedded within a larger system of state and national policy and programs and large bureaucratic institutions.

These systems come together to form a larger multidimensional ecology in which the family system is the center (see Figure 13.1). To varying degrees, and in a variety of ways, the child and family exist within and interact with all these informal and formal support systems and their various components. Although much of the literature on family support focuses on services and supports to families, in practice it is very much an interactive relationship.

The Changing Environment

As Graves and Gargiulo (1993) point out, the enormous changes that have taken place in our society are experienced most dramatically in the family system. Numerous factors, including changes in the structure of the family, the entry of more mothers into the work force, poverty, race, and ethnic and language differences between community and school members, have contributed to difficulties in managing the multiple extrafamilial system roles required to meet obligations to children.

The context within which the family and other primary support systems is functioning has undergone dramatic changes. Changes in family composition, expected roles, patterns of work, and other circumstances of families have all

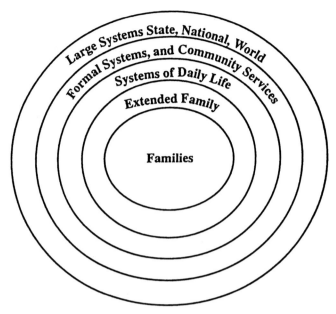

Figure 13.1 An Ecology of Nested Family Support Systems.

contributed to this phenomenon. All families, both urban and rural, are often faced with lack of time, energy, and money; inadequate housing and schools; lack of community support; difficult family relations; innumerable social problems; and barriers related to race, class, culture, and language (Liontos, 1992). The increasingly complex problems and challenges most modern families experience present many hurdles in the upbringing of their children. All families increasingly need to develop or access new and greater networks of support to deal with social stresses that interfere with their ability to function on behalf of their children.

Poor families generally face an even greater challenge due to lack of services, amenities, barriers to accessing services, and minimal purchasing power to buy supports on the market, such as tutoring, high-quality day care, and transportation. In distressed communities the combination of negative influences on a child's learning and development can ricochet through the system in ways that do damage to key supports in children's lives. For instance, a fear of violence in many neighborhoods diminishes use of public spaces such as parks and other neighborhood gathering places.

One can easily surmise that holes and gaps in the web of institutions that support children have a complex impact not only on the availability of supports, but also on the roles that families play within those systems. Extended families, neighborhood institutions, and schools undergo tremendous institutional stress in their attempts to meet the multiple needs of poor and at-risk children.

Roles of Families in Systems

Families potentially intersect with systems in a variety of ways. As Melson (1980) suggests, "Each family appears in a special relationship to its environment...[even though] the life of the family often seems to be organically part of its setting, to have grown up out of its nurturing soil.... [The] family exercises all the power and creativity in its grasp to fashion an environment that will reflect its aspirations, values, and means" (p. 1). Parent roles can include client, communicator, advisor, provider, co-designer, evaluator, advocate, and change agent. The taking up of these roles can vary with the skills, capacity, and desires of parents and other family members; the resource base of a family; the stages of a family life cycle; and the willingness or capacity of a system to allow and encourage parents to take on these roles.

Since the 1970s, there has been a growing recognition of the important role of parent involvement in systems of education, service, and support. With the passage of PL 94-142, the Education for All Handicapped Children Act (reauthorized as the Individuals with Disabilities Education Act) the federal government codified the essential role of families of children with disabilities in their children's education process.

A primary rationale for parent involvement has historically been based on a belief that the family members who spend the most time with young children are uniquely positioned to have the greatest potential impact on the children's development (Shriver & Kramer, 1993). Yet family policy and modes of family intervention vary greatly. Using a synthesis of classification schemes Dunst et al (1988) arrive at four broad classes of family-oriented prevention/early intervention programs:

1. Professional-centered models in which the locus of expertise and problem definition rest solely with the professional. Interventions are generally based on the notion of family deficits or pathology.
2. Family-allied models in which families are enlisted to assist professionals in implementing interventions that have been defined as necessary. Although families are viewed as having some role in meeting needs, they are seen as necessitating frequent and intense interventions from professionals.
3. Family-focused models in which families and professionals collaborate to define the needs and interventions for the family. Families are viewed as having the capability to meet needs and have an impact on their own lives, but require the guidance and advice of professionals.
4. Family-centered models in which the families' needs and wants determine service delivery and resource provision. Professionals become the agents and instruments of families as they work with them to identify resources and support that will further strengthen the capacity of families to build their informal and formal resource networks. Interventions are nearly always strength- and competency-based.

There are many systemic barriers that can limit families' ability to work in tandem with professionals. Systems of support that are based in professional-centered or family-allied models are often difficult and time-consuming for families to access and often unintentionally impose conflicting agendas on family members. Even when attempting to use family-focused and family-centered approaches, environmental factors within the family system can make the link between professionals and families even more difficult to establish. Numerous studies of the past 10 years have shown that parent participation is often complicated by a combination of the following: a lack of awareness of rights, services, and procedures associated with special education; isolation and stressful life circumstances; logistical barriers to accessing services; low self-confidence; and culturally based unqualified trust in authority (Harry, 1992. See also Lynch & Stein, 1987; Smith & Ryan, 1987 cited in Harry 1992). Differences can develop into obstacles and barriers when formal systems of service and support do not maximize opportunities to use family-centered approaches.

Families of children with special needs can face even greater challenges, for a family system can be greatly affected by having a member who is a child with disabilities. As Graves and Gargiulo (1993) argue, given the enormous stress often experienced by many modern families, sustaining viable support systems for families is even more difficult when coupled with the challenges of raising special-needs children. In addition to its profound impact on family environment, having a child with special needs also has an impact on how the family relates to all the systems that support the child's growth and development. Having a child with special needs requires the family to intersect with informal and formal systems in many unique ways, often necessitating greater amounts of individual involvement and advocacy on behalf of the child, and frequently requiring access to more, and broader, arrays of services.

Both the modes of intervention and the nature and extensiveness of the roles that families play are subjects of ongoing analysis and debate in research and practice arenas. Most children's and families' systems have to some degree accepted the notion that the observations, insights, and ideas of parents can be powerful tools for enhancing the quality of services (Hanson & Vosler-Hunter, 1991). Nevertheless, the degree and nature of both the recognition and practice of involving families in multiple roles vary across formal systems of service and support for children and families. For instance, although strongly embraced as a value, involving parents at the policy and planning level is a relatively new component of traditional child-welfare services (Family Impact Seminar, 1994). Since the 1980s there has been growing interest in the role of organized advocacy, with the idea that parent advocacy to improve education and services contributes new skills and perspectives that ultimately can position parents to work more effectively with educators, schools, and child-serving agencies to improve those systems.

The mandate for parental participation in the special education process is one of the hallmarks of the Individuals with Disabilities Education Act of 1991. However, rapidly changing demographics of school populations, as well as the fact that children from minority groups still tend to be overrepresented in special education programs (Artiles & Trent, 1994), means that school personnel are continually faced with the need to ask whether the communication patterns designed for mainstream parents will be adequate to the task at hand (Harry, 1992). Harry (1992) identifies examples of how the power of the accumulated knowledge of a group of professionals, manipulation arising from differential knowledge and authority between parents and professionals, and the parents' sense of risk associated with challenging individuals providing needed services who are often quite kind in their presentation all contribute to the parents subordinate position in what is essentially a power dynamic.

In a special education system that includes a disproportionate number of minority and poor children, parents often regard the system of education with some degree of distrust and fear. As Winters (1993) so aptly observed, insufficient time is often given to the process of engaging parents, which can be lengthy and absorb what appears to be an inordinate amount of scarce professional time. Most of these parents cannot process the data and concentrate on the extent and range of implications contained in the schools' findings. As a result, the focus of the exchange is lost, diminishing expectation of positive outcomes. Winters states that although procedural changes at the local level could modify factors that exacerbate the distance between parents and the school, wider systemic changes are required to alter the "outcome of desperation and powerlessness these parents experience" (p. 47).

A systemic understanding of partnerships takes into account these and other forces that come into play as families attempt to fulfill their roles and responsibilities to their children. In prevention and early intervention efforts there has been a movement toward the family-centered model of intervention and practice, based partly on the growing body of evidence indicating that interventions that are family centered are more likely to have broad-based positive influences on a number of aspects of child, parent, and family functioning (Dunst et al., 1991. Also see Kagan, Powell, Weissbord, & Zigler, 1987; Powell, 1988; Weiss & Jacobs, 1988; Weissbord & Kagan, 1989 cited in Dunst et al 1991). Many of the most effective parent-school partnership approaches involve parents in multiple roles as supporters of children's education at home and in school, as partners with staff, as advocates, and as decision makers. Effective programs also entail the utilization of multiple strategies on the part of schools to build on the strengths of family systems (U.S. Department of Education, 1994).

An ecological systems approach is informed by an understanding of the inherent symbiosis between formal and informal systems that have an impact on the child. Research has confirmed that, regardless of the population, informal support from the primary personal support networks of families is not only a powerful stress-buffering and health-promoting influence on families, but also bears more strongly on outcomes than many formal system interventions (Dunst et al., 1988). A systemic approach also recognizes the implications of viewing families as environments within environments. Both principle and practice point to the adoption of approaches that maximize the ability of families to act as consumers and definers of resource supports in the larger environment, while at the same time work in partnership with families to take into account the family environment itself. A systemic understanding of working with families does not lay blame on parents, schools, or other systems of support, but takes into account the variety of forces that are at play in families and that often create additional stress in families with children with special needs. Policymakers need to understand the extraordinary complexity of managing the family environment within a larger environment in which advocacy is required to sustain a viable system of supports for the child with special needs. "Parents involved in systems change...are asked to play both [public and private] roles simultaneously. They are asked to be their public selves, following the usual expectations for people who sit on boards and participate in meetings, while also drawing upon the deepest emotional experiences of their private lives and self" (Gilkerson, 1994, p. 23). Practitioners and educators also need to listen to parents' own assessment of the breadth and depth of the roles they can play in larger systems. In recent years, there has been growing evidence that placing parent involvement in a holistic context that responds to the changing, fluid nature of families; recognizing diversity within family structures; and attempting to respond to the variety of cultures represented within the education system is the most effective approach (Swap, 1993). Partnerships will require a new form of *"reciprocal enculturation* [author's emphasis]...a process whereby new cultural patterns are acquired by both systems...as they develop and mature, and each can be endowed with a new energy that changes its configuration" (Winters, 1993, p. 3).

Guiding Principles for Systemic Approaches to Family Support

The principles of family support outlined in Table 13.2 are drawn from the fields of family support and preservation (Family Impact Seminar, 1994) and family-centered prevention and intervention (Dunst, et al 1991). For practical purposes, one must understand and embrace both perspectives, for in combination they form an entire ecology of family support that encompasses interventions within the family environment and family-centered practices and policies in the environment in which the family functions (see Table 13.2).

Table 13.2 Guiding Principles of a Multidimensional Family Ecology Perspective

Principles	Description
1. Enhance community by promoting shared values and common needs in ways that create the coming together of people around mutually beneficial interdependencies.	Interventions need to emphasize the common needs of all people, and base interventions around those commonalities, they should focus on building interdependencies between members of the community and the family unit.
2. Mobilize resources and supports. Contribute to building systems of support that enhance the flow of resources in ways that assist families with parenting responsibilities.	Intervention must be conceptualized as the aggregation of the many different types of support and assistance provided by families, informal and formal support networks. Utilize, reinforce, and support informal sources of family support, including specific efforts to strengthen family's personal support networks. Resources and supports should be made available to families in ways that are flexible, individualized, and responsive to the needs of the entire family.
3. Share responsibility and collaborate: share ideas and skills between parents and professionals in ways that build and strengthen collaborative arrangements.	Interventions should employ partnerships between parents and professionals as a primary mechanism for supporting and strengthening family functioning. Resources and support mobilization interactions should be based on mutual respect, and non-biased sharing of information and perspectives.
4. Acknowledge the strength and persistence of family ties, even when they are problematic.	Policies and programs that attempt to assist or change a member of the family must take into account the influence of the family on that person and how the desired change or service may affect the family system as a whole.
5. Treat parents as full partners in meeting the needs of children. If families are to fulfill the various functions society expects of them, policies and programs must treat them as partners.	The well-being of society is promoted by the well-being of families. Hence, it is in society's best interest to insure that public policies and programs support family stability.
6. Support and supplement family functioning, rather than substitute for family functioning.	A primary objective of family policies and programs should be the strengthening of families' own abilities to manage and fulfill their own functions, including care of their own members. Services that substitute for families should be provided only in those situations where it is clear that the family will not be able to function sufficiently even with support or when the burden on the family is excessive.

Table 13.2 Continued

Principles	Description
7. Acknowledge the diversity of family life and protect family integrity.	Families come in many forms and configurations. Programs and policies must take into account their different effects on different types of families, since an objective of a policy or a program may be beneficial for one type of family but may harm another. Interventions should be conducted in ways that accept, value, and protect a family's personal and cultural values and beliefs.
8. Address pressing needs as families define them. Families in greatest economic and social need, and those determined to be most vulnerable to breakdown should have priority in government policies and programs.	Targeting the vulnerable has long been a tradition in the government's responsibility to promote the general welfare. In the absence of adequate support systems, the needs of children and families are essential to address. Nevertheless, these need to be understood as stop-gap measures that will negatively impact the family support system in the absence of systemic approaches.
9. Adopt a consumer driven human service delivery model and practices that support and strengthen family functioning.	Service-delivery programs should emphasize promotional approaches Resources and support mobilization should be consumer driven rather than service provider driven or professionally prescribed. Families need to be provided with information and resources that will empower them, and given rights to make or share in decisions. Policies and programs need to acknowledge that there is a continuum of family participation.

It is increasingly clear that in today's complex world, families, neighborhoods, schools, service providers and others must explore and be open to new ways of joining together in strategic relationships that benefit children. Schools and family support providers have a responsibility–and an interest–in developing relationships that position and train parents to maximize the potential of familial and informal support systems to serve as mediating forces in the growth and development of these children. Approaching family support in terms of an environment within an environment allows us to understand and take into account the variability among family systems, while at the same time maximizing the potential of families to develop and impact the systems of support and services that are vital to family functioning on behalf of children with special needs.

References

Artiles, A.J., & Trent, S.C. (1994). Overrepresentation of minority students in special education: a continuing debate. *Journal of Special Education 27(4)*, 410–437.

Barber, P., Turnbull, A., Behr, S., & Kerns, G. (1989). A family systems perspective on early childhood special education. In B. Hanft, (Ed.), *Family centered care: An early intervention resource manual* (pp. 1–13). Rockville, MD: American Occupational Therapy Association.

Beal, E., & Chertkov, L.(1992) Family school intervention: A family systems perspective. In M.Fine & C.Carlson, (Eds.), *The handbook of family-school intervention: A systems perspective.*(pp. 288–301) MA: Allyn & Bacon.

Carter, E.A., & McGoldrick, M. (Eds.). (1980). *The family life cycle: A framework for family therapy*. New York: Gardner Press.

Davies, D. (April 1988). *Poor families, poor schools: An exploratory study of the perspectives of low income parents and teachers in Boston, Liverpool, and Portugal.* Presentation at the annual conference of the American Educational Research Association, New Orleans, LA.

Dunst, C.J., Johanson, C., Trivette, C.M., & Hamby, D. (1991). Family-oriented early intervention policies and practices: family centered or not? *Exceptional Children 58, (2)*, 115–126.

Dunst, C.J., Trivette, C., & Deal, A. (1988). *Enabling and empowering families: Principles and guidelines for practice*. Cambridge, MA: Brookline Press.

Falicov, C. (1988). *Family transitions: Continuity and change over the life-cycle*. New York, NY: Guilford.

Family Impact Seminar. (1995). *Disconnected dads: Strategies for promoting responsible fatherhood*. Washington, DC: Author.

Family Impact Seminar. (1994). Families and partners in FP/FS planning. *Roundtables on the Implementation of the Family Support and Preservation Program*. Washington, DC: Author

Family Impact Seminar. (1993). *Changing the paradigm: Strategies for family-centered systems reform*. Washington, DC: Author.

Federal Interagency Coordinating Council. A briefing paper on Part H of the Individuals with Disabilities Education Act (IDEA). Report submitted April 20–21, 1995. Chapel Hill, NC: NEC*TAS.

Gaith, A. (1977). The impact of an abnormal child upon parents. *British Journal of Psychiatry, 130*, 405–410.

Germain, C. (1994). Emerging conceptions of family development over the life course. *Families in Society, 75(5)*, 259–269.

Gilkerson, L. (1994). Putting parents in leadership roles. *Zero to Three, 14(4)*, 23.

Graves, S.B., & Gargiulo, R.M. (Winter 1993). Early childhood special education: strategies for supporting families. *Day Care and Early Education*, pp. 47–48.

Hanft, B., (Ed.). (1989). *Family centered care: An early intervention resource manual*. Rockville, MD: American Occupational Therapy Association.

Hanson, S., & Vosler-Hunter, R. (Fall 1991/Winter 1992). Parents as policymakers: challenges for collaboration. *Focal Point, 6(1)*, 1–5.

Harry, B. (1992). An ethnographic study of cross-cultural communication with Puerto Rican-American families in the special education system. *American Educational Research Journal, 29*, 471–494.

Harry, B., Allen, N., & McLaughlin, M. (1995). Communication versus compliance: African American parents' involvement in special education. *Exceptional Children, 61(4)*, 364–377.

Liontos, L.B. (1992). *At risk families and schools*. Washington, DC: ERIC Clearinghouse on Education Management.

Lynch, E.W., & Stein, R. (1987). Parent participation by ethnicity: A comparison of Hispanic, Black, and Anglo families. *Exceptional Children, 54*, 105–111.

McGoldrick, H.M., & Carter, B. (1993). The changing family life cycle: A perspective on normalcy. In F. Walsh (Ed), *Normal family processes*. New York: Guilford.

McGoldrick, M., Pearce, J., & Giordano, J. (Eds.). (1982). *Ethnicity and family therapy*. New York: Guilford.

Melson, G. (1980). *Family and environment: An ecosystem perspective*. Minneapolis, MN: Burgess.

National Commission on Children. (1993). *Strengthening and supporting families*. (Implementation Guides Series). Washington, DC: Author.

Ooms, T., & Preister, S. (Eds.). (1988). *A strategy for strengthening families: Using family criteria in policymaking and program evaluation*. Washington, DC: Family Impact Seminar.

Schorr, L.B. (1989). *Within our reach*. New York: Doubleday.

Shriver, M.D., & Kramer, J.J. (1993). Parent involvement in an early childhood special education program: A descriptive analysis of parent demographics and level of involvement. *Psychology in the Schools, 30*, 255–263.

Snyder, W., & Ooms, T. (1992). *Empowering families, helping adolescents: Family centered treatment of adolescents with alcohol, drug abuse and mental health problems*. Rockville, MD: U.S. Department of Health and Human Services.

Staton, J., Ooms, T., & Owen, T. (1991). *Family resource, support, and parent education programs: The power of a preventive approach*. Washington, DC: Family Impact Seminar. 1991.

Stephens, S.A., Leiderman, S.A., Wolf, W.C., & McCarthy, P.T. (1994). *Building capacity for system reform*. Philadelphia, PA: Center for Assessment and Policy Development.

Swap, S.M. (1993). *Developing home-school partnerships: From concepts to practice*. New York: Teachers College Press.

Tseng, W., & Hsu, J. (1991). *Culture and early problems and therapy*. New York: Hayworth.

Terkelson, K. (198). Toward a theory of the family life cycle. In. E. Carter & M. McGoldrick, *The Family Life Cycle* (pp. 21–52). New Jersey: Halset Press.

Turnbull, A.P., & Turnbull, H.R. (1986). Stepping back from early intervention: An ethical perspective. *Journal of the Division for Early Childhood, 10*, 106–117.

United Nations. (1994). *The concept of family health*. Vienna, Austria: Author.

United Nations. (1993). *Family as an environment: An ecosystem perspective on family life*. Occasional Papers Series, No. 5. Vienna, Austria: Author.

United Nations. (1992). *Family: Forms and functions*. Occasional Papers Series, No. 2. Vienna: Author.

United States Department of Education (1994). *Strong families, strong schools: Building community partnerships for learning*. Washington, DC, Government Printing Office

Visher, J., & Visher, E.B. (1982). Step-families and step-parenting.In F. Walsh, Normal Family Processes,(pp. 331–353. New York: Guilford.

Walsh, F. (1993). *Normal family processes*.New York: Guilford.

Weiss, H.B., & Halpern, R. (1990). *Community-based family support and education programs: Something old or something new?* New Haven, CT: Yale University School of Public Health.

Winters, W. (1993). *African American mothers and urban schools: The power of participation*, Washington, DC: Howard University.

Zill, N., & Nord, C.W. (1994). *Running in place. How American families are faring in a changing economy and an individualistic society*. Washington, DC: Child Trends.

Zill, N. (Winter 1993). The changing realities of family life. *Aspen Institute Quarterly, 5*, 27–51.

14

Teacher Education

PAUL T. SINDELAR and KAREN L. KILGORE
University of Florida, USA

In this chapter, we review contemporary teacher education as it relates to the provision of services to students with disabilities. We argue that there is more reason than ever for collaboration in teacher education between special and general educators, that the general and special education teacher education reform agendas remain largely distinct, and that collaboration is beginning to emerge nonetheless. We describe the common ground that general and special education teacher educators might forge and consider contemporary teacher education in special education and the issues that impinge on it.

Introduction

There is greater diversity in U.S. school population today than ever before. Diversity has resulted primarily from the changing demographics of school-aged children, although other factors, such as the movement to include students with disabilities in less restrictive environments, have contributed. One clear implication of the growing diversity of school children is the need to prepare teachers who are skilled in accommodating diversity, whether it results from racial, cultural, or ethnic differences, poverty, or disability. Teacher preparation of this sort will require collaboration among teacher educators who heretofore have worked independently and who have developed distinct professional cultures, but the call for collaborative teacher education is not new. In fact, nearly two decades ago, Reynolds and Birch (1977) argued that

> changes in the roles of both regular and special education personnel are inevitable. . . and require major shifts in the training and retraining of personnel. We are in a period of major renegotiations of the relation between regular and special education. . . . Adaptations are being made in colleges and universities where tough renegotiations are underway. (p. 13)

Children with limited proficiency in English, children from cultures in which education is not strongly valued, children living in poverty, children who are homeless, and children at-risk for school failure for yet other reasons constitute the most rapidly growing sector of the school-age population and are changing the character of contemporary education, particularly in urban areas. Given contemporary school practice, these demographic changes are likely to result in increased demand for compensatory, remedial, and special education services–systems already taxed by high demand and limited resources. Moreover, our public schools also are contending with more students with disabilities.

In 1992–93, nearly 5.2 million infants, toddlers, children, and youth through age 21 received special education services in the United States (U. S. Department of Education, 1994), a 3.7% increase from the year before and a 28% increase over 1976–77 special education enrollments. The growth in special education enrollments cannot be accounted for fully by growth in the school-aged population. In 1992–93, students with disabilities (served under Part B of the Individuals with Disabilities Education Act (IDEA) or Chapter 1) made up 6.4% of the birth-to-21 population; in 1976–77, they constituted 4.5% of it (U. S. Department of Education, 1994).

Growth in special education populations has been limited to students classified as learning disabled. These students, who once represented less than a fourth of all students with disabilities, now represent more than half. Students with LD exist at the interface of general and special education, and nearly 80% of them spend at least 40% of the school day in general education classrooms. Although whether they all have true disabilities may be arguable, that they all challenge their teachers is a certainty. Classroom teachers are expected to work effectively with them along with the growing numbers of students at-risk for school failure—and the other students in their classes. These same classroom teachers are likely to have been trained to refer students with problems for special education services of one sort or another (Pugach, 1988).

To meet the needs of a changing and challenging student population, schools must become flexible and responsive institutions (Schorr, 1986), for as Goodlad (1986) noted, "we are running out of organizational and special grouping types of solutions" (p. xi). Both general and special educators have called for reforms of the nation's public schools system to meet the needs of our changing society. General and special educators differ, however, in the ways they characterize the problems facing schools and the solutions they propose, as we describe in the following section.

Calls for School Reform: The General Education Agenda

Since the early 1980s, reformers have questioned the extent to which schools adequately prepare students to meet changing economic and social conditions. Beginning with *A Nation at Risk* (National Commission on

Excellence in Education, 1983), various national commissions issued numerous reports sounding the alarm that the nation's educational system was in serious decline. These initial reform efforts emphasized the schools' role in preparing a competent work force for the nation's economy. These reforms included top-down regulation systems characterized by centralized management and extensive mandates resulting in changes in curriculum, higher academic standards for students, and more rigorous assessment procedures. The "success" of these reform efforts included an increase in school attendance, a decrease in dropout rates, slightly improved SAT scores, and more students enrolled in higher education (Finn, 1991; Webb & Sherman, 1989). The traditional organization of the school system and classroom practices remained relatively unchanged.

Many scholars and practitioners characterized these reforms as insufficient and called for significant restructuring of the educational system (Cuban, 1984). Publications such as *A Nation Prepared: Teachers for the 21st Century* (Carnegie Forum on Education and the Economy, 1986) and *Tomorrow's Teachers: A Report of the Holmes Group* (Holmes Group, 1986) called for changes in teacher preparation and the conditions necessary for teachers to work effectively. Underlying these calls for reform was the assumption that those who are involved in and most fully understand the educational needs of their students—faculty and staff in schools, not district, state, or federal bureaucrats—should be responsible for making decisions concerning student learning. Rather than implementing top-down mandates, teachers were asked to be problem-solvers, to collaborate with other professionals, and to be accountable for student learning. These reform efforts, however, did not specifically address the changing demographics of the student population or the need for creating an educational system capable of meeting these students' needs. In fact, none of the major reports calling for educational reform specifically referred to students with disabilities (Lilly, 1987; Pugach & Sapon-Shevin, 1987).

The most recent calls for reform, however, have begun to focus on the needs of at-risk or disadvantaged students (Little, 1993; Paul & Rosselli, 1995; Ysseldyke, Algozzine, & Thurlow, 1992). Reformers have noted the rising rates of special education and alternative education placements and interpreted these trends as indicative of the educational system's failure to provide appropriate educational programs for a culturally diverse and increasingly at-risk student population. These concerns were highlighted in the President's education summit in 1989 and culminated in a set of national goals (Task Force on Education, 1990). Educational researchers argued that disparities between home and schools, communities and school systems, and teachers and students accounted for the difficulties school personnel faced as they attempted to provide an appropriate education for all children (Cochran-Smith & Lytle, 1992; Cummins, 1986; Fordham, 1988; Ogbu, 1978).

Reformers have called for the development of culturally responsive curricula and programmatic changes for linking schools with homes and communities (Comer, 1989; Heath, 1982). The "full service school" was proposed as a means to promote meaningful family involvement in the school and to provide school-based family support services (Comer & Haynes, 1991). To benefit families most in need, full-service schools collaborate with other helping agencies (social services, health care, employment) to establish and operate cross-disciplinary prevention and intervention programs. Family services centers and outreach activities based at the full-service school provide social and economic services (e.g. AFDC, WIC, food stamps, child support enforcement), health services (e.g., referrals to appropriate subspecialty clinics and therapies), educational programs (e.g., adult basic education, GED classes) and employability training (e.g., career counseling). Teachers in full service schools must necessarily engage in collaborative work involving other educators, social workers, nurses, and other human services professionals (Duchnowski, Dunlap, Berg, & Adiegbola, 1995; Lawson, 1994). The reforms aimed at providing educational programs for at-risk students, like earlier reforms, did not specifically address the needs of students with disabilities.

The Inclusion Movement: The Special Education Agenda

Since the publication of Will's (1986) paper on what has come to be known as the Regular Education Initiative (REI), the movement to include students with disabilities in less restrictive environments has commanded much of the attention of special education scholars. Compelling arguments for the inclusion of individuals with both mild (Gartner & Lipsky, 1987; Reynolds, Wang, & Walberg, 1987; Wang, Reynolds, & Walberg, 1986, 1988) and severe disabilities (Halvorsen & Sailor, 1990) have been put forward, as have equally compelling rejoinders (Braaten, Kauffman, Braaten, Polsgrove, & Nelson, 1988; Fuchs & Fuchs, 1994; Kauffman, 1989, 1993; Kauffman, Gerber, & Semmel, 1988). The effect of this debate on educational practice has not been fully realized, and probably won't be for some years. To date, changes in placement practices have not been dramatic: Whereas 27% of all students with disabilities, aged 3–21, were served in general education classes in 1986–1987 (U.S. Department of Education, 1989), 36% were so served in 1991–1992, the most recent year for which such data are available (U.S. Department of Education, 1994). The percentage of students with learning disabilities served in general education classes also increased over these years, from 16% to 24%.

Interpretations of U.S. Department of Education placement data have varied. Sawyer, McLaughlin, and Winglee (1994), for example, studied placements of students with disabilities from 1977–1978 to 1989–1990. They reported little change in the percentages of students placed in general public school placements, by which they meant general education classrooms, resource rooms, and separate classes. There were exceptions, however: Students with serious emotional disturbances, mental retardation, and multiple disabilities were more likely to be excluded from general public schools in 1989–1990, and students with orthopedic impairments were more likely to be included in general public schools in 1989–1990. Within public schools, a reciprocal pattern of increased general education classroom placement and decreased resource room placement was observed.

McLeskey and Pacchiano (1994), on the other hand, focused their study of Department of Education placement data on students with learning disabilities and included in their analyses annual data from 1979 to 1989. They reported a decreasing trend in general class and resource room placement (taken together), and an increasing trend in separate class placement for students with learning disabilities. It is not surprising that the latter authors were less sanguine than the former about the effects of the REI on placement practices. Whereas McLeskey and Pacchiano (1994) concluded that "little progress has been made. . . toward mainstreaming students with learning disabilities," (p. 514), Sawyer, McLaughlin, and Winglee (1994) asserted that "integration efforts, including the REI, appear to have impacted school practice, particularly with respect to the percentages of students with disabilities who are now receiving a majority of their education in general education classrooms" (p. 213).

Obscured in this debate about changes in placement practices is the fact that large numbers of students with disabilities already are placed and educated in general education classrooms. In 1991–1992, over 1.7 million students with disabilities were placed there (for at least 79% of the school day), and an additional 1.6 million were placed in resource room programs, for which classroom placement constitutes at least 40% and at most 79% of the school day (U.S. Department of Education, 1994). These data make clear that, independent of the issue of inclusion and its effects on placement practices, classroom teachers need to be prepared to work students with disabilities. Given the changing demographics we described previously, they also need to be prepared to work with the growing number of students in their classes who have problems learning and comporting themselves in socially acceptable ways. Teachers no longer have the luxury of referring students with problems out of their classrooms.

Current Status of Teacher Education

The changing character of the U.S. population of school children, the large number of students with disabilities served primarily in general education classrooms, and the potential of inclusive education policy to increase this number have brought questions about the content of teacher education and requirements for professional licensure to center stage. The available evidence on the content and quality of the preparation of classroom teachers suggests that not all of them are being prepared to work with challenging students. It is significant to note, however, that teacher educators in general education and special education have responded differently to the needs of preparing teachers to educate a diverse student population. In the following section, we summarize the different ways that teacher educators in general education and special education have called for teacher education reform.

General Education: Preparing Teachers to Work in Restructured Schools

Teacher educators in general education have emphasized the increased professionalization of teachers and have considered ways to prepare teachers to play significant roles in school reform, engage in shared decision making, and advocate for structural changes that would improve student learning. Teacher educators have developed programs around the notion of teachers as reflective practitioners—decision makers capable of interpreting and applying research to problems of practice, studying their own teaching as a means to improve practice and promote student learning, and understanding the educational and ethical implications of their decision making (Clift, Houston, & Pugach, 1990; Valli, 1989; Zeichner & Liston, 1987). Teacher educators also began to view teachers as creators rather than merely consumers of knowledge and advocated preparing teachers to engage in action research in the classroom (Ross, 1989). Teacher educators have noted the importance of assisting preservice students in developing the ability to collaborate with others (Johnson & Pugach, 1992; Lieberman, 1988) in recognition of the body of research that demonstrated the powerful role of collegiality and professional collaboration in creating positive learning environments for students and teachers (Goodlad, 1984; Little, 1982). Preparation in organizational theory, human relations, and the politics of education have also been viewed as essential for teachers to participate effectively in restructured schools (Murphy, 1990).

General Education: Preparing Teachers in Professional Development Schools

The 1980s brought with it calls not only for school reform but also for significant restructuring of teacher education. In fact, the traditional structure of teacher education—a collection of loosely connected courses across departments and colleges without programmatic coherence and lacking significant connections with schools—has been criticized repeatedly (Association of Teacher Educators, 1991; Carnegie Forum on Education and the

Economy, 1986; Holmes Group, 1986). Goodlad (1990), in noting the weak link between schools and the institutions that prepare teachers, argued that teacher preparation must be integrally linked with school reform. He stated, "The education of teachers must be driven by a clear and careful conception of the educating we expect our schools to do, the conditions most conducive to this education (as well as the conditions that get in the way), and the kinds of expectations that teachers must be prepared to meet. Further, the renewal of schools, teachers, and the programs that educate teachers must proceed simultaneously" (p. 4). The Holmes Group (1990), Goodlad (1990), and others (Levine, 1992) called for a new organizational structure, the Professional Development School, as a means of integrating school reform and teacher preparation. Professional Development Schools were thought of as collaborative school/university partnerships to support student learning, provide for the professional education of teachers, and facilitate inquiry directed at the improvement of practice. The design of Professional Development Schools offered the opportunity for colleges and schools of education to reconceptualize teacher education and to forge new relationships with schools.

General Education: Preparing Teachers for a Diverse Student Population

In preparing teachers for a diverse student population, teachers educators in general education have focused their efforts on preparing preservice teachers to address equity issues related to differences in the student population in class, ethnicity, race, and gender. Teacher educators have advocated strategies to challenge preservice students' assumptions about children and families based on race, class, and ethnicity (Taylor & Dorsey-Gaines, 1988). They've stressed the importance of preparing preservice teachers to understand the potential for oppositional relationships to develop between educators and students from different cultural backgrounds, to explore limitations of teachers' knowledge about values and cultures different from their own, and to develop culturally relevant and responsive curricula (Asante, 1991; Cochran-Smith & Lytle, 1992; Foster, 1990; Ladson-Billings, 1990). Teacher educators also proposed ways to prepare teachers to create stronger links between schools, homes, and communities (Zeichner, 1989). Adler and Goodman (1986) and others (Zeichner & Liston, 1990) have argued that teacher education must become a means of awakening the social consciousness of teachers and preparing them to work towards a more just society.

General Education: Preparing Teachers to Educate Students with Disabilities

As teacher educators in general education proposed reforms in teacher preparation to deal with a diverse student population, issues related to the education of children with disabilities have been notably absent. Relatively few publications have appeared in the general education literature in which the need to prepare teachers to educate students with disabilities was emphasized. For example, a review of publications in the *Journal of Teacher Education* from 1985 to 1995 revealed very few articles (Pugach, 1988; Stephens, 1988) that dealt with preparing teachers to work with students with disabilities. In general education programs, preparing teachers to educate students with disabilities remains on the periphery.

In their survey of state departments of education, Swartz, Hildalgo, and Hays (1991–92) found that 15 states required a special education course for elementary education certification. An additional 24 states required only the coverage of special education content, a requirement most commonly fulfilled by infusing it in general education courses. Previously, Patton and Brathwaite (1990), in a 1987 survey of certification directors in 52 states and territories, found that (a) 37 states required special education content for initial certification, (b) 12 (of the 37) states required a special education course, and (c) 9 states required special education content for recertification.

This pattern of growth may be traced even further. Patton and Braithwaite's survey was a follow-up to one the same authors conducted a decade before (Patton & Braithwaite, 1980). In the original, they found that only 11 states required special education content for initial certification. At about the same time, Smith and Schindler (1980) reported that 10 states had a course requirement (of which 9 required one course and 1 required two courses) and 5 had a competency requirement. Four years later, Ganschow, Weber, and Davis (1984) reported that 17 states required one course, 2 states required two courses, and 15 states had nonmandated requirements, such as competency guidelines.

Less information is available about pre-school and secondary teacher education program requirements. Swartz et al. (1991–1992) found that requirements for secondary certification were similar to elementary requirements; 14 states required a course, and an additional 23 required content coverage. Twenty-five states required special education coursework for certification in early childhood education. Few states (5–8) required experience with students with disabilities for any level of certification. About special education teacher education programs, Swartz et al. reported that 24 states required special education teachers to hold general education certification, and that 38 states required a course on consultation.

The picture painted by surveys of university teacher education programs offer no more cause for optimism. Fender and Fiedler (1990), for example, found that only 80% of the education departments they surveyed required preservice teachers to take a course on students with disabilities, and only 19% required students to take more than one course. The courses were described as having a traditional lecture format and seldom requiring field experience. Their content emphasized characteristics of individuals and was organized categorically. Jones and Messenheimer-Young (1989) also studied course content; they found that when two courses were required, the content of the two often was redundant. They contrasted introductory special education courses with those they described as "mainstreaming" courses and found considerable overlap in content of the two, at both the knowledge (9 of 10 topics) and application (7 of 10 topics) levels. (Swartz et al., however, found the topic of "instructional methods" to be a more common certification requirement than "mainstreaming," which they equated with survey classes.)

In a survey of education departments in 35 institutions in New York, Kearney and Durand (1992) found that (a) only a third of the programs either were NCATE-accredited (and thereby held to the NCATE standards about accommodation), offered dual special/general education preparation, or offered training in consultation or collaboration; (b) over half required no more than one course in special education; and (c) nearly two-thirds required their students to spend no more than 16 hours in an integrated mainstream practicum setting. In California, where general certification is a prerequisite for special education certification, elementary and secondary teacher education programs were more likely to require special education content coverage, through both separate courses (88%) and infusion (62%) (Swartz et al., 1991–92).

These studies fail to address the sufficiency of a course or two as a means for preparing teachers for work with students with disabilities, and even the most optimistic data failed to impress the researchers who collected them. Swartz et al. concluded, for example, that "it is clear that training regular educators to teach special needs learners is not yet a priority" (p. 60). A more fundamental concern is whether classroom teachers are being prepared adequately for the children now populating their classrooms. On the basis of their "Study of the Education of Educators," Goodlad and Field (1993) expressed their doubts. They asserted that "more attention needs to be focused on helping [preservice classroom teachers] build the capabilities they need to meet the diverse learning needs of the wide range of students who are, or should be, in their classes" (p. 240). Others (Fullan, 1991; Reynolds, 1992) have echoed the same concern. If there is reason to believe that teachers are not being prepared well to work with students assigned to general education classrooms, how can we expect them to work effectively with students with disabilities?

It has been argued that the most serious teacher education problem confronting special education today is preparing classroom teachers with the knowledge, skills, and dispositions to work effectively with students with disabilities (Sindelar, Ross, Brownell, Griffin, & Rennells, 1994). A larger contingent of more effective classroom teachers would (a) allow for more expeditious reintegration of students in special education programs, (b) reduce the number of referrals of students with disabilities, and (c) stem the rising tide of referrals of students whose difficulties in school are not the result of disabilities. Shortages of certified special education teachers have been chronic and severe. It is conceivable that the intractable, unmet demand for special education teachers could be allayed by having more, and more effective, classroom teachers.

The importance of well-prepared classroom teachers to the success of special education practice has long been recognized (Reynolds & Birch, 1977). It was first construed as a natural consequence of the passage of P.L. 94–142, but, in spite of a substantial federal investment, early efforts failed. It is now understood as a means to achieve two ends: the reintegration of students with disabilities into general education classrooms and the accommodation there of students with learning and behavioral problems, whether they have disabilities or not.

Collaborative Teacher Education Programs

Korinek and Laycock (1988) analyzed the 20 top-rated federally funded teacher education programs in 1986 and 1987 and found a high and growing proportion that emphasized the Regular Education Initiative. Yet in spite of nearly a decade of federally funded projects, strong sentiment for collaborative teacher education, and some sentiment for the unification of general and special education programs (Sindelar, Pugach, Griffin, & Seidl, 1994; Stainback & Stainback, 1987), descriptions or studies of such programs are scarce in the literature. Our search yielded eleven. In this section, we review this literature with particular attention to the scope of the program. We found more examples of descriptions of program parts (Gerlach, Clawson, & Noll, 1994; Heller, Spooner, Spooner, & Algozzine, 1992; Heller et al., 1991–92; Welch & Sheridan, 1993) or structures (Affleck & Lowenbraun, 1995; Pasch, Pugach, & Beckum, 1991; Pugach, 1992) than full program descriptions (Ellis et al., 1995; Feden & Clabaugh, 1986; Kemple, Hartle, Correa, & Fox, 1994; Maheady, Harper, Mallette, & Karnes, 1993; Valero-Figueira, 1986). Given the enormity of the task of redesigning teacher education programs, it's probably not surprising that few full program descriptions are available in the literature. Before we consider the current literature on collaborative programs, however, we look back on the Dean's Grant program, an early and broad scale effort to promote the development of teacher education programs with themes of mainstreaming and accommodation.

Deans' Grants

Deans' Grants—a 1974 initiative of what was then called the Bureau of the Education for the Handicapped (BEH)—"resulted from the growing demand for support to prepare teachers at the elementary and secondary levels to accommodate handicapped pupils in regular classrooms" (Behrens & Grosenick, 1978, p. 1) and were intended to promote integration of teacher education programs in general and special education. Although this outcome was never fully realized (Askamit, 1990; National Support Systems Project, 1980; Pugach, 1992), it was not for lack of investment or effort. The Deans' Grants Program was funded from 1975 to 1983, and special education courses (or content) were added (or was infused) to teacher education curricula nationwide.

Furthermore, the language used by scholars who wrote about Deans' Grants projects has a remarkably contemporary ring. Corrigan (1978), for example, contended that "the most vivid truth that emerges from any analysis of the goals and concepts included in P.L. 94-142 and a review of teacher education as it exists in this country today is that we will not succeed if we continue to conceptualize special education in a framework separate from regular education at any level of the educational spectrum" (p. 21). Although it may be impossible to relate the particulars of current programs to the work conducted under the auspices of the Dean's Grants program, it is simple to see in statements like this the philosophical roots of collaborative teacher education.

Field Experience and Supporting Course Approaches

The four examples we found (Gerlach et al., 1994; Heller et al., 1992; Heller et al. 1991–92; Welch & Sheridan, 1993) of programs with collaborative elements all took roughly the same shape: Elementary education majors complete field experiences while taking a related course or set of courses. The emphasis on field experience seems entirely appropriate (Buck, Morsink, Griffin, Hines, & Lenk, 1992), even though experience may not be sufficient to improve teachers' attitudes about mainstreaming students with disabilities and their ability to do so effectively. Although it is widely believed that experience with students with disabilities improves the attitudes of teachers, the research is equivocal. In a recent synthesis of this literature, Scruggs and Mastropieri (1995) found that willingness to accept mainstreamed students covaried more directly with the severity of students' disabilities and the degree of integration than with teachers' experience with students with disabilities. Teachers judge students with mild disabilities to be better candidates for mainstreaming, and indeed, the collaborative programs we found all focused on this population.

For example, Heller and his colleagues at the University of North Carolina at Charlotte (Heller et al., 1992; Heller et al., 1991–1992) offered a summer practicum for elementary education teachers in general education settings with at-risk students. Project REACH (Regular Educators Accommodating Children with Handicaps), as it was called, also involved classroom instruction in the afternoon and pre-practicum orientation. The program was evaluated by requiring students to complete a self-evaluation. On it, REACH participants expressed more willingness after the practicum than before to have students with disabilities assigned to their classes. They also expressed more confidence in their abilities to (a) use instructional strategies, (b) make instructional modifications, (c) provide instruction in various groups, (d) work with other teachers, and (e) participate in building-based decision making.

It's regrettable that Heller et al. used only self-report measures to evaluate their program, because their findings would have been more convincing had they collected other data, such as supervisors' ratings or direct observations of classroom performance. The problems of using self-report measures are as well-known as they are thorny. We believe that teacher education research could be improved greatly with more attention to outcome measures and more serious commitment to evaluating the outcomes (Sindelar, Ross, et al., 1994). The problem is not unique to the work of Heller and his colleagues, as we shall see. Teacher education research is difficult to conduct well, and, in our opinion, research in special education to date has not been done with the sophistication that has characterized the best research in general teacher education.

Gerlach, Clawson, and Noll (1994) described a similar program for preservice elementary and secondary education students at Indiana University of Pennsylvania. The preservice students were prepped for student teaching with a 1-credit course that covered among other topics (a) instructional support teams, (b) professional collaboration, and (c) using supports for students in inclusive classrooms. Student teachers co-taught with their supervising teachers and visited other classrooms. The students rated the course topics as important to very important and offered other generally favorable evaluations of the project. The participants claimed to have met their goals, but no data were offered to corroborate the questionnaire data or establish that they were indeed more effective in their work with difficult students.

Finally, Welch and Sheridan (1993) described a federally funded program at the University of Utah in which students in Educational Psychology, Educational Administration, Elementary and Secondary Education, and Special Education participated together during practicum in multidisciplinary teams. Students were required to take two seminars and two courses as well. The courses were entitled "Educational Partnerships Serving Exceptional Students" and "Collaborative Educational Problem-Solving and Conflict Management." The authors offered no evaluation data in this program description.

Aside from the obvious limitations of these studies and our reluctance to draw conclusions from them, we'd question the assumption implicit in this approach that the solution to the problem of teacher education reform lies in effecting changes solely in what classroom teachers do. However admirable, these efforts apparently are not based on a vision of inclusive education that involves a reconceptualization of school practice and of what both classroom and special education teachers do. Classroom teachers who complete such programs might be better able to accommodate students with problems as a result of their participation, and to do so might make school practice somewhat more inclusive. However, that these teachers would regard the accommodation of differences as a defining feature of their work seems less likely.

College Organization

The organization of a college or school of education can be a barrier to the development of collaborative teacher programs, if only for logistical reasons (Kemple, Hartle, Correa, & Fox, 1994). If the division of the teacher education effort between departments of special education and curriculum and instruction is dysfunctional (Pugach, 1992), then organizational changes might also provide a means to achieve collaboration. The Center for Teacher Education (CTE) at the University of Wisconsin-Milwaukee (Pasch, Pugach, & Fox, 1991; Pugach, 1992) serves just such a function. Its creation sanctioned an ongoing dialogue among teacher educators there about the redesign and governance of teacher education.

The CTE is an organizational unit in which faculty from all certification programs are represented. Pugach (1992) noted that although initially "building trust. . . constituted the de facto agenda of the CTE" (p. 260), the establishment of trust set the occasion for more substantive consideration of teacher education reform. Affleck and Lowenbraun (1995) described a similar process in their case study of teacher education reform at the University of Washington. There the building of trust began with the team-teaching of a 12-credit block, a collaboration prompted by a new college dean for the purpose of initiating a process to define a shared vision for the college. The participants in the team-teaching met as a group and communicated to the faculty in the departments a sense of the vision they created together. If the experience of the special education group is any indication, area faculties debated the implications of the vision for their own work and ultimately reconceptualized their work to be consistent with it.

These two examples illustrate the importance of time for discussion to the process of teacher education reform; it also illustrates the importance of strong leadership to provide it. Opportunities for faculty to come together to talk were created at both institutions as part of the participants' regular loads. Reform is not likely to prevail when the work of reform is added to existing faculty loads. In special education, we struggle to address the problem of chronic and severe teacher shortages. In efforts to meet schools' need for teachers, little time is available to reflect on the quality or nature of the work that's being done. It's likely that many special education teacher education programs across the country suffer from such overcommitment to the problem of shortages; as a result, few resources are available to invest in curricular reform.

Program Descriptions

We found five papers with sufficient detail about integrated programs (Ellis et al., 1995; Feden & Clabaugh, 1986; Kemple et al., 1994; Maheady et al., 1993; Valero-Figueira, 1986) for us to consider them full program descriptions. Feden and Clabaugh's work at La Salle University predated the Regular Education Initiative, but foreshadowed in several ways its focus on the classroom accommodation of students with disabilities. The development of this program was fostered by the authors' dissatisfaction with categorical special education and teacher education, both of which are based on the assumption that effective practice differed as a function of "stereotypical conceptions of what it means to be mentally retarded, emotionally disturbed, or learning disabled" (p. 182). The authors believed that recent advances in the knowledge base of cognitive psychology—the emergence of the concept of active learners—permitted teachers to treat students as individuals, and cognitive psychology became the organizing theme for this program. Because there is no benefit to differentiating on the basis of disability, they argued, the distinctions between general and special education were meaningless. In fact, Feden and Clabaugh likened dual certification programs to Siamese twins, "two truly different organisms joined together through an unnatural connection" (p. 184), and argued instead for the integrated preparation of educators who would be expected to work effectively with all children. About the only thing that separates Feden and Clabaugh's thinking from that of contemporary teacher educators is the absence from the former of a full vision of inclusive school practice.

It is disappointing to have heard nothing from Feden and Clabaugh in the decade since the publication of this paper and to know nothing more about the success of their program and the abilities of their graduates to deal with students in the individualized manner they described. Yet, like Heller et al., Feden and Clabaugh cannot be singled out for their failure to conduct or disseminate program evaluations. Either serious program evaluation does not get published, or the work is not being done. We hope that the former is the case, because renegotiating standards of acceptability for the publication of program evaluations offers a solution to the problem, whereas lack of commitment to the process of program evaluation does not.

The work of Valero-Figueira (1986) is of the same vintage as Feden and

Clabaugh's, but the program she described, a bilingual special education teacher education program at George Mason University, lacks the contemporary flavor of the LaSalle program. In it, general education, special education, and bilingual education faculty contribute to the preparation of bilingual special education teachers, and bilingual and multicultural content is infused into existing courses. In this program, unlike Feden and Clabaugh's, distinctions among specialists are maintained. However, more clearly than Feden and Clabaugh, Valero-Figueira analysed the supports required to develop and institutionalize the teacher education program, and enumerated as most important administrative support, faculty support and participation, financial support from the institution, and compatibility with institutional priorities.

Maheady et al. (1993) described the Reflective and Responsive Educator or RARE program at the State University of New York, College at Fredonia (SUNY-Fredonia). RARE is a general education teacher education program in which classroom teachers are prepared to accommodate students whose learning needs differ as a result of cultural diversity or disability. The program has five major components, including (a) continuous field experience requirements from freshman to senior years, (b) systematic training in the application of established teaching and management practices, (c) training in collaboration, (d) systematic evaluation, and (e) a conceptual framework that emphasizes teacher reflection. It should be noted that Maheady and his colleagues do not use reflection in the same sense that it is commonly used in the teacher education literature. Their use is more closely akin to data-based decision making, and the general philosophy of the program is more eclectic than its title would imply. Thus, prominent among the established practices that RARE students learn is curriculum-based measurement; others include cooperative learning, classwide peer tutoring, and the K-W-L comprehension strategy.

This program is notable for a number of reasons. For one, it is a general education program at an institution at which no special education program exists. It seems extraordinary that a general teacher education faculty working to restructure its program would choose as a theme the accommodation of students with disabilities. Their doing so also attests to the appeal of the concept of accommodation and the potential for using it to bring together the interests of groups focused on diversity and disability. Also notable is the quality of the conceptual work that went into the program. It may be the best thought-out program yet to be described in the literature (with the possible exception of the two papers that follow). It is described in detail and completely, and the SUNY-Fredonia faculty seems to have achieved an enviable level of integrity in its curriculum. Perhaps most important from our point of view is the clear focus on program evaluation. The RARE faculty has assumed that their work has a discernible effect on their students, and that the work of their students has a discernible effect on the learning of the school children with whom they work. Only rarely does the evaluation of a teacher education program extend to the demonstration of graduates' effects on school students. It's also important to note that the restructuring that resulted in the RARE program was stimulated by a state department of education edict that required teacher education programs at public institutions across New York to restructure. Maheady et al. and their colleagues elsewhere in the SUNY system were charged with the responsibility of preparing elementary and secondary classroom teachers to work with students with disabilities and students who are culturally or linguistically different. This mandate was issued in 1989. If its effect at other institutions was similar to its effect at SUNY-Fredonia, then descriptions of other teacher education programs with themes of accommodation should be appearing in the literature soon.

Like the RARE program, the Multiple Abilities Program or MAP at the University of Alabama (Ellis et al., 1993) is complete, carefully thought out, and innovative. Graduates of the MAP are recommended for an experimental "Multiple Abilities" certificate, approved by the Alabama State Board of Education, which comprises two traditional certificates: Mild Learning and Behavior Disabilities (K-6) and Early Childhood and Elementary Education (K-6). There are 33 students currently enrolled in the program and being prepared to "meet the needs of students without regard for the labels students have or the settings in which elementary instruction takes place" (p. 38). Ultimately, the goal of the program is to alter school practice.

Both the program content and the teacher education methods are innovative. Teaching competencies, initially generated by a focus group of teachers and teacher educators, are organized into four domains, including professionalism, the learner, facilitating learning, and communication and collaboration. Cutting across these domains are three major themes: understanding child development and diversity, facilitating empowerment, and using authentic assessment and instructional practices. Courses are taught over the 5-semester duration of the program in the manner of a spiraling curriculum. Students are able to develop complex understandings that the time restraints of a semester and disassociation from practice (typical of many teacher education courses) preclude. There also is heavy emphasis on field experiences, which include three semester-long apprenticeships in elementary education, special education, and, where possible, inclusion classrooms.

Kemple, Hartle, Correa, and Fox (1994) described the Unified Program, an early childhood education/early childhood special education (ECE/ECSE) teacher education program they developed at the University of Florida. In their paper, they described in some detail how they established their collaboration and the barriers they had to overcome in the process. They also described how they reconciled their initially disparate world views and identified the rather substantial literature base in ECE/ECSE that supports collaboration. The participants shared commitments to inclusive educational practice, a family-focused approach to early intervention, and developmentally appropriate practice; and sensitivity to multicultural issues.

The program faculty had strong support from their home departments and the college, and cite the support of a college administrator (the assistant dean for student services) as being instrumental to their success at institutionalization. Their program is as an integrated junior/senior/master's program designed to prepare teachers to work with children with and without disabilities, from birth to 8 years of age. Graduates receive certification in preschool and primary education and Florida's Pre-Kindergarten Handicapped Endorsement.

The faculty's commitment to inclusive education brought the issue of program integrity to the forefront. They were concerned about the availability of practicum and student teaching placements where practice was inclusive. This issue may prove to be an even more difficult barrier for collaborative teacher education programs for elementary teachers, because the acceptance of inclusive practice is far less widespread in general education than early childhood education. Indeed, inclusion is essentially absent from the education and teacher education reform literatures. Collaborative teacher education will flourish only to the extent that inclusive settings are available for field experiences; where they're not available, teacher educators may have to work with school colleagues to create them. The professional development school work at the University of South Florida (Paul, Duchnowski, & Danforth, 1993) illustrates how that concept may be applied to the creation of inclusive school environments and the education of professionals to work in them.

The work of Kemple and her colleagues is built upon the belief that early intervention improves the long-term academic and social efficacy of children with disabilities and children at-risk (Berrueta-Clement, Schweinhardt, Barnett, Epstein, & Weikart, 1984; Lazar, 1988), a belief shared commonly by early childhood and early childhood special educators. By contrast, beliefs about the effectiveness of interventions for school-aged students with disability are varied and often in conflict. Lavely and McCarthy (1995) posited three reasons why early childhood educators and early childhood special educators are more likely to be successful at collaboration than educators serving older students: (a) differences among young children are less pronounced, (b) measures of success in early childhood programs are not limited to cognitive functioning or academic performance but include a strong focus on socialization of young children, and (c) family involvement is typical of programs in both early childhood and special education. This convergence of beliefs— and the divergence of beliefs about school-aged children—strongly suggest that early childhood teacher education ought to move more rapidly into collaborative teacher preparation than programs for teachers of school-aged children.

Collaborative Teacher Education: An International Perspective

Special education practices and the preparation of special education teachers in other countries are shaped by different conceptions of the purposes of general and special education, varied structures of educational systems, and diverse beliefs about the roles of people with disabilities in society (Jones, 1993). Consequently, teacher preparation programs in other countries vary widely. However, the inclusion of individuals with disabilities is as prominent an issue in Europe as it is in the United States, and inclusion has had an impact on the practice of teacher education there (Bladini, 1992; Dens, 1991; Diniz, 1991; Fernandes & Pinto, 1991; Metcalfe & Diniz, 1991; Preuss & Hofsass 1991; Watson, 1991).

In the Scandinavian countries, in particular, children are no longer described as having "learning difficulties." Rather, schools are described as having "teaching difficulties" (Bladini, 1992). Attending to special educational needs, therefore, is the responsibility of the whole school. As Dens (1991) explained, "This [viewpoint] promotes a problem solving attitude which is more directed towards negotiation and situation analysis than to the quick referral of the problem to someone else, perceived as a specialist in the matter" (p. 129). Consequently, teacher educators in Europe are rethinking the preparation of classroom teachers for the range of diversity they will encounter in student ability, behavior, and ethnic and cultural background and to prepare special education teachers to provide support, consultation services, and training to their [regular education] colleagues (Dens, 1991; Upton, 1990).

Teacher educators in European countries have defined the role of special educators as change agency (Askeroi, 1988; Bladini, 1992; Dens, 1991; Fernandes & Pinto, 1991; Metcalfe & Diniz, 1991; Preuss & Hofsass, 1991). The role of a change agent is not to diagnose and treat individual children but rather to appraise the effectiveness of the school's efforts to educate the child. Teacher educators in Sweden (Bladini, 1992), England (Metcalfe & Diniz, 1991), West Germany (Preuss & Hofsass, 1991), and Portugal (Fernandes & Pinto, 1991) have described their efforts to prepare teachers to be change agents.

As other countries expand and modify their special education programs and as the American educational system assumes a more multicultural and international orientation, opportunities for international collaborative work among teacher educators are increasing (Jones, 1993). It would appear that Europeans have addressed many of the same issues facing special educators in the United States. We have much to offer one another and much to gain by developing mutually beneficial relationships.

Obstacles to Collaborative Teacher Education

As we noted in our introduction, the need for collaborative teacher education has been recognized and articulated in the literature for the past two decades. Yet, as is clear from the scarcity in the literature of program descriptions, progress has been slow. The obstacles to restructuring teacher education are as numerous and challenging as those facing the restructuring of public schools. In the following section, we briefly review several obstacles that must be overcome in order to create collaborative teacher education programs.

Bureaucratic Constraints

Schools of education, like other public sector bureaucracies, have a stake in stability and predictability (Schön, 1983). Professionals working within bureaucracies are hampered by systems of rules and regulations and often have difficulty focusing on the core function for which the organization was developed. Program reform in colleges and schools of education is often constrained by complex and cumbersome state certification and licensure regulations as well as academic requirements within the institution itself (Lilly, 1992; Rude, 1994). Certification and institutional requirements are even more burdensome for teacher educators attempting to develop collaborative programs. Existing bureaucratic structures within colleges and schools of education also work against communication and collaboration among teacher educators across departments (Pugach, 1992), but as described by Maheady et al. (1993), Ellis et al. (1995), and Kemple et al. (1994), teacher educators can overcome significant bureaucratic challenges in developing collaborative programs.

Lack of a Common Culture

Collaborative efforts among general and special educators are hampered by what Erickson (1987) referred to as a lack of common culture. The cultures of special education and general education conflict in terms of terminology, values, beliefs, and practices (Lilly, 1989; Pugach, 1992). Besides, general and special educators do not have a history of collaboration (Bondy, Ross, Sindelar, & Griffin, in press). Communication and collaboration between special educators and general educators often becomes problematic due to the differences in the ways they define teaching and learning and the professional vocabulary they use to describe educational phenomena (Bondy et al., in press). In the following paragraphs, we present an overview of the differences between special and general educators that have made collaboration difficult.

Special Education. Special education is a complex field, comprising programs ranging from neonatal intensive care to supported employment. Special education professionals tend to define their work by dimensions of disability category, degree of severity, and age. Despite the diversity of the field, special education practice has a clear positivistic (if no longer behavioral) perspective on teaching and learning. Effective and practical approaches for working with infants, toddlers, children, and youth with disabilities, and their families have roots in applied behavior analysis. Special educators believe that "some children, including many of those at risk and those with disabilities, require more extensive, structured, and explicit instruction" (Harris & Graham, 1994, p. 238). The preparation of special education teachers typically emphasizes mastery of competencies related to individual differences and instructional adaptation. Methodology focuses on individuals, not groups, so special education methods are likely to have only limited utility in general education classrooms.

General Education. General education is defined by distinct content areas (reading, math, language arts) and age group domains (early childhood, elementary, middle, and secondary school). Despite the range of age groups served and the content areas covered, one can argue reasonably that a constructivist perspective on teaching and learning (Bredekamp, 1987; Calkins, 1986; Dewey, 1990; Goodman, 1990; Kamii, 1984; Zemelman, Daniels, & Hyde, 1993) informs much of contemporary general education research and practice. Professional education groups such as the National Association for the Education of Young Children and the National Council of Teachers of Mathematics have revised their standards of practice to reflect constructivist theory. Constructivism has a long history with general education and is rooted in the educational theories of Dewey, Piaget, Vygotsky, and others. Constructivists view children as active learners who construct knowledge through interaction with and immersion in meaningful contexts and authentic learning tasks. Children's prior understandings shape new learning. The notion that children passively respond to the environment and learn through the transmission of knowledge by others is rejected as is the idea that demonstrated success on basic skills is a necessary prerequisite to more advanced learning and higher-order thinking. Constructivists view the role of the teacher as that of facilitating student learning rather than explicitly providing knowledge and information.

Stereotypical Perceptions. The lack of a common culture between special and general education has led to stereotypical perceptions among teacher educators about each others' educational philosophy and practices (Bondy et al., in press; Harris & Graham, 1994). General educators have characterized the behavioral approaches associated with special education as mechanistic attempts to "pour knowledge into children's minds with little consideration for children's understandings and use of this knowledge or even for the children themselves" (Harris & Graham, 1994, p. 242). General educators interpret phrases such as "diagnosis of skills and competencies" and "remediation of deficiencies" as evidence of an assembly line model of teaching. Similarly, special educators hold stereotypical views of general educators.

We've characterized constructivist approaches as creating chaotic classrooms that ignore the needs of students with disabilities and render them hopelessly confused (Mather, 1992). Special educators question whether anyone is teaching or learning in classrooms where "whole language" and "discovery learning" are practiced. These stereotypical views of colleagues' perspectives impede the open communication essential to developing shared understandings and common goals (Fullan, 1991; Goodman, 1988; Rudduck, 1990).

Who Is Responsible for Educating Children with Special Needs?

A belief shared by general and special educators is the idea that special educators possess special skills to educate exceptional children and, therefore, must bear special responsibility for educating them. This belief is pervasive and has profoundly impacted practices in schools and teacher education (Sindelar, Pugach, et al., 1994; Welch & Sheridan, 1993). Pugach (1988) has argued that the existence of a separate sytem of special education has led general educators to relinquish responsibility for educating children who experience difficulty in school. Teacher educators in general education prepare their students to teach typically developing children, but not children with special needs. Consequently, general educators in the schools refer students experiencing academic and behavioral problems to special educators. This shared belief in the efficacy of a separate system of special education is an additional barrier to the development of both integrated services within the schools and collaborative teacher education programs.

Overcoming the Obstacles

Despite these considerable obstacles, some teacher educators have been successful in their efforts to create collaborative programs. In the following section, we describe strategies some have used in overcoming barriers and establishing a thematic focus on accommodation.

Managing the Beauracracy

Negotiating a bureacracy is a daunting task. Initiating and sustaining change requires time, energy, and administrative support—resources often in short supply. Reward structures within schools and colleges of education must be altered to motivate faculty to engage in the hard work of teacher education reform. Teacher educators who have been successful in their collaborative efforts have garnered the support of administrators in their institutions and in state departments of education (Affleck & Lowenbraun, 1995; Ellis et al., 1995; Kemple, et al., 1994; Maheady et al., 1993; Pugach, 1992). Kemple et al. described the support provided by a college administrator and the head of early childhood programs in the state department of education. Affleck and Lowenbraun (1995) and Pugach (1992) described ways in which deans of colleges of education created infrastructures to provide forums for teacher educators to come together to discuss and plan for reform. Maheady and his colleagues restructured their program in response to a state agency mandate, while Ellis et al. had an experimental certificate approved by their SEA. Clearly, the prompting or assistance of key administrators can spur the development of collaborative programs.

Visionary leadership is needed to help create the institutional conditions necessary for collaborative endeavors (Sindelar, Ross, et al., 1995). Howey (1990) described the need for special leadership and new networks among stakeholders to support the development of innovative teacher education programs. Fundamental changes in the ways in which we prepare teachers will require strong leadership from within schools of education and the public schools.

Creating a Common Culture

Earlier, we argued that special education and general education lack a common culture. Nevertheless, some teacher educators have been able to find common ground and build a foundation for collaboration. Affleck and Lowenbraun (1995) and Pugach (1992) described the importance of creating a time and place to come together and share ideas. Conversations among special educators and general educators, however, can be marked by conflict or avoidance due to the differences in professional language and stereotypical thinking (Bondy et al., in press; Kemple et al., 1994; Maheady et al. 1993). Kemple and her colleagues (1994), however, asserted that differences in language may be easily resolved. Unified faculty "made concerted efforts to use general terms. . . while educating one another in the specific terminology of their respective areas" (p.46). Others (Bondy et al., in press; Maheady et al., 1993) have noted the importance of exploring colleagues' perspectives rather than reacting negatively to differences in terminology. By communicating clearly and attempting to understand others' perspectives, reformers have found common values and have developed a shared commitment to a vision and the collaboration required to achieve it.

Although conflict between positivistic and constructivist world views continue to be a source of contention between special and general educators, special educators have begun to investigate constructivist theory as a potential means for understanding and addressing the needs of children with disabilities. The Fall, 1994, issue of the *Journal of Special Education* was devoted to the application of constructivist pedagogy to the education of students with disabilities. It was hoped that special educators would consider "the strengths, potential limitations, and issues represented by the constructivist approach for students with disabilities and those at risk for school failure" (Harris & Graham, 1994, p. 233). A growing number of special educators have argued that the field of special education has been constrained by a disgnostic prescriptive model of teaching and that alternative perspectives on teaching and learning need to be considered (Blanton, 1992; Skrtic, 1991). Graham and Harris (1994) proposed a future in which teachers and teacher educators "construct a stance based on pragmatic, sincere, and productive rapprochment among multiple paradigms" (p. 245). Special educators appear to be developing an appreciation for diverse perspectives, a significant step toward developing shared understandings among special and general educators.

At the same time, general educators have begun to understand both the complexities posed by student heterogeneity and the limitations of traditional instructional methods (Boyer, 1983; Goodlad, 1984, 1986; Sizer, 1984; Slavin, 1990). Boyer (1983) and Goodlad (1984) emphasized the monotonous landscapes of general education classrooms and called for instructional innovation and experimentation. Slavin (1990) and others (Goodlad, 1986) have argued persuasively for heterogeneous grouping and adapting instruction for individual needs. Sailor (1991) agreed; he noted that "sufficient parallels exist between the general and special education reform agendas to suggest that the time may be at hand for a shared educational agenda" (p. 8).

Are Differences a Barrier or a Source of Strength?

Although we have described the differences between special education and general education as barriers to collaboration and the development of integrated teacher education programs, we believe that these differences do not have to be reconciled in order to initiate collaboration. As Graham and Harris (1994) stated, no single "intervention or approach can address the complex nature of school success or failure" (p. 245). Faculty in collaborative teacher education programs (Bondy et al., in press; Kemple et al., 1994) have commented that recognition and respect for colleagues' expertise is an essential part of the foundation for preparing teachers to teach all children. In fact, professional respect for and recognition of the benefits of multiple perspectives may be a cornerstone of collaboration. Differences in perspectives may enable preservice students to grasp the complexities of teaching and learning and provide them with a range of possible avenues to prevent learning and behavioral problems and to accommodate all of their students.

Sharing the Responsibility for Educating Children with Special Needs

It is perhaps this issue that poses the most compelling reason for collaboration. The challenges posed by students in the public schools demand a collaborative response from all school professionals. As long as children experiencing academic and behavioral difficulties are viewed as someone else's responsibility, their education will suffer. The problems these students face in succeeding in school require the commitment of the entire school system: administrators, teachers, and support personnel alike. Learning to accept and share responsibility for students with special needs should begin during preservice preparation and continue during teachers' inservice years (Pugach, 1992).

Sharing responsibility for students with special needs has been incorporated into early childhood programs faster than K-12 programs. Kemple et al. (1994) suggested that the success of their collaboration depended, in part, on their shared commitments to inclusiveness, developmentally appropriate practices, and the use of least intrusive placements and most normal strategies for intervention. Commitment to these themes could also serve as a foundation for collaborative programs serving older children as well.

Uniting General Education and Special Education Reform

In their relationships to one another, general and special education are like two people sitting back to back. They abut, but have no easy way to communicate. They have different perspectives on the world. Thus, in their concern for students who do not succeed in schools, general educators focus on children from diverse cultural and linguistic backgrounds. As they engage in the restructuring of schools and talk about accommodating diversity, they need only to expand their definition of diversity to connect with the special education reform agenda. We must agree that diversity may result from disability in the same sense it results from cultural and linguistic difference.

On the other hand, as special educators advocate restructuring for the inclusion of special education students, they need to understand the full range of issues that schools must address. Inclusion will not be successful in the current political context of schools if special educators are not involved in all aspects of reform (Stainback & Stainback, 1987). We believe that general educators have much to contribute to the conversation about including students with disabilities successfully. We also believe that special educators have much to contribute to broader conversations of school reform. The problem seems to boil down to defining the issue in a way that inspires both groups to commit to it.

Contemporary Trends and Issues in Special Education Teacher Education

Our argument for collaborative teacher education programs is based on the assumption that what classroom teachers need to know and to be able to do, and the dispositions they must have about their work and the students they teach, match the skills, knowledge, and dispositions essential for effective special education teaching. This is a fairly common (Feden & Clabaugh, 1986; Pugach, 1992) and long-held (Reynolds & Birch, 1977) assumption, sometimes expressed tautologically as "good teaching is good teaching is good teaching." Yet the generalization does not apply to work with all students with disabilities. It's been argued elsewhere (Pugach, 1992) that teachers who work with students with sensory impairments, severe behavioral disorders, and severe and profound developmental disabilities, for example, require knowledge and skills that a generalist would not have mastered.

The most obvious examples of distinct skills are those of teachers of students who are visually impaired or hearing impaired. Teaching braille reading or sign language, for example, are skills that teachers of visually impaired and hearing impaired students need. Likewise, the sophisticated crisis management and intervention skills of teachers of students with severe emotional or behavioral disorders differentiate their repertoires from that of a generalist. Even the accommodation of students with severe or profound disabilities in general education classrooms requires specialized skills and knowledge that not all teachers need to have.

Because of these distinct repertoires and the needs they serve in schools, teacher education programs in special education are, and are likely to remain, alive and well. We do not intend to suggest that collaborative work with general education colleagues will replace special education teacher education. Rather, general teacher education as we envision it will require the involvement of special education teacher educators. We must rethink our work and recognize both our role in the preparation of classroom teachers and the role of our general education colleagues in the preparation of special education teachers. But we can't wait to be asked into the conversation on reform (Pugach, 1992; Sapon-Shevin, 1988). The work at the University of Washington described by Affleck and Lowenbraun (1995) illustrates vividly how a special education faculty participated in a process of reform and created for itself a role in the general teacher education program.

In the sections that follow, we consider some contemporary issues in special education and their implications for teacher education and the preparation of school specialists. Our list of issues is selective and not exhaustive; it includes (a) categorical versus noncategorical certification, (b) critical teacher shortages, (c) attrition and retention, and (d) professional certification.

Categorical Versus Noncategorical Certification

Using the term *versus* in this heading may suggest more contention than has appeared in the literature. Beginning with Reynolds's 1979 paper, the issue has surfaced periodically in the special education literature, almost invariably in the form of a proposal for noncategorical certification (Sindelar, McCray, & Westling, 1993). What's more, the research that's focused on the issue has confirmed that for categories of mild disabilities, at least, differentiated preparation does not produce differentiated outcomes. For example, Gable, Hendrickson, Shores, and Young (1983) demonstrated that teachers don't plan or conduct themselves differently based on the disability classifications of their students. Furthermore, Marston and his colleagues (Marston, 1987; O'Sullivan, Marston, & Magnusson, 1989) have shown teachers to be as effective with students they are not certified to teach as they are with students they are certified to teach.

This body of research has had little effect on states' certification practices (and perhaps rightfully so given the small number of studies), and categorical certification is alive and well. Three surveys conducted over the past two decades (Belch, 1979; Chapey, Pyszkoski, & Trimarco, 1985; Cranston-Gingras & Mauser, 1992) demonstrate this point. The results of these studies have shown that (a) there has been little change in the years separating the first and last of these surveys in the proportion of states with strictly noncategorical certification structures (22% to 27%), (b) there has been a dramatic growth in the proportion of states offering a noncategorical certification option, and, (c) in spite of (b), there has been little change in the proportions of states maintaining categorical certification (Sindelar, Pugach, et al., 1994).

Progress toward the noncategorical certification of special education teachers, such as it is, may be stemmed by an action of the Council for Exceptional Children (CEC), which, in promulgating knowledge and skills standards for teachers, has maintained distinctions based on categories of disability ("CEC Finalizes," 1995; Swan & Sirvis, 1992). We recognize that professional standards and self-regulation are essential for the professionalization of teaching, of course, and applaud CEC's effort to develop them. At the same time, we regret the continued reliance on categories of disability for defining professional specialization. Disability categories simply are not functional in this sense. Knowing that children are classified as learning disabled, for example, tells us little about the particular nature of their learning problems and less about how to remedy them. That the knowledge and skills essential for effective teaching (and teacher education) can be differentiated by disability categories seems even less probable. Furthermore, the notion of teacher as generalist, as we have advocated in this chapter for elementary teachers and teachers of students with mild disabilities, might help link special education reform with broader programs of education reform.

It seems to us that categorical certification in special education, having survived a couple of decades of strong, consistent criticism, is alive and kicking. Furthermore, if we are correct about the potential effects of CEC's adoption of disability specific standards for teachers, it may be stronger now than it has been over the past 20 years.

Critical Teacher Shortages

Before we discuss shortages of special education teachers, we'd like to elaborate on a point we made earlier in this chapter about the potential of collaborative teacher education to ameliorate them. When the problem of shortages is couched in terms of whether schools have enough qualified special education teachers, only one solution is presented: to increase the supply of special education teachers. Such thinking may lead to emergency credentialing as a solution, for example, because it increases supply (albeit in a questionable manner). When the problem is posed in a slightly different way and whether schools have enough professional expertise to accommodate special education students successfully is asked, the number of possible solutions is expanded to include the development in classroom teachers of expertise in dealing with students with disabilities. However, the issue has been and continues to be (Simpson, Whelan, & Zabel, 1993) framed as shortages of special education teachers, not shortages of expertise in dealing with students with disabilities.

In 1991–92, the most recent year for which such data are available, 27,282 special education teachers were needed in the states and territories to fill vacancies and replace less-than-fully certified teachers (U.S. Department of Education, 1994). This total represented nearly 9% of all special education teachers employed, and an additional 2,288 teachers were needed for positions in programs for children with disabilities aged 3 to 5. Five years earlier, 26,798 teachers were needed and the ratio of vacancies to teachers employed (expressed as a percentage) was just over 9% (U.S. Department of Education, 1989).

Teacher shortages are not consistent across states, regions within states, and disability areas. U.S. Department of Education data (1994) demonstrate this first point. In 1991–1992, the magnitude of shortages (vacancies as a percentage of employed teachers) ranged from less than 1% in Wyoming to 28% in New York. Shortages did not seem to vary with state size. As measured by total number of teachers employed, some larger states (e.g., California, Illinois, Pennsylvania, and Texas) had below average shortages (5.8%, 2.1%, 1.9% and 5.8%), while some smaller states (Idaho, Montana, New Hampshire, and New Mexico) had above average shortages (26.4%, 12.6%, 15.8%, and 17.6%).

Shortages are not uniform within state boundaries either. In their studies of teacher supply and demand in Wisconsin, Lauritzen and Friedman (1993) illustrated this point explicitly. Shortages of special education teachers ranged from "extreme" to "slight" and occurred in 9 of 12 regions of the state and in Milwaukee. In three east central regions, by contrast, supply was considered "normal to demand." This supply and demand picture differed dramatically from that of elementary education. For this field, in all but the extreme southwestern region of the state, where oversupply was "slight," and in Milwaukee, where shortages were "extreme," oversupply was judged to be "extreme" throughout the state.

Shortages also vary by disability category. In 1991–1992, shortages ranged from 6.9% for cross-categorical to 51.5% for traumatic brain-injury (U.S. Department of Education, 1994). Whereas shortages in larger categories like learning disabilities, speech, and mental retardation were relatively small (8.2%, 9.0%, and 7.1%), the shortage of teachers of students classified as seriously emotionally disturbed, the fourth largest category, was nearly twice the overall average (16.0%).

The concept of "reserve pool" may help illuminate the issue of shortages, however. In her studies of the career paths of special education teachers, Singer (1993a, 1993b) has shown that from one-third to one-fourth of all special education teachers who leave teaching return to teaching within 5 years of their departure. With transfers from other schools, teachers from the reserve pool constitute nearly three-fourths of all new hires (Bobbitt, Faupel, & Burns, 1991). Graduates of teacher education programs probably constitute little more than one-fourth of all new hires (E. E. Boe, personal communication, April 28, 1995).

One of the strategies states have adopted to address the problem of less than fully certified teachers is the issuance of temporary or emergency teaching credentials and the offering of alternative certification programs. The data base on the effectiveness of alternative certification in special education is limited and inconclusive (Sindelar & Marks, 1993). More at issue, however, is the appropriateness of what is clearly a secondary, subject matter approach to teacher education (i.e., alternative certification) to a discipline in which teaching methods play a more prominent role than teaching content. Although it may make some sense for a biologist, for example, to become a biology teacher via an alternative route in which pedagogical training is abbreviated, the same does not hold true in a field in which basic skill instruction is fundamental. For this reason, and in spite of evidence to suggest that alternative programs can work well when done right (Rosenberg & Rock, 1994), it seems regrettable that alternative routes to special education certification are proliferating (Buck, Polloway, & Robb, in press).

Among the categories of professionals who work with students with disabilities, shortages are not limited to teachers (U.S. Department of Education, 1994). Occupational (15.8%) and physical therapists (19.8%), for example, are in high demand by schools, at least in part as a result of the higher salaries available to OTs and PTs from private sector and medical employers. Among speech pathologists and audiologists, too, employment in schools may not be preferred, and as a result, schools experience shortages of both. In 1991–92, 3,907 (9.0%) teachers for students with language impairments were needed along with 97 (9.7%) audiologists. The shortage of speech pathologists has probably been exacerbated by the recent promulgation by the American Speech, Hearing, and Language Association (ASHA) of a requirement for a master's degree for ASHA certification and entry into the profession. School districts are decrying the requirement because it has further restricted the supply of speech professionals available to them.

On the other hand, shortages of district (3.8%) and state (5.4%) administrators and supervisors are modest, as are those of school psychologists (5.9%) and counselors (6.6%), perhaps because school administrators, school psychologists, and school counselors have fewer alternative employment opportunities available to them. Nonetheless, with the downward extension of IDEA mandates to age 3 and the implementation of Part H of the law, the importance of establishing and maintaining adequate supplies of related service professionals has never been greater. These two events have also served to highlight the importance of collaborative training among the professions identified in Part H.

Attrition and Retention

Teacher attrition and retention captured the attention of the Office of Special Education Programs (OSEP) some years ago when they were first considered critical aspects of the supply and demand problem and what was known about them seemed confused and inconclusive. At that time, estimates of special education teacher attrition ranged from 7.3% nationally to 19% for the state of Illinois (Brownell & Smith, 1993), and differences in the attrition rates of special and general education teachers were estimated to be as large as 8% in Wisconsin. Given the small amount of available data, its lack of clarity, and the importance of reducing attrition, OSEP funded a research priority to support studies of these phenomena in urban schools. Three projects were funded in this directed priority and a fourth the next year as a field initiated study. The work of these four projects is nearing completion, and a sense of the size and shape of the problem of attrition is beginning to emerge. Estimates from these projects have varied widely but range from 7.1% to 13.1%, and average 9.1% (Pyecha & Levine, 1995). Although this estimate is surprisingly small, attrition among special educators was reliably greater than attrition among general educators.

These projects promise to have a pronounced impact on the problem of retention, because in addition to estimating the amount of attrition in urban school districts, the researchers were expected to identify factors related to attrition and develop school-based strategies to promote retention. A great deal is known about factors related to attrition in general education, and in two recent literature syntheses (Billingsley, 1993; Brownell & Smith, 1992), the implications of this body of research for special education were educed.

These papers helped to dispell the notion that higher attrition among special education teachers could be attributed to stress associated with working with students with disabilities, handling the paperwork requirements of special education, or dealing with such external constituents as parents and general education teachers. Billingsley (1993) argued that although the attrition rate of special education teachers was reliably greater than the attrition of classroom teachers, job stress does not reliably differentiate between the two groups. Nonetheless, the difficulty of working with challenging students has been cited by teachers leaving special education employment as a factor in their decision to leave (Brownell & Smith, 1992).

Age, gender, and academic ability are teacher characteristics that relate to attrition. That younger teachers (and older teachers—the relationship is U-shaped), more capable teachers, and women are more likely to leave special education teaching is especially problematic because teachers in our field are predominately female and, on the average, younger than general education teachers. Also, teachers' sense of professional efficacy is associated with reduced rates of attrition. Certain student variables also pertain; attrition is known to be elevated among teachers of students with behavioral disorders and among teachers employed in multicategorical programs. Among the workplace variables that also relate consistently to the probability of attrition are (a) role ambiguity or role conflict; (b) support, especially support from administrators; (c) opportunities for collaboration; and (d) participation in decision-making.

It is certainly true, as both Brownell and Smith (1992) and Billingsley (1993) suggest, that the relationships among these variables are complex and that the effects of individual factors may be mediated by intervening variables, such as job satisfaction. And although direct, functional correspondences seem improbable, this research, along with the studies being done with OSEP sponsorship, will serve to identify for employing districts the kinds of factors to which their retention efforts may be best directed.

National Certification

As we mentioned previously, CEC has promulgated essential knowledge and skill standards for beginning special education teachers. Because it also serves as the special education proxy for NCATE, CEC is now well-positioned to establish a process of national professional certification. There would seem to be little reason to oppose CEC's move into the national arena. In fact, given the current assault on the professionalization of teaching, its timing could not be better.

However, that special educators can agree on a certification structure to codify in standards is doubtful. Furthermore, the fact that special education's relationship to general education is being redefined is certain to have important implications for the way we certify teachers. For example, some have argued for a complete unification of general and special education teacher education (Stainback & Stainback, 1987), and others have called for the common training of classroom teachers and teachers of students with mild disabilities (Sindelar, Pugach, et al., 1994). To presume that we have a clear, agreed-upon structure upon which to build a national certification process seems wrong. Given that neither the nature of special education practice in the future nor how best to educate and certify teachers can be foretold, and given the potential for the ossification of standards, it seems premature to initiate this process today.

In "Who Will Teach? Who Will Serve?" (National Clearinghouse on Professions in Special Education, 1993), it was assumed that "rigorous national standards will assure that all professionals in special education, related services, and early intervention have the knowledge and skills to provide competent high quality services to all children with disabilities" (p. 14). Yet all teacher educators have known students who complete programs satisfactorily but fail to develop into competent professionals. We also all know gifted "teachers," often experienced paraprofessionals, who have no formal education in teaching students with disabilities. So completing a teacher education program, however well-crafted and well-aligned with a template of standards, is no guarantee that an individual will become a competent teacher, as this assumption implies. In fact, to assume that teaching competence results automatically from the satisfaction of requirements contradicts the experience of teacher educators, oversimplifies the process of becoming a teacher, and is indefensible for these reasons.

In "Who Will Teach? Who Will Serve?", concern also is expressed about "tremendous state-to-state discrepancies in the official standards that define a 'qualified' special educator" (p. 6). Tremendous state-to-state discrepancies in the competence of the teaching workforce would be greater cause for concern, but no evidence of such discrepancies exists. In fact, the entire area of certification and accreditation is devoid of research findings, a fact that led Lilly (1992) to conclude that "licensure and state program approval. . . are predominantly political functions, and those who engage in such activities . . . depend little on research to inform their planning and actions" (pp. 157–158).

Proponents argue correctly that the promulgation of national certification standards will promote the professionalization of teaching. Professions are defined, in part, by control over entry and shared standards of practice. However, national standards alone are not sufficient for professionalization, and we wonder whether teachers themselves would realize any benefit from national certification. Teachers everywhere are poorly paid, overworked, and underappreciated. Their caseloads are so large as to prevent them from achieving the individualization of instruction that defines special education.

Teachers spend their own money to purchase basic supplies for their classrooms. Often, their sense of professional isolation is redoubled by the physical isolation of working in portable classrooms apart from their colleagues in the school building.

Will teachers operating in such a context feel a greater sense of professionalism because they have a national certificate hanging on the wall? We think not. In the professionalization of teaching, there are more serious and fundamental problems to overcome than the lack of national certification. To pursue it without vigorous efforts to improve the working conditions of teachers, for example, seems fruitless.

Proponents also argue that national certification will promote reciprocity among states, thereby facilitating the movement of teachers to areas of high demand (Swan & Sirvis, 1992). This assertion belies the facts that the vast majority of teachers are white, middle class women, who are unlikely to move from state to state (Lauritzen 1990a, 1990b; Nicholas, 1990). It also presumes that the same state bureaucracies that have bollixed the implementation of existing reciprocity agreements will implement national certification with aplomb. We are not optimistic about this possibility.

Because the American public has questioned the importance of teacher education, the goal of increased professionalism might be more readily achieved by engaging in systematic evaluation of our graduates, thereby demonstrating that we do make an important difference (Sindelar, Ross, et al., 1994). Judgments about the adequacy of programs then could be based on the competence of graduates rather than on how well their requirements fulfill standards. Lilly stated this argument persuasively: "Education is moving rapidly toward models of accountability that will be more outcomes-driven than process driven. Regulatory systems will give way to considerable flexibility in how things are done, and strong emphasis on outcomes assessment" (p. 158). If Lilly is correct, then wouldn't we be smart to be thinking about this issue more earnestly?

Summary

Regardless of where one sits on the issue of inclusion, the need to prepare classroom teachers for work with students with disabilities must be evident. Growing numbers of the students are experiencing increasingly difficult problems in school, and classroom teachers must be capable of helping them. Special education and other compensatory programs—their capacities already stretched thin—cannot deal with numbers as large as these. We agree with Goodlad when he argues that we seem to have reached the end of our ability to solve educational problems with solutions that involve the organization or grouping children.

We believe that the preparation of classroom teachers with skills and knowledge about, and positive dispositions toward, teaching students with special needs will require collaboration between general and special education teacher educators. Historically, these groups have not worked together in any sort of collaborative fashion, and such barriers as the lack of a common professional culture and distinct reform agendas must be overcome to initiate the process. The existence of collaborative programs demonstrates that these barriers can be overcome, and a common ground–concern for students experiencing difficulty in schools—is beginning to emerge. We are optimistic about the prospects of increased collaborative effort in teacher education.

We also believe that the creation of a better classroom teacher would help address the chronic problem of teacher shortages in special education. Certainly, more expertise would be available in schools for work with students with disabilities, even if more certified special education teachers weren't. The issue of critical teacher shortages is only one of several that confront special education teacher education today. About another, noncategorical certification, we've concluded that categorical labeling, service delivery, teacher education, and certification are probably here to stay. Also, the movement toward national professional certification may cast in stone a structure for preparing and certifying teachers at the same time special education is in the process of redefining itself as a profession, and our continued reliance on specialization may be antithetical to elements of the more general reform agenda.

References

Adler, S., & Goodman, J. (1986). Critical theory as a foundation for methods courses. *Journal of Teacher Education, 37*(4), 2–8.

Affleck, J. Q., & Lowenbraun, S. (1995). *Managing change in a research university special education program.* Manuscript submitted for publication.

Asante, M. K. (1991). The Afrocentric idea in education. *Journal of Negro Education, 62,* 170–180.

Askamit, D. L. (1990). Practicing teachers' perceptions of their preservice preparation for mainstreaming. *Teacher Education and Special Education, 13,* 21–29.

Askeroi, E. (1988). In-service training in special education as research and innovation work. *European Journal of Teacher Education, 11,* 157–167.

Association of Teacher Educators. (1991). *Restructuring the education of teachers.* Reston, VA: Author.

Behrens, T., & Grosenick, J. K. (1978). Deans' grants projects; supporting innovations in teacher-education programs. In J. K. Grosenick & M. C. Reynolds (Eds.), *Teacher education: Renegotiating roles for mainstreaming* (pp. 1–5). Reston, VA: Council for Exceptional Children.

Belch, P. (1979). Toward noncategorical teacher certification in special education: Myth or reality? *Exceptional Children, 46,* 129–131.

Berrueta-Clement, J. Schweinhart, L., Barnett, W. S., Epstein, A., & Weikart, D. (1984). *Changed lives: The effects of the Perry Preschool Program on youths through age 19.* Ypsilanti, MI: High/Scope Press.

Billingsley, B. S. (1993). Teacher retention and attrition in special and general education: A critical review of the literature. *Journal of Special Education, 27,* 137–174.

Bladini, U. B. (1992). From teachers of backward children to agents for educational change. *Teacher Education and Special Education, 15,* 307–313.

Blanton, L. P. (1992). Preservice education: Essential knowledge for the effective special education teacher. *Teacher Education and Special Education, 15,* 88–96.

Bobbitt, S. A., Faupel, E., & Burns, S. (1991). *Characteristics of stayers, leavers, and movers: Results from the teacher follow-up survey of 1988–89.* Washington, DC: U.S. Department of Education, National Center for Education Statistics.

Bondy, E., Ross, D. D., Sindelar, P. T., & Griffin, C. C. (in press). Planning to prepare teachers for inclusive education: The purposes and processes of team building. *Teacher Education and Special Education.*

Boyer, E. (1983). *High school: A report on secondary education in America.* New York: Harper & Row.

Braaten, S. R., Kauffman, J. M., Braaten, B., Polsgrove, L., & Nelson, C. M. (1988). The Regular Education Initiative: Patent medicine for behavioral disorders. *Exceptional Children, 55,* 21–27.

Bredekamp, S. (Ed.). (1987). *Developmentally appropriate practices in early childhood programs serving children from birth through age 8.* Washington, DC: National Association for the Education of Young Children.

Brownell, M. T., & Smith, S. W. (1992). Attrition/retention of special education teachers: Critique of current research and recommendations for retention efforts. *Teacher Education and Special Education, 15,* 229–248.

Brownell, M. T., & Smith, S. W. (1993). Understanding special education teacher attrition: A conceptual model and implications for teacher educators. *Teacher Education and Special Education, 16,* 270–282.

Buck, G., Morsink, C., Griffin, C., Hines, T., & Lenk, L. (1992). Perservice training: The role of field-based experiences in the preparation of effective special educators. *Teacher Education and Special Education, 15,* 108–123.

Buck, G. H., Polloway, E. A., & Robb, S. M. (in press). Alternative certification programs: A national survey. *Teacher Education and Special Education.*

Calkins, L. (1986). *The art of teaching writing.* Portsmouth, NH: Heinemann.

Carnegie Forum on Education and the Economy Task Force on Teaching as a Profession. (1986). *A nation prepared: Teachers for the 21st century.* New York: Author.

CEC finalizes national certification standards for special educators. *CEC Today, 1*(11), 1,5.

Chapey, G. D., Pyszkoski, I. S., Trimarco, T. A. (1985). National trends for certification and training of special education teachers. *Teacher Education and Special Education, 8,* 203–208.

Clift, R. T., Houston, W. R. & Pugach, M.C. (1990). (Eds.). *Encouraging reflective practice in education.* New York: Teachers College Press.

Cochran-Smith, M., & Lytle, S. L. (1992). Interrogating cultural diversity: Inquiry and action. *Journal of Teacher Education, 43*(2), 104–115.

Comer, J. P. (1989). Racism and the education of young children. *Teachers College Record, 90,* 352–361.

Comer, J. P. & Haynes, N. M. (1991). Parent involvement in schools: An ecological approach. *Elementary School Journal, 91,* 271–277.

Corrigan, D. C. (1978). Public law 94–142: A matter of human rights; a call for change in schools and colleges of education. In J. K. Grosenick & M. C. Reynolds (Eds.), *Teacher education: Renegotiating roles for mainstreaming* (pp. 17–29). Reston, VA: Council for Exceptional Children.

Cranston-Gingras, A., & Mauser, A. J. (1992). Categorical and noncategorical teacher certification in special education: How wide is the gap? *Remedial and Special Education, 13*(4), 6–9.

Cuban, L. (1984). School reform by remote control. *Phi Delta Kappan, 66*(3), 213–215.

Cummins, J. (1986). Empowering minority students: A framework for intervention. *Harvard Educational Review, 56,* 280–298.

Dens, A. (1991). Integration of disabled pupils in the regular school system in the Dutch speaking community of Belgium. *European Journal of Teacher Education, 14,* 117–130.

Dewey, J. (1990). *The school and society and the child and the curriculum.* Chicago: University of Chicago Press.

Diniz, F. A. (1991). Special education: An overview of recent changes. *European Journal of Teacher Education, 14,* 107–116.

Duchnowski, A., Dunlap, G., Berg, K., Adiegbola, M. (1995). Rethinking the participation of families in the education of children: Clinical and policy issues. In J. L. Paul, H. Rosselli, & D. Evans (Eds.), *Integrating school restructuring and special education reform* (pp. 188–213). Fort Worth, TX: Harcourt Brace College Publishers.

Ellis, E. S., Rountree, B., Casareno, A. B., Gregg, M., Schlichter, C. L., Larkin, M. J., & Colvert, G. C. (1995). The multiple abilities program (MAP): Integrated general and special education teacher preparation. *Alabama Council for Exceptional Children Journal, 12*(1), 37–46.

Erickson, F. (1987). Conceptions of school culture: An overview. *Educational Administration Quarterly, 23*(4), 11–24.

Feden, P. D. & Clabaugh, G. K. (1986). The "new breed" educator: A rationale and program for combining elementary and special education teacher preparation. *Teacher Education and Special Education, 10,* 58–64.

Fender, M. J., & Fiedler, C. (1990). Preservice preparation of regular educators: A national survey of curricular content in introductory exceptional children and youth courses. *Teacher Education and Special Education, 13,* 203–209.

Fernandes, M. T. & Pinto, P. R. (1991). Teachers as change-agents for young people with special needs: An approach to teacher training in Portugal. *European Journal of Teacher Education, 14,* 149–154.

Finn, C. (1991). *We must take charge: Our schools and our future.* New York: The Free Press.

Fordham, S. (1988). Racelessness as a factor in black students' school success: Pragmatic victory or phyrrhic victory? *Harvard Educational Review, 58,* 54–84.

Foster, M. (1990). The politics of race: Through the eyes of African-American teachers. *Journal of Education, 172*(3), 123–141.

Fuchs, D., & Fuchs, L. S. (1994). Inclusive schools movement and the radicalization of special education reform. *Exceptional Children, 60,* 294–309.

Fullan, M. G. (1991). *The new meaning of educational change.* New York: Teachers College Press.

Gable, R. A., Hendrickson, J. M., Shores, R. E., & Young, C. C. (1983). Teacher-handicapped child classroom interactions. *Teacher Education and Special Education, 6,* 88–95.

Ganschow, L., Weber, D. B., & Davis, M. (1984). Preservice teacher preparation for mainstreaming. *Exceptional Children, 51,* 74–76.

Gartner, A., & Lipsky, D. K. (1987). Beyond special education: Toward a quality system for all students. *Harvard Educational Review, 57,* 367–395.

Gerlach, G., Clawson, D., & Noll, M. B. (1994). A restructured collaborative student teaching experience to prepare teachers for inclusion. *Action in Teacher Education, 16*(3), 58–65.

Goodlad, J. (1984). *A place called school.* New York: McGraw-Hill.

Goodlad, J. (1986). Foreword. In E. Cohen, *Designing group work: Strategies for the heterogeneous classroom.* New York: Teachers College Press.

Goodlad, J. (1990). *Teachers for our nation's schools.* San Francisco: Howwey-Bass.

Goodlad, J. I., & Field, S. (1993). Teachers for renewing schools. In J. I. Goodlad & T. C. Lovitt (Eds.), *Integrating general and special education.* (pp. 229–251). San Francisco: Jossey Bass.

Goodman, J. (1988). University culture and the problem of reforming field experiences in teacher education. *Journal of Teacher Education, 39*(5), 45–53.

Goodman, Y. M. (Ed.). (1990). *How children construct literacy: Piagetian perspectives.* Newark, DE: International Reading Association.

Halvorsen, A., & Sailor, W. (1990). Integration of students with severe and profound disabilities: A review of research. In R. Gaylord-Ross (Ed.), *Issues and research in special education. Vol. 1.* (pp. 110–172). New York: Teachers College Press.

Harris, K. R., & Graham, S. (1994). Constructivism: Principles, paradigms, and integration. *Journal of Special Education, 28,* 233–247.

Heath, S. B. (1982). What no bedtime story means: Narrative skills at home and school. *Language in Society, 11*(1), 49–76.

Heller, H. W., Spooner, M., Spooner, F., Algozzine, B. (1992). Helping general educators accommodate students with disabilities. *Teacher Education and Special Education, 15,* 269–274.

Heller, H. W., Spooner, M., Spooner, F., Algozzine, B., Harrison, A., & Enright, B. (1991–2). Meeting the needs of students with handicaps: Helping regular teachers meet the challenge. *Action in Teacher Education, 13*(4), 44–54.

Holmes Group. (1986). *Tomorrow's teachers.* East Lansing, MI: Author.

Holmes Group. (1990). *Tomorrow's schools.* East Lansing, MI: Author.

Howey, K. R. (1990). Changes in teacher education: Needed leadership and new networks. *Journal of Teacher Education, 41*(1), 3–9.

Johnson, L. J., & Pugach, M. (1992). Continuing the dialogue: Embracing a more expansive understanding of collaborative relationships. In W. Stainback & S. Stainback (Eds.), *Controversial issues confronting special education* (pp. 215–222). Boston: Allyn & Bacon.

Jones, S. D., & Messenheimer-Young, T. (1990). Content of special education courses for preservice regular education teachers. *Teacher Education and Special Education, 12,* 154–159.

Jones, T. W. (1993). International special education inservice training: Challenges and solutions. *Teacher Education and Special Education, 16,* 297–302.

Kamii, C. (1985). *Young children reinvent arithmetic.* New York: Teachers College Press.

Kauffman, J. M. (1989). The Regular Education Initiative as Reagan-Bush education policy: A trickle-down theory of education of the hard-to-teach. *Journal of Special Education, 23,* 256–278.

Kauffman, J. M. (1993). How we might achieve the radical reform of special education. *Exceptional Children, 60,* 6–16.

Kauffman, J. M., Gerber, M. M., & Semmel, M. I. (1988). Arguable assumptions underlying the Regular Education Initiative. *Journal of Learning Disabilities, 21,* 6–11.

Kearney, C. A., & Durand, V. M. (1992). How prepared are our teachers for mainstreamed classroom settings? A survey of postsecondary schools of education in New York state. *Exceptional Children, 59,* 60–11.

Kemple, K. M., Hartle, L. C., Correa, V. I., & Fox, L. (1994). Preparing teachers for inclusive education: The development of a unified teacher education program in early childhood and early childhood special education. *Teacher Education and Special Education, 17,* 38–51.

Korinek, L. A., & Laycock, V. K. (1988). Evidence of the Regular Education Initiative in federally funded personnel preparation programs. *Teacher Education and Special Education, 11,* 95–102.

Ladson-Billings, G. (1990). *Making a little magic: Teachers talk about successful teaching strategies for black children.* Paper presented at the American Educational Research Association, Boston.

Lauritzen, P. (1990a). *Comprehensive assessment of service needs for special education in Wisconsin, 1990.* Madison: Wisconsin Department of Public Instruction.

Lauritzen, P. (1990b). *Wisconsin teacher supply and demand: An examination of data and trands, 1990.* Madison: Wisconsin Department of Public Instruction.

Lauritzen, P., & Friedman, S. J. (1993). Meeting the supply/demand requirements of the Individuals with Disabilities Education Act. *Teacher Education and Special Education, 16*, 221–229.

Lavely, L. & McCarthy, J. (1995). Early intervention in the context of school reform and inclusion. In J. L. Paul, H. Rosselli, & D. Evans (Eds.), *Integrating school restructuring and special education reform* (pp. 188–213). Fort Worth, TX: Harcourt Brace College Publishers.

Lawson, H. (1994). Toward healthy learners, schools, and communities. *Journal of Teacher Education, 45*(1), 62-70.

Lazar, I. (1988). Measuring the effects of early childhood programs. *Community Education Journal, 18*, 8–11.

Levine, M. (Ed.). (1992). *Professional practice schools.* New York: Teachers College Press.

Lieberman, A. (Ed.). (1988). *Building a professional culture in schools.* New York: Teachers College Press.

Lilly, M. S. (1987). Lack of focus on special education in literature on educational reform. *Exceptional Children, 53*, 325–326.

Lilly, M. S. (1989). Teacher preparation. In D. K. Lipsky & A. Gartner (Eds.), *Beyond separate education* (pp. 143–157). Baltimore, MD: Brookes.

Lilly, M. S. (1992). Research on teacher licensure and state approval of teacher education programs. *Teacher Education and Special Education, 15*, 148–160.

Little, J. W. (1982). Norms of collegiality and experimentation: Workplace conditions of school success. *American Educational Research Journal, 19*, 325–340.

Little, J. W. (1993). *Teachers' professional development in a climate of educational reform.* New York: Teachers College Press.

Maheady, L., Harper, G. F., Mallette, B., & Karnes, M. (1993). The reflective and responsive educator (RARE): A preservice training program to prepare general education teachers to instruct children and youth with disabilities. *Education and Treatment of Children, 16*, 474–506.

Marston, D. (1987). Does categorical certification benefit the mildly handicapped child? *Exceptional Children, 53*, 423–431.

Mather, N. (1992). Whole language reading instruction for students with learning disabilities: Caught in the cross fire. *Learning Disabilities Research and Practice, 7*, 89–95.

McLeskey, J. & Pacchiano, D. (1994). Mainstreaming students with learning disabilities: Are we making progress? *Exceptional Children, 60*, 508–517.

Metcalfe, J. A., & Diniz, F. A. (1991). Teachers as change-agents in special educational needs: Experience in England. *European Journal of Teacher Education, 14*, 139–148.

Murphy, J. (1990). Helping teachers to teach in restructured schools. *Journal of Teacher Education, 41*(4), 50–56.

National Clearinghouse on Professions in Special Education. (1993). *Who will teach? Who will serve?* Alexandria, VA: Author.

National Commission on Excellence in Education. (1983). *A nation at risk: The imperative for educational reform.* Washington, DC: U.S. Government Printing Office.

National Support Systems Project. (1980). *The dean's grant projects: A descriptive analysis and evaluation.* Minneapolis: Author.

Nicholas, G. (1990). *ASCUS research report: Teacher supply and demand in the United States, 1990.* Evanston, IL: Association for School, College, and University Staffing.

Ogbu, J. (1978). *Minority education and caste.* New York: Academic.

O'Sullivan, P. J., Marston, D., & Magnusson, D. (1987). Categorical special education teacher certification: Does it affect instruction of mildly handicapped pupils? *Remedial and Special Education, 8*(5), 13–18.

Pasch, S. H., Pugach, M. C., & Fox, R. G. (1991). Case one: A collaborative structure for institutional change in teacher education. In H. L. Barnes, M. C. Pugach, & Beckum, L. C. (Eds.), *Changing the practice of teacher education: The role of the knowledge base* (pp. 109–138). Washington, DC: American Association of Colleges for Teacher Education.

Patton, J. M., & Braithwaite, R. L. (1980). P. L. 94–142 and the changing status of teacher certification/recertification: A survey of state education agencies. *Teacher Education and Special Education, 3*(2), 43–47.

Patton, J. M., & Braithwaite, R. L. (1990). Special education certification/recertification for regular educators. *Journal of Special Education, 24,* 117–124.

Paul, J. L., Duchnowski, A. J., & Danforth, S. (1993). Changing the way we do our business: One department's story of collaboration with public schools. *Teacher Education and Special Education, 16,* 95–109.

Paul, J. L., & Roselli, H. (1995). Integrating the parallel reforms in general and special education. In J. L. Paul, H. Rosselli, & D. Evans (Eds.), *Integrating school restructuring and special education reform* (pp. 188–213). Fort Worth, TX: Harcourt Brace College Publishers.

Preuss, E. & Hofsass, T. (1991). Integration in the Federal Republic of Germany: Experiences related to professional identity and strategies of teacher training in Berlin. *European Journal of Teacher Education, 14*(2), 131–138.

Pugach, M. (1988). Special education as a constraint on teacher education reform. *Journal of Teacher Education, 39*(3), 52–59.

Pugach, M. (1992). Unifying the preservice preparation of teachers. In W. Stainback & S. Stainback (Eds.), *Controversial issues confronting special education: Divergent perspectives* (pp. 255–270). Needham, MA: Allyn & Bacon.

Pugach, M., & Sapon-Shevin, M. (1987). New agendas for special education policy: What the national reports haven't said. *Exceptional Children, 53,* 295–299.

Pyecha, J., & Levine, R. (1995, May). *The attrition picture: Lessons from three research projects.* Paper presented at the National Dissemination Forum on Special Education Teacher Satisfaction, Retention, and Attrition, Washington.

Reynolds, A. (1992). What is competent beginning teaching? A review of the literature. *Review of Educational Research, 72,* 1–35.

Reynolds, M. C. (1979). Categorical vs. noncategorical teacher training. *Teacher Education and Special Education, 2*(3), 5–8.

Reynolds, M. C., & Birch, J. W. (1977). The interface between regular and special education. *Teacher Education and Special Education, 2*(1), 12–27.

Reynolds, M. C., Wang, M. C., &Walberg, H. J. (1987). The necessary restructuring of special and regular education. *Exceptional Children, 53,* 391–398.

Rosenberg, M. S., & Rock, E. E. (1994). Alternative certification in special education: Efficacy of a collaborative field-based teacher preparation program. *Teacher Education and Special Education, 17,* 141–153.

Ross, D. D. (1989). Action research for preservice teachers: A description of why and how. *Peabody Journal of Education, 64*(3), 131–150.

Rude, H. (1994, September). *Professional licensure of educators.* Paper presented at the Vision 2000 Conference, Tampa.

Rudduck, J. (1990). *Innovation and change.* Bristol, PA: Open University Press.

Sailor, W. (1991). Special education in the restructured school. *Remedial and Special Education, 12*(6), 8–22.

Sapon-Shevin, M. (1988). Working toward merger together: Seeing beyond distrust and fear. *Teacher Education and Special Education, 11,* 103–110.

Sawyer, R. J., McLaughlin, M. J., & Winglee, M. (1994). Is integration of students with disabilities happening? An analysis of national data trends over time. *Remedial and Special Education, 15,* 204–215.

Schön, D. A. (1983). *The reflective practitioner: How professionals think in action.* New York: Basic Books.

Schorr, L. B. (1988). *Within our reach: Breaking the cycle of disadvantage.* New York: Doubleday.

Scruggs, T. E., & Mastropieri, M. A. (1995). *Teacher perceptions of mainstreaming: A research synthesis.* Manuscript submitted for publication.

Simpson, R. L., Whelan, R. J., & Zabel, R. H. (1993). *Remedial and special education, 14*(2), 7–22.

Sindelar, P. T., & Marks, L. J. (1993). Alternative route programs: Implications for elementary and special education. *Teacher Education and Special Education, 16*, 146–154.

Sindelar, P. T., McCray, A. D., & Westling, D. L. (1992). A proposed certification model for special education. *Remedial and Special Education, 13*(4), 10–13.

Sindelar, P., Pugach, M., Griffin, C., & Seidl, B. (1994). Contemporary teacher education: Changing philosophy and practice. In J. L. Paul, H. Rosselli, & D. Evans, (Eds.), *Integrating school restructuring and special education reform* (pp. 140–166). Fort Worth: Harcourt Brace College Publishers.

Sindelar, P. T., Ross, D. D., Brownell, M. T., Griffin, C. C., & Rennells, M. S. (1994, September). *Teacher education for Florida's 21st century.* Paper presented at the Vision 2000 Conference, Tampa.

Singer, J. D. (1993a). Are special educators' career paths special? *Exceptional Children, 59*, 262–279.

Singer, J. D. (1993b). Once is not enough: Former special educators who return to teaching. *Exceptional Children, 60*, 58–72.

Sizer, T. (1984). *Horace's compromise: The dilemma of the American high school.* Boston: Houghton Mifflin.

Skrtic, T. M. (1991). *Beyond special education: A critical analysis of professional culture and school organization.* Denver: Love.

Slavin, R. E. (1990). General education under the regular education initiative: How must it change? *Remedial and Special Education, 11*(3), 40–50.

Smith, J. E., Jr., & Schindler, W. J. (1980). Certification requirements of general educators concerning exceptional pupils. *Exceptional Children, 46*, 394–396.

Stainback, S., & Stainback, W. (1987). Facilitating merger through personnel preparation. *Teacher Education and Special Education, 10*, 185–190.

Stephens, T. M. (1988). Eliminating special education: Is this the solution? *Journal of Teacher Education, 39*(3), 60–64.

Swan, W. W., & Sirvis, B. (1992). The CEC common core of knowledge and skills essential for all beginning special education teachers. *Teaching Exceptional Children, 25*(10), 16–20.

Swartz, S. L., Hildalgo, J. F., & Hays, P. A. (1991–92). Teacher preparation and the regular education initiative. *Action in Teacher Education, 56*(4), 55–61.

Task Force on Education. (1990). *Educating America: State strategies for achieving the national education goals.* Washington, DC: National Governors' Association.

Taylor, D., & Dorsey-Gaines, C. (1988). *Growing up literate: Learning from inner city families.* Portsmouth, NH: Heinemann.

Upton, G. (1990). Teacher training and special education alternative systems of delivery and their implications for educational development. *International Journal of Special Education, 5*(3), 301–310.

U.S. Department of Education. (1989). *Eleventh annual report to Congress on the implementation of the individuals with disabilities education act.* Washington, DC: Author.

U.S. Department of Education. (1994). *Sixteenth annual report to Congress on the implementation of the individuals with disabilities education act.* Washington, DC: Author.

Valero-Figueira, E. (1987). Bilingual special education personnel preparation: An integrated model. *Teacher Education and Special Education, 9*, 82–88.

Valli, L. (1989) Collaboration for transfer of learning: Preparing preservice teachers. *Teacher Education Quarterly, 16*(1), 85–95.

Wang, M. C., Reynolds, M. C., & Walberg, H. J. (1986). Rethinking special education. *Educational Leadership, 44*(1), 26–31.

Wang, M. C., Reynolds, M. C., & Walberg, H. J. (1988). Integrating the children of the second system. *Phi Delta Kappan, 71*, 64–67.

Watson, J. (1991). Current developments in Scotland in inservice courses for teachers of pupils with special educational needs. *European Journal of Teacher Education, 14*(2), 155–162.

Webb, R. B., & Sherman, R. R. (1989). *Schooling and society.* New York: Macmillan.

Welch, M. & Sheridan, S. M. (1993). Educational partnerships in teacher education: Reconceptualizing how teacher candidates are prepared for teaching students with disabilities. *Action in Teacher Education, 15*(3), 35–45.

Will, M. C. (1986). Educating children with learning problems: A shared responsibility. *Exceptional Children, 52*, 411–415.

Ysseldyke, J., Algozzine, B., & Thurlow, M. (1992). *Critical issues in special education.* Boston: Houghton Mifflin.

Zeichner, K. M. (1989). Preparing teachers for democratic schools. *Action in Teacher Education, 11*(1), 5–10.

Zeichner, K. M., & Liston, D. (1987). Teaching student teachers to reflect. *Harvard Educational Review, 57*, 23–48.

Zeichner, K. M., & Liston, D. (1990). Traditions of reform in U. S. teacher education. *Journal of Teacher Education, 41*(2), 3–20.

Zemelman, S., & Daniels, H., Hyde, A. (1993). *Best practice: New standards for teaching and learning in America's schools.* Portsmouth, NH: Heinemann.

15

School–Community Linkages

ANDREA G. ZETLIN
School of Education, California State University, Los Angeles, USA

WILLIAM L. BOYD
College of Education, The Pennsylvania State University, USA

This chapter examines the use of schools as coordinating organizations for health and human services to insure more beneficial outcomes for students. As argued since the 1890's, unless barriers to learning, development, and health are removed, children and youth cannot succeed in school. Past social reform efforts have been mostly peripheral to the school's mission. The more recent school-linked services movement is aimed at creating and implementing an integrated care and educational system which includes a reconceptualization of the relationship between the school, community, and larger society. Variations in collaboration and interagency partnerships as practiced today are described and outcome data from currently operating programs are presented. For special education students in particular, it is argued that many of their medical and psychological needs can be best served through a school-linked services program which avoids fragmentation and duplication.

Introduction

Today, we very much recognize that the schooling of many children is significantly compromised by health and social problems that require services beyond what parents and schools are able to provide (Behrman, 1992). We find students falling along a continuum of those ready, healthy, and able to achieve at school to those who experience an array of barriers to learning including deficiencies in necessary prerequisite skills, dysfunctional home situations, peers who are negative influences, and inadequate health and social support services (Adelman, 1993). To eliminate or minimize the effect of these barriers, schools have sought alliances with other agencies to ensure that nonschool issues that affect the performance of students are addressed.

These noninstructional services should not to be viewed as a diversion from the main task of school. Schools are already affected by the consequences of noneducational problems among students and their families, and they often deal with such problems with few resources and little expertise. Growing numbers of students, especially those from urban areas, are requiring increasing amounts of support before they can benefit fully from classroom instruction. For these students a comprehensive set of enabling services must accompany their educational program if we are to assure access to learning (see Figure 15.1). Consequently, closer ties are being sought between schools and other agencies to ensure that children receive the full array of services they need to enable them to pursue their education (Boyd & Crowson, 1992; Gray, 1992). By joining with social and health agencies to provide nonaca-demic services, schools can concentrate on the educational performance of students—the function schools are best suited to handle—and escape criticism that the school's academic mission is being derailed (Wang, Haertel, & Walberg, 1995).

Presently, school-linked or school-based service integration programs are emerging in which public and private community agencies work collab-oratively with schools toward a shared goal of attending to the whole child's needs. A comprehensive and integrated array of services are made available as needed to promote healthy development and minimize barriers to learning in ways that address the whole family and its multiple problems and needs (Adelman, 1993). Depending on the extent of a child and/or family's need, these collaborative partnerships are able to make accessible a broad spectrum of enabling services such as medical and dental care, social service advocacy, mental health counseling, provision of basic needs (e.g., food, clothing, shelter referral), legal aid, before-and after-school programs, and English classes and job training for parents.

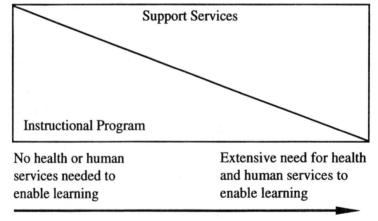

Figure 15.1 Hypothesized Relationship between Need for Health and Human Services and Instructional Efficiency.

Historical Trends

Tyack (1992), an educational historian, reminds us that for over 100 years, reformers have advocated using schools, with their captive populations, as a base to address a myriad of social ills. In successive reform movements since the 1890s, schools have acted as instruments of social change by taking on more and different responsibilities (Cremin, 1976). School lunches, kindergartens, medical and dental care, home visits, vocational guidance and counseling programs, services for wayward youth, and summer programs are but a few noneducation services which were introduced into the schools beginning at the turn of the century in response to the social and economic problems of our society (Tyack, 1992).

These early reform efforts were based on a cultural deficit model in which poor and immigrant parents were viewed as incapable of providing for all their children's needs (Tyack, 1992). Parents were seldom consulted about the new social and health programs offered and while most programs were welcomed, some were seen as intrusive and a threat to parents' authority. School officials were divided in their reactions to the programs as well. Progressive educators embraced the broadening of the school's agenda while conservatives viewed the new services as noneducational and worried about how they would be supported. Despite opposition, by 1940 an array of health and social services were readily available and experiencing steady growth in our public schools (Tyack, 1992).

Even during the Great Depression, budgets and staff increased for services, particularly for health services. Almost all cities with a population over 30,000 offered some form of public health service, and physicians, nurses, dentists, and dental hygienists were employed directly by the school districts. After World War II, school lunch and mental health programs took firm root to combat poverty and address the school drop-out problem. For the most part, these health and social services were recognized as addressing the real needs of children and were supported by educators as well as the public.

But as these programs became institutionalized in the schools, they were transformed to become part of the system. For example, the original visiting teachers, who were volunteers or charity-supported, made home visits to see why children were having difficulty in school and to help impoverished families use resources in the community. Once they became school employees, their time was directed to enforcing compulsory attendance.

The 1950s and 1960s ushered in other changes. During the 1950s, services were made available to the whole student population rather than being focused primarily on the poor. Guidance counselors shifted goals from finding jobs for poor youths to advising students about courses and appropriate tracks, and assisting in school discipline. Vacation (summer) schools expanded from places to repeat failed work to places where advanced work could also be done. In prosperous school districts, with substantial tax bases, such programs thrived while disadvantaged, inner city districts struggled to offer even minimal services (Tyack, 1992). In the 1960s, federal programs from the War on Poverty (i.e., Head Start and Title 1), targeted poor and minority groups specifically. These compensatory education programs shifted focus to assist children and families, and were more client sensitive, encouraging community participation. For schools and other public agencies accustomed to running health and social services without consulting clients, this sometimes resulted in conflict (Tyack, 1992).

Since the 1970s, concern about academic standards and maintaining the nation's competitiveness has overtaken concern about poverty and equality (Kagan, 1994). Combined with significant budget cutbacks, this shift away from a broadened school mission has resulted in reductions in nonacademic services throughout the public schools.

Present Day Social Reform Efforts

Renewed interest in school-linked services is due to large-scale social changes (i.e., increased poverty, teenage and single parents, the immigration of ethnic minorities to major cities) which have resulted in multiple responsibilities being placed on public schools (Holtzman, 1992). Current initiatives at the local, state, and federal levels, however, have drawn lessons from the long history of school reform. Past efforts, despite their regularity and intensity throughout the century, remained peripheral to the regular school program and were vulnerable to retrenchment or elimination when funds were scarce. Today's reformers are engaged in creating and implementing an integrated care and educational system which includes a dramatic reconceptualization and restructuring of the relationship between the school, the community, and the larger society (Guthrie & Guthrie, 1991).

The prevailing system of human service delivery—which includes education, health, and social services as separate entities—is characterized as a large unwieldy bureaucracy in which services are fragmented, overlapping, and often inaccessible for those who need them most (Guthrie & Guthrie, 1991; Morrill, 1992). Preventive action is rare (i.e., problems must become acute before services are brought to bear), and programs are implemented in isolation without consideration for the overall condition of the child and family.

The current emphasis on interagency collaboration is seen by many as crucial to reconfiguring the nature and structural alignment of mainstream institutions (Kagan, 1994). By combining a wealth of expertise and a variety of perspectives in interagency partnerships, systems can be reoriented away from the narrowness of single-agency mandates toward attending to the multiple problems of children and families in a comprehensive, meaningful way (Melaville & Blank, 1991).

Interagency collaboration is based on the belief that no one agency can provide all the necessary services for children and families (Tonelson & Waters, 1993). In a collaborative effort, all contributing parties must see the necessity and value of collaboration in order to achieve successful service delivery. Integral to improving service coordination is strengthening the ability of agencies to work together, share scarce resources, and take advantage of each other's respective disciplinary knowledge. The collaboration must include a broad cross-section of people and agencies who are in close communication, engaged in joint planning and policy development, and focused on accountability (Chang, 1993; Gardner, 1989).

School-linked Service Integration Models

Whether community services are located physically at the school site, or bridges are built between the school and a wide range of public and private community-based agencies, the intent of the school-linked service integration movement is to develop effective connections between the school and community service agencies (Gardner, 1994). Together, schools and community agencies can redefine their responsibilities, share decision-making, and jointly develop a comprehensive system to promote child growth and development. The overall goal of school-linked services is to ensure that all children are equally able to succeed by addressing their multiple needs in a coordinated manner (Chang, 1993).

There is no single model for school-linked service integration which predominates. Presently, many different types of collaborative programs have been initiated that vary in the composition and intensity of services delivered, skill of staff and mode of delivery, and target group served (Morrill, 1992). They range from single, one component partnerships between a school and an outside agency or business to sophisticated, complex, multicomponent, multiagency collaborations (Dryfoos, 1994). In most cases, services are joined to the schools via informal agreements, contractual agreements, established systems of referral, and sometimes mechanisms that enable staff members of various community agencies to be "outposted" or shared. While the approaches are diverse, what they all have in common is the intent of ensuring access to and continuity of health and social services to students and their families (Kagan, 1994).

There are also key differences in the process of collaboration practiced. While most centers have moved beyond simple cooperation toward more coordinated activity (i.e., defined by degree of institutional autonomy of partners), they differ in the "negotiated order" among participating agencies (Gray, 1992).

Thus, there is wide variety and creativity in children's services coordination to date, and no "one best way" to proceed. Nevertheless, as experimentation proceeds, and indeed as the pace of program development increases, the pros and cons of comparative approaches to services-coordination are beginning to become clear. For example, differences in impact may be associated with variation in the locus of service-provision. A school-based approach benefits from the school's position as a dominant neighborhood institution but can suffer from excessive control by schools. A school-linked approach can more effectively balance school and nonschool contributions but may still be too heavily "institutions" oriented. A community-based model can incorporate a wider diversity of resources and facilities (e.g., churches, community organizations, clubs) but may lose some focus and "sharpness" in its dispersion of stakeholders.

One important issue an analysis of integrated services programs raises is, *Just how much coordination among services is necessary and desirable?* The literature on coordinated services tends to be ambivalent on this issue. For example, while distinguishing between "cooperation" and "collaboration," Hord (1986) says that both are "valued models, but each serves a unique purpose and yields a different return" (p. 22). But she then contradicts this by saying that "collaboration is highly recommended as the most appropriate mode for interorganizational relationships" (p. 26).

The idea of alternative models for coordinated ventures has been advanced not only by Hord (1986), but also by Intriligator (1992), who suggests that interagency interactions can be usefully examined along a continuum of cooperation to coordination to collaboration. In cooperation the independence of individual agencies may be little affected, changes in institutional policy and structure are minimal, and "turf" is not a serious issue. Under collaboration (at the other end of the continuum), however, there will be a loss of institutional autonomy, *interagency* policy making in place of agency independence, and a need to go beyond "turf" toward consensus and well-established trust. Experience thus far nationally suggests that, rather than either cooperative, coordinative, or collaborative, some efforts have tended simply to be "co-located." However, even in co-location, difficult issues can arise over shared facilities usage, managerial control, resource allocation, professions protection, and information flow.

In general, then, the state-of-the-art in children's services collaboration has typically not progressed to an "idealized" point where participating organizations in projects share completely in the delivery of services, agree fully on goals and outcomes, contribute resources equally, share control and leadership, communicate and interact smoothly, and operate as "we" rather than "us/them." Rather, it is far more likely thus far that projects will be struggling with problems in blending other services into the institutional dominion of the school, in reaching a shared sense of mission and shared leadership/control in collaborative ventures, and in building effective communicative linkages between the project's array of service-providers (Crowson & Boyd, 1993).

In *Together We Can*—a very helpful guide to collaboration developed jointly by the U.S. Department of Health and Human Services, and the U.S. Department of Education—a five-stage process of building collaboration for comprehensive family services is laid out. The steps include: (1) getting together, (2) building trust, (3) developing a strategic plan, (4) taking action, and (5) going to scale (Melaville, Blank, & Asayesh, 1993, p. 20). The ultimate goal of "going to scale" (i.e., applying the principles of coordination widely across an entire jurisdiction, rather than narrowly in one limited pilot project) raises the issue of how ambitious and comprehensive coordinated service ventures should try to be, especially at the outset.

Indeed, one way of comparing coordinated services is according to their differing styles of administrative implementation (Boyd & Crowson, 1992). Projects are frequently initiated as *strategic* interventions—pragmatically and iteratively moving toward a goal of coordination and problem-solving as the project unfolds. The alternative, often recommended, models a strategy of *systemic* reform, where key institutional constraints (e.g., agencies' functional boundaries, conflicting reward systems, differing norms and conventions, professional training differences, and the like) are identified and a comprehensive overall coordination and implementation plan is developed before proceeding further.

As a practical matter, there are advantages in starting with less ambitious projects, but also some significant hazards. Such ventures can get under way faster since they can avoid the complex negotiations and transaction costs of trying to work out all the details of complicated interagency agreements. Rather than requiring elaborate formal agreements, they can rely in part on a more informal approach, for example, building on positive personal networks among cooperating agency and school personnel. By contrast, large comprehensive reform efforts require long and complex planning processes involving many actors and agencies. The practical advantages of the less ambitious approach are reflected in the conclusions of a GAO report entitled, "Integrating Human Services: Linking At-Risk Families with Services More Successful Than System Reform Efforts" (Government Accounting Office, 1992).

The hazards in the less ambitious approach are that such ventures can easily succumb to what Sid Gardner (1992) calls "projectitis," that is, limited and temporary projects which ultimately do little to change the overall character of fragmented delivery systems for children's services. Thus, the long-term challenge of the school-linked service integration movement is to reconfigure relationships between the school community and public service agencies (Kagan, 1994).

First and foremost, school-linked services should not simply be "add-ons" to the school program. As Gardner (1992) cautions, additive projects do not change institutions because they operate as new activities grafted on top of the existing system. Rather, the ultimate goal is formation of a new kind of community-oriented school, a "seamless institution" with some kind of joint or shared governance structure. Second, service delivery must shift emphasis from being program-centered to being family-centered. This implies acknowledgement of the central role that families play in their children's well-being and in the mobilization/coordination of community supports to assist families in carrying out their roles (Family Resource Coalition, 1993). More intensive intervention is called for which is comprehensive, promptly delivered, and cuts across professional and programmatic categories (Morrill, 1992). Lastly, maximum responsiveness to the community must be assured through changes in the working relationship between service providers and the people they serve. As Chang (1993) notes, communities must be given the opportunity to participate in the design and implementation of program and policies.

Collaborative Service Delivery

Social problems rarely exist in isolation. Children suffering from child abuse, for example, are likely to experience other problems in their homes, such as family involvement in substance abuse and inadequate parental supervision (Holtzman, 1992). Rather than referring families to various agencies, usually in different locations, school-linked integrated service programs offer many services, typically through a system of case management. Case managers, from the school or community agencies, assess, treat, or refer families to a variety of services and then track the referrals and outcomes (Gardner, 1992). This coordinated approach avoids the bureaucratic pitfalls which often prevent families from accessing needed services (i.e., difficulty comprehending eligibility requirements, incomplete knowledge of available services, transportation/childcare problems, language barriers, etc.), and spares families from involvement in inefficient and ineffective "programs" which address social problems in isolated and compartmentalized ways (i.e., teen pregnancy, substance abuse, gang involvement, school dropout, and low self-esteem.)

A list of the components that might be incorporated into a program that shares responsibility for education among families, schools, and the community was suggested by Joy Dryfoos (1994) and is presented in Table 15.1. Dryfoos argues that for a program to be effective, it must encompass both quality education and comprehensive support services. She notes that "no single component, no magic bullet, can significantly change the lives of disadvantaged children, youth, and families. Rather it is the cumulative impact of a package of interventions that will result in measurable changes in life scripts" (p. 12).

Probably the two most recognized models of school-linked services are Zigler's (Zigler & Lang, 1991) Schools of the 21st Century and Comer's

Table 15.1 Components of Comprehensive School-linked Service Programs.

Quality Education Provided by Schools	Support Services Provided by Community Agencies	Provided by Schools or Community Agencies
Effective basic skills	Health screening and services	Comprehensive health education
Individualized instruction	Dental services	Health promotion
Team teaching	Family planning	Social skills training
Cooperative learning	Individual counseling	Preparation for the world of work (life planning)
School-based management	Substance abuse treatment	
Healthy school climate	Mental health services	
Alternatives to tracking	Nutrition/weight management	
Parent involvement	Referral with follow-up	
Effective discipline	Basic services: housing, food, clothes	
	Recreation, sports, culture	
	Mentoring	
	Family welfare services	
	Parent eduction, literacy	
	Child care	
	Employment training/jobs	
	Case management	
	Crisis intervention	
	Community policing	

SOURCE: Adapted from Dryfoos (1994)

(1985) School Development Program. Both programs promote schools that function as community centers and have in common: (1) the mobilization and integration of community expertise and resources; (2) emphasis on community renewal, family preservation, and child development; and (3) the active involvement of all stakeholders in the identification and development of policies and procedures. In Zigler's model, family support systems are linked with child care systems. Program components include full-day child care for preschool and school-age children, parent education and family support services, literacy training, training and support for family day-care providers, and teen pregnancy prevention services (Zigler & Lang, 1991).

The School Development Program, in operation in over 165 schools, emphasizes the social context of teaching and learning. The program is a school-based management approach to making school a more productive environment for poor, minority children. Within the model, heavy emphasis is placed on mental health services, and strengthening and redefining the relationships between school staff, parents, and students. Four major components comprise the main thrust of the program: a governance and management team, a mental health team, a parent participation program, and a program for curriculum and staff development. The basic goal is to create schools that offer children stability as well as role models to nurture them and increase their chances of academic success (Comer, 1985).

School-Linked Services and Special Education

Schools are presently the only long-term care and support system for chronically vulnerable children and youth (Hooper-Briar & Lawson, 1994). Many of the medical and psychological services required by students in special education can be served through the school-linked services program. For medically fragile children, health care services such as suctioning mucous from the airways of children, inserting feeding tubes, or administering insulin and other injections or medications can be done by medical personnel in the center rather than by teachers and aides (Dryfoos, 1994). For psychosocial problems, student study teams comprised of center practitioners, school personnel, and special education staff can review referrals from teachers and parents and develop comprehensive action plans which detail how best to serve students' needs and who will do what.

For families with children with disabilities, negotiation through the patch-work of disjointed service agencies and programs can be nearly impossible. In some cases, the resources of many agencies must be activated in order to best address the existing needs of the child and family. For example, a child with emotional problems typically receives special education services from the school as well as counseling services from a mental health agency. If there is a health problem, it might be attended by the Department of Health Services. If his or her mother is a single parent receiving AFDC, then the Department of Social Services is involved with the family. If the child or another family member is caught up in the court system, the Department of Probation or even Child Welfare Services may be brought in. When there is interagency collaboration, then feedback and mutual exchange of ideas can occur and the number of overlaps and/or gaps in service can be reduced. Further, agencies that share ideas and information and coordinate efforts in structured collaboration can avoid the misinterpretation of responsibility that often occurs when agencies operate independently (i.e., one agency believes that another is providing for needs that end up going unattended). Not only can interagency collaboration offer a clearer understanding of each agency's goals and purposes, the collaborative process more clearly outlines the needs of the individual or family as they relate to the service providers (Tonelson & Waters, 1993).

The following case study illustrates how the service integration center at a school can serve as the primary case manager, advocating for the family and facilitating comprehensive services within a reasonable time frame. Sammy, a first grader, was referred by his teacher to the school-based service center because of serious behavior problems. A case manager followed up and learned that Sammy's mother, a drug user, had abandoned him to the care of grandparents who were having a difficult time managing him. Workers from the County Departments of Mental Health and Child Welfare Services, both on-site service providers in the center, worked with the school to locate his mother, and obtained her consent for a psychoeducational assessment of Sammy. The evaluation confirmed that Sammy had attention deficit hyperactivity disorder (ADHD) and medication was prescribed. The case manager and child welfare worker continued to work closely with the grandparents to obtain physical custody of Sammy and to transfer AFDC benefits from Sammy's mother to his grandparents. Finally, a meeting of the student study team, attended by Sammy's grandparents and the case manager, resulted in Sammy's placement in a special education classroom where his academic program would be modified and counseling would be provided. The process took five months during which time the school, in collaboration with two public agencies, developed a joint service plan to address the needs of Sammy and his family in a holistic fashion (Zetlin, Ramos, & Valdez, 1995).

Prerequisites for Setting Up School-Linked Programs

The following discussion summarizes emerging principles for interagency collaboration and points out pitfalls that social service and educational administrators should avoid. *First*, quality leadership is essential. There must be a top level catalyst who (a) recognizes that the current delivery of education, health, and human services is not meeting the needs of at least some of the population served and (b) has a vision for interagency collaboration, as well as the authority to facilitate it (i.e., doing business "differently" and more effectively).

Second, we must understand the commitment of asking for parent involvement in the planning and implementation of a school-linked services center. Inherent in this commitment must be a willingness: (a) of administrators and professionals to relinquish some of their power as decisions are made as to how business is to be conducted, what services and agencies to recruit and support, what needs are to be addressed and in what order; and (b) for school/center staff to "teach" parents how to be involved (i.e., to nurture the development of their "voice").

Third, we must be committed to ensuring that policies and practices are culturally compatible. This goes beyond translating letters in the language of the home or assuring that a translator is present at meetings. For example, at one Los Angeles school-based center where ESL classes are offered for parents, parents attend with their younger children because of lack of money for child care. Since the center runs two ESL classes on alternate days, center staff helped parents set up a reciprocal child-care program in a nearby classroom where parents serve as sitters on the days they are not in class (Zetlin, Campbell, Lujan, & Lujan, 1994).

Fourth, we must make long-term commitments to program development since it may be 5 or 10 years before we see the kinds of outcome data which society will applaud. Such long-term commitment includes: (a) a willingness to persevere as we struggle to work out issues of turf, leadership, and mission; (b) acceptance of the dynamic nature of the process and the need to make changes in response to evaluation data and community input; and (c) commitment from school districts to forego their policy of transferring site administrators every three to five years and allow a principal to remain in place during the initial period of growth and development.

Fifth, we must be committed to the "nuts and bolts" needs of the project and to seeking stable funding for operating costs. This includes: (1) providing adequate space for the project (which may be difficult in some overcrowded school districts, but critical to the identity of the project) and also providing funding for a Center Coordinator who is available for interagency networking, case management, troubleshooting, and operations management (i.e., the coordinator is the "glue" that holds the pieces together); (2) providing training and cross-training opportunities for participating school and community agency workers. Support for training and cross-training is critical so participants can (a) learn one another's language and programs, (b) negotiate the necessary new roles and relationships between educators and other client service personnel (thus overcoming turf protection), and (c) tackle such issues as communication, confidentiality, and liability; and (3) incorporating the school-linked program into the regular budget so that when start-up monies—demonstration grant monies—diminish or disappear, the program does not disappear too. Only when the program becomes central to the operation of the school and community will powerful supportive constituencies, parents, educators, service providers, be committed to fight for its continued existence.

Sixth, there need to be variations in the models we develop so that programs are individualized to the particular needs and concerns of the school and community (i.e., schools with large immigrant populations, highly transient populations, or large homeless populations; communities in need of child care or after-school care, job training and employment, or those struggling with high drop-out rates, gang membership, substance abuse, or teenage pregnancy.) No one model fits all settings and works well in all cases. Variations of the model need to be available to suit differing local needs and concerns. And most important, detailed evaluations of all models must take place to yield a much needed knowledge base on how to provide school-linked service integration that is both feasible and cost-effective.

Seventh, these integrated service projects must develop partnerships with local universities to provide the technical assistance for program development and evaluation. University faculty must also be involved for the purposes of interprofessional education. The school-linked service integration center provides a collaborative setting for training educators and service workers so that they develop skills for coordinating efforts with workers from related fields (Adler & Gardner, 1993). Until now most university training of professionals in children's services inadvertently impeded collaborative efforts and interprofessional relations. Such training is heavily constrained by a separation of knowledge bases by discipline and certification systems. We need to begin building collaboration skills into undergraduate and graduate programs by restructuring our training programs.

Conclusion

Most of the exemplary school-linked programs are still in the development phase, so their effectiveness—whether they can substantially change the lives of high-risk children and families—is largely unknown. From the limited evaluation data available thus far, Dryfoos (1994) has identified the following patterns of outcomes:

- programs are located in communities and schools with the greatest need and are being utilized most by the highest risk students
- availability of school health services has led to a decrease in absences for minor illnesses and less use of emergency rooms in areas with school clinics
- substantial numbers of students and families are accessing mental health counseling that was not available in their communities before
- student behavior is being influenced by the provision of health education in classrooms and group counseling covering a range of problems (i.e., substance use, family relations, sexuality, peer relationships, etc.)
- students, families, and teachers report improvement in the school environment and a high level of satisfaction with the accessibility, convenience, and support offered in the centers.

In terms of the five types of school outcome data commonly measured—attendance, achievement data, reduced behavior problems, self-esteem and dropout rates—overwhelmingly positive results were identified by 44 collaborative school-linked programs reported in the literature (Wang, Haertel, & Walberg, 1995).

To sum up, the current reality is that schools cannot, on their own, provide for all that today's students need. New kinds of arrangements of community resources have to be brought together to ensure that children can grow up to be responsible, productive, and fully participating members of our society. School-linked services are strategies and resources to insure more beneficial outcomes for children and youth (Dryfoos, 1994; Hooper-Briar & Lawson, 1994).

While we have seen the supply of services within the schools turned on and off over the past century, current practices have schools joining with health and social service systems to develop powerful new institutions. Supported by a combination of federal, state, and local initiatives, the school-linked service integration movement is growing rapidly, and promises cutting-edge reform in the ways schools and public service agencies interact and respond to the needs of students, families, and communities.

The fundamental goal of the service integration movement is to improve the conditions of teaching and learning within schools by attending to the personal and social problems that interfere with success in school. By providing the necessary family and social supports essential for child growth and development, and improving the school climate within which learning takes place, the needs of high-risk children and their families can be addressed and access to future opportunities can be equalized. As Dryfoos (1994) dramatically states, "Without a concerted effort, millions of young people will continue to fail and will have no hope of growing into responsible and productive adults."

References

Adelman, H.S. (1993). School-linked mental health interventions: Toward mechanisms for service coordination and integration. *Journal of Community Psychology, 21,* 309–319.

Adler, L., & Gardner, S. (Eds.). (1993). *The politics of linking schools and social services. The 1993 yearbook of the Politics of Education Association.* Washington, DC: Falmer Press.

Behrman, R.E. (1992). Introduction: School linked services. *The Future of Children, 2*(1), 6–7.

Boyd, W.L., & Crowson, R. (1992, October). *Integration of services for children: A political economy of institutions perspective.* Paper presented at an Invitational Conference on School/Community Connections. Washington, DC: National Center on Education in the Inner Cities.

Chang, H.N. (1993). Diversity: The essential link in collaborative services. *Education and Urban Society, 25,* 221–230.

Comer, J.P. (1985, September). *The school development program: A nine step guide to school improvement.* New Haven, CT: Yale Child Study Center.

Cremin, L. (1976). *Public education.* New York: Basic Books.

Crowson, R.L., & Boyd, W.L. (1993, February). *Structures and strategies: Toward an understanding of alternative models for coordinated children's services.* Paper prepared for the National Center on Education in the Inner Cities, Temple University, Philadelphia, PA.

Dryfoos, J.G. (1994). *Full-service schools: A revolution in health and social services for children, youth, and families.* San Francisco: Jossey-Bass Publishers.

Family Resource Coalition. (1993). Family support and school-linked services. *Family Resource Coalition Report, 12,* 3–4.

Gardner, S.L. (1992). Key issues in developing school-linked integrated services. *The Future of Children*, 2(1), 85–94.

Gardner, S.L. (1989). Failure by fragmentation. *Equity and Choice*, 2, 4–12.

Government Accounting Office. (1992, September). *Integrating human services: Linking at-risk families with services more successful than system reform efforts.* Washington, DC: U.S. Government Printing Office.

Gray, B. (1992, October). *Obstacles to success in educational collaboration.* Paper presented at the Conference on School/Community Connections: Exploring Issues for Research and Practice, National Center on Education in the Inner Cities, Leesburg, VA.

Guthrie, G.P., & Guthrie, L.F. (1991). Streamlining interagency collaboration for youth at risk. *Educational Leadership*, 49, 17–22.

Hooper-Briar, K., & Lawson, H.A. (1994). *Serving children, youth, and families through interprofessional collaboration and service integration: A framework for action.* Oxford, OH: The Danforth Foundation and the Institute for Educational Renewal at Miami University.

Holtzman, W. (1992). Community renewal, family preservation, and child development through the School of the Future. In W. Holtzman (Ed.), *School of the Future* (pp. 3–18). Austin, TX: American Psychological Association and Hogg Foundation for Mental Health.

Hord, S.M. (1986). A synthesis of research on organizational collaboration. *Educational Leadership*, 43, 22–26.

Intriligator, B.A. (1992, October). *Designing effective interorganizational networks.* Paper presented at the Annual Meeting of the University Council for Educational Administration (UCEA), Minneapolis, MN.

Kagan, S.L. (1994). Readying schools for young children: Polemics and priorities. *Phi Delta Kappan*, 76, 226–233.

Melaville, A.I., Blank, M.J., & Asayesh, G. (1993). *Together we can: A guide for crafting a profamily system of education and human services.* U.S. Department of Education, U.S. Department of Health and Human Services. Washington, DC: U.S. Government Printing Office.

Melaville, A.I., & Blank, M.J. (1991). *What it takes: Structuring interagency partnerhsips to connect children and families with comprehensive services.* Washington, DC: Education and Human Services Consortium.

Morrill, W.A. (1992). Overview of service delivery to children. *The Future of Children*, 2(1), 32–43.

Tonelson, S.W., & Waters, R. (1993). Interagency collaboration. In B.S. Billingsley (Ed.), *Program leadership for serving students with disabilities* (pp. 55–66). Richmond, VA: Virginia Department of Education.

Tyack, D. (1992). Health and social services in public schools: Historical perspectives. *The Future of Children*, 2(1), 19–31.

Wang, M.C., Haertel, G.D., & Walberg, H.J. (1995). The effectiveness of collaborative school-linked services. In L.C. Rigsby, M.C. Reynolds, & M.C. Wang (Eds.). *School/community connections: Exploring issues for research and practice* (pp.283–310). San Francisco: Jossey-Bass Publishers.

Zetlin, A.G., Campbell, B., Lujan, M., & Lujan, R. (1994). Schools and families working together for children. *Equity and Choice*, 10, 10–15.

Zetlin, A.G., Ramos, C., & Valdez, A. (1995). *Integrating services in a school-based center: An example of a school-community collaboration.* Unpublished manuscript. Los Angeles, CA.

Zigler, E.F., & Lang, M.L. (1991). *Child-care choices: Balancing the needs of children, families, and society.* New York: The Free Press.

Epilogue

MARGARET C. WANG, MAYNARD C. REYNOLDS, and
HERBERT J. WALBERG

We, the delegates of the World Conference on Special Needs Education
representing 92 governments and 25 international organizations. . .
hereby reaffirm our commitment to Education for All, recognizing the
necessity and urgency of providing education for children, youth and
adults with special educational needs within the regular education system.
(United Nations, 1994)

An early step in creating this volume was the preparation of a prospectus
in which we, the editors, offered a vision of the work to be performed. The
introductory pages of this book reveal much of that original design, which
includes a commitment in belief to education for all persons and to efforts
for delivering education in the most inclusive ways possible.

Now, having worked through the writing of the dozens of scholars who
contributed to the book, we have sharpened and reshaped some of our
perspectives. In this final chapter, we reflect briefly on what we now make of
the situation of special and remedial education. Just as the prospectus was a
prediction and a plan, this chapter provides a bit of retrospective rumination.
It is totally a product of the editors; none of the others who contributed to
the book are responsible for it.

In general, we think the fields of special and remedial education have
succeeded remarkably well (at least in parts of the so-called "developed"
world) in achieving the initial enrollment of nearly all children and youth in
school programs. But, in terms of outcomes there is much less reason to be
satisfied. A candid review of programs, we believe, tells a mixed story of
positives and negatives; of worthy ideas and practices, but also of wasteful
and sometimes demeaning misdirection. By its candor, we hope that what is
revealed in this volume is some glimpse of "what works" and also of "what
does not work" in programs for students with special needs. We agree with
the many authors of earlier chapters that we are in a period of critical appraisal
and revision of programs.

What follows here is organized into three main sections. First, we acknowledge pioneering work in many countries around the world in the scientific and professional development of special and remedial education. Second, we propose a tri-partite structuring of special and remedial education programs and present ideas about progress, problems, and prospects in each of the three areas. In a third and penultimate section, we offer a series of general observations which have been developed or sharply reaffirmed as our work on the book proceeded.

International Perspective

As Katherine Butler mentions in her chapter, educators throughout the world increasingly recognize the need for effective programs for students who are at risk of falling behind or who have already done so. It is fitting that we here acknowledge the long-standing tradition of international information sharing, inspiration, and collaboration. Perhaps the most notable early example is the French educator Louis Braille, blinded at age three, who developed the system of dot printing and writing named after him and used by the blind in many parts of the world. Perhaps influenced by Braille was the Englishman William Moon, who invented in 1845 the system of raised Roman letter outlines that, while less prevalent, is more easily learned by those blinded after childhood. In the field of deaf education, the American Thomas Hopkins Gallaudet studied in England and France where he learned from the abbé Sicard the sign method of communication which he promulgated widely in the United States.

In recent times, Norway and other Scandinavian countries pioneered the inclusion of special-needs children in regular classes, a movement that has vastly expanded throughout the world. In this respect, perhaps Italy has taken the greatest strides in the last several years. International organizations, too, are playing a larger role in information sharing, innovation, and international collaboration. Since 1978, for example, the Organization for Economic Cooperation and Development (OECD) has carried out work on the integration of the disabled into mainstream schools. In that effort, Peter Evans (1993) of the OECD edited a thematic issue of the *European Journal of Special Needs Education* devoted to the subject with chapters by scholars from Australia, Germany, Sweden, the United Kingdom, and the United States. Similarly, the United Nations Educational, Scientific, and Cultural Organization (UNESCO) convened a conference in Salamanca, Spain for educators from many parts of the world to consider issues of inclusion. In a related UNESCO report, Hegarty (1993) noted that about 80% of the estimated 200 million students with disabilities in the world live in developing countries. In making recommendations to better serve them, he drew on American, Canadian, French, English, Indian, UNESCO, and Spanish studies.

Special and remedial educators throughout the world need to acknowledge their debts to such pioneers and continuing innovators in countries other than their own. Such important influences are easily forgotten, but then how many scientists and professionals have had time to master the histories of their disciplines. When we pass along insights and techniques to the next generation, we often fail to point out their origins. If our students have not learned them, they can hardly be expected to remember. Still, we as editors can at least point out the many known and unknown, acknowledged and unacknowledged debts of ours and undoubtedly the chapter authors to foreign scholars and educators of the past and present.

A Tri-Partite View

Much of special and remedial education is not truly distinctive or "special" so much as it is a more intensive version of what is good education for all students. Often it involves going one-on-one (teacher and student) or into intensive small group instruction to meet needs of selected students. Included are those who have shown limited profit from initial tries at ordinary classroom instruction or those who are considered to be "at risk" for academic failure because they are members of groups for which the base rate of academic failure is especially high. In ATI (aptitude-treatment-interaction) terms, these programs illustrate ordinal interactions. The instruction offered is not different in "type," but is offered in magnified or intensified form for the selected students. We discuss programs of this area below, under the heading "Ordinal Adaptations."

In the second set of programs, there are distinct types of instructional practices that are offered to selected students; for example, braille for students who are blind, nonaural communication systems for those who are deaf. In ATI terms these are situations that illustrate disordinal interactions. That is, they are characteristics of individuals that can be used (with validity) in allocating students to different types of instruction for at least parts of their school program. What is distinct about these programs also goes beyond the competencies most regular teachers can be expected to offer (and sustain); thus, the teacher preparation here has some truly "special" aspects. We discuss this area under the heading "Disordinal Adaptations."

A third area consists of practices in service to students who have very great needs but in domains where the knowledge base for school organization and instruction is exceedingly thin. The problems are untractable under most conditions. For example, we include here students who are extremely disturbed, and sometimes dangerous; also, those who may be extremely withdrawn and unresponsive. There is no clear and promising approach that teachers, working alone, can use with such students with expectations for problem solution. Closely coordinated work with community agencies, such as mental health clinics or correctional agencies, and almost always with parents, may be helpful and must be tried. But frequently these situations result in frustration and possibly in suspension or expulsions from school. We discuss this area under the tentative rubric "Unremitting Conditions."

Ordinal Adaptations

This is the big and recently developed aspect of categorical school programs. We include here the very large programs now offered in the United States and Canada for so-called learning disabled (LD) students, plus those for students labeled as mildly retarded, emotionally disturbed (mild levels only), speech impaired (some), or "at risk." This encompasses at least 75% of students now enrolled in special education programs and nearly all economically disadvantaged students identified for compensatory programs.

In our view, most of the classification and labeling practices in this domain are faulty. Indeed, the LD identification practice which requires demonstration of a large negative discrepancy between expectations for learning and actual learning achievements leaves that field virtually immobilized for services to children below about third-grade level. This reduces LD staff to the role of mere speculators in the most promising developments relating to learning problems, those that operate at first grade or earlier levels. Changes in legislation relating to such programs create more flexibility, and such programs can be expected to have increasing impacts, especially in schools that serve children of economically poor families.

We feel reaffirmed in our view that there is no truly "special" education in this domain. But clearly there are many students who are not achieving well in the basic aspects of the school curriculum. They need intensive, aggressive help. The best results are achieved when interventions are offered early, as in preschool programs, in adaptive forms of primary level programs, and sustained throughout the school years.

This is a domain in which all of the related programs—in the schools, in teacher preparation, and in research operations—need to be reformed. There is, for example, no clear reason for separate teacher preparation tracks in fields of learning disabilities or mild mental retardation, and there is no reason for separate school programs for students in those same categories. It is notable, in this connection, that the most rapidly developing "category" of special education these days is "cross-categorical." For example, the 16th U.S. annual report on implementation of federal special education laws (U.S. Department of Education, 1994) reports that 69,919 teachers (about 23% of the total number of special education teachers) were employed in cross-categorical programs.

We hope to see more cross-categorical programs in the future, or better still more *non*categorical programs in the schools and in the colleges. They should be closely linked with regular school programs, involve no labels for students, and focus on children at the earliest possible stages of schooling. This large area of service requires affirmations that are deep enough to permit major changes.

All is not clear about an appropriate future for ordinal adaptations. Continuing innovation and research are important. A specific need is for researchers to provide clear descriptions (marker variables) relating to all aspects of their work (students, methods, context, etc.) so that appropriate generalizations can be considered.

Disordinal Adaptations

Here are the truly "special" areas, including programs for students who are blind, deaf, severely retarded, and some speech impaired. They meet the ATI test for distinctiveness of approach as required for some students. The beginnings of special education were in these areas more than a century ago and they remain the most distinct. But this does not mean that things are well settled in them. The following are some of the noteworthy concerns that emerged for us in the process of working through chapters on these several topics.

- *The high frequency of co-morbidity.* This is to say that very often the child who is blind also has other significant disabilities. Similarly, the person who is severely limited in cognitive abilities often is also limited in hearing, motor abilities, or in other ways. It may be that the ways we classify and report on these disabilities ("unduplicated counts by category") encourage too simple a view of the human needs involved. One senses also that preparation programs for teachers and other helpers are constricted, giving too little attention to the co-morbidity data. This is territory for concern and work.
- *Full inclusion.* Much recent debate has focused on full inclusion in the schools and classrooms of students who are severely retarded or multi-disabled. Writers for this volume have reminded us of a broader concern through the voice given to students who are deaf and others who, for themselves, are not always enthusiasts for full inclusion. Indeed, in the case of some deaf students—especially those enrolled in specialized schools and Gallaudet University—there is a claim for honor and appreciation of what is distinct in every way about their special schools and broader culture. There is, for us, some shock associated with learning how early in life some children with vision problems might be placed in special school settings. It may be that the discussions about full inclusion have too often bypassed considerations of children who do not see or hear.
- *Innovation and research.* We complete this volume with a sense that far too little systematic research and creative work is focused on programs

and issues in these areas—especially as related to visual and hearing impairments. Somehow, sets of capable and persistent researchers have been engaged in work with severely retarded persons and in speech-language fields. They have impressive specialized journals in which to report their research. But in areas of vision and hearing there is reason for much concern for lack of clear progress in instructional programs and for well-sustained research programs. These could well be areas for special planning efforts, perhaps with leadership through national or international bodies.

Unremitting Conditions

Sadly, the "special" programs of the schools have higher rates for school dropouts, suspensions, and expulsions than do regular school programs. The characteristics of the students on whom everyone "gives up" are well known: males more than females, minority students more than others, low achievers more than average achievers. Often the students we refer to here are labeled as severely emotionally disturbed or by one of the recent versions of that rubric. Special educators provide relief to regular educators by taking these students out of mainstream programs, at least for limited periods. Without doubt some students are helped by extraordinary teachers in the special programs. But often the programs are oriented mainly to problem minimizing and fall far short of problem solving.

We have felt awkward about a heading for this section of our discussion, just as special educators have had unusual and persisting difficulties in finding labels for the students involved. We've chosen—very tentatively—the term *Unremitting Conditions*, partly to convey quite honestly the disheartening results of special school programs to serve children who show highly disturbing behavior or who present currently intractable problems.

Above all others, the students of concern here require more than school services. The rising numbers of communities in which schools are joining efforts with families, mental health clinics, correctional agencies, drug addiction treatment centers, and family service agencies give some hope that better opportunities can be provided in the future to these children and youth. But it is in consideration of this domain of unremitting problems that we conclude that truly broad and radical reformation is required in our society, touching upon the most basic aspects of family and community life. It is cruel to children and a disgrace for all of us to permit such neglect and abuse of children that teachers observe during the school year; and it is a cruelty and hoax upon teachers even to imagine that they alone can serve the needs of these children. The diagnosis and the treatment of students with major unremitting problems run appropriately not just to the students, their families, or their schools but to the broader community and all of its root structures.

Some General, Concluding Observations

In brief summary form, these are concluding thoughts and observations. Mostly, they are actions we think deserve to be taken.

- *Not just schools.* We feel reaffirmed in support of efforts to coordinate programs of schools, families, and community agencies. Children will be healthy only when families and communities are healthy, and children will learn well in school only when they are well supported in learning activities by their parents and neighbors. Students will improve in post-school behavior and accomplishments when transitions from school to adult life are carefully planned and broadly supported by the total community. These are matters for patient, long-term effort. There is good reason for special and remedial educators to help lead the way.
- *Students as key figures in planning.* There is a trend toward a more explicit and specific *futures* orientation in planning individualized student programs. This can be highly paternalistic, or it can give strong, growing voice to students themselves. Students, we think, should learn to set purposes for their own lives, to set objectives for their own learning, to evaluate for themselves their progress in learning, and to choose activities they find rewarding in their lives—all with help and encouragement by parents and teachers. In many parts of this volume we have sensed trends and commitments in these ways, which we find encouraging and worthy of strong support.
- *Paraprofessionals.* Much of what exceptional students need in today's complex world is not the stuff of professional work. Important parts of their needs run to the unpredictables of daily life, to genuinely caring attention, to occasional respite from academic demands, to patient listening and talking through issues and plans, etc. We may have overestimated the need for constant professional involvements in all of these matters and underestimated the potential contributions of well-selected paraprofessionals. There is a general rise in distrust of professionals these days, a fact wisely taken into account as we staff the schools and other agencies and recognize that, for many children, parents are less often available than we desire. Carefully selected paraprofessionals can contribute much to programs for exceptional students.
- *Integration of professionals.* Clearly the schools are making more inclusive arrangements in organizing programs to serve the special needs of students, and collaborative arrangements among schools and other agencies are emerging rapidly. But the special teachers and regular teachers, the school psychologists, nurses, and social workers are still meeting one another as strangers when they leave college and enter

community work. There is great need for replication of the trends toward integration in the infrastructures of the field; that is, in the colleges, in administrative offices (at all levels), in the advocacy and professional groups, in public policies, and in the supporting literature of the field. The several chapters of this book tell a mainly disappointing story in these affairs. We believe that leaders in every facet of the infrastructure should feel challenged to take steps toward integration.

• *Resilience*. The fields of special and remedial education seem often to focus on defects and deficiencies—of children and programs, as if these were the proper starting points for positive efforts. But as Louis Braille demonstrated long ago, one cannot design programs for better reading by pupils who are blind just by analyzing their defective vision. The solution was to look for what was intact or positive and available for use in reading—in the case of braille that meant tactile sensory processes. There is a general turn in research to what offers strengths in the lives of children, to what generates trust, protection, and supports in their lives. This turn to the resilience side and the advances in research and practices that are emerging there is very promising; again, special and remedial educators have every reason to help lead the way to this positive side of the life of children.

Money

A great deal of money flows (in the developed nations) in special categories of support for programs considered in this volume. We concur strongly with the idea that categorical funding systems should be designed for leadership. Programs follow dollars and that means dollars should lead.

There are major problems in present funding systems. For example, schools often receive funds for having problems, but not for solving them. A kind of "bounty hunt" mentality follows. Disincentives are created for problem solutions. That can and should be changed so that resources follow both problems and solutions, with a shift in balance toward the success side.

Also, ways must be found to bring categorical funds into sharing arrangements with general education. Money must be shared just as responsibilities for programs are shared. Similarly, steps must be taken to bring about shared research endeavors between categorical areas and general education. Changing the flow patterns for money will be difficult because people who are now well paid in segregated arrangements will not be anxious to enter complicated and more general arrangements. Their ownership and rewards may be diminished. It seems likely, however, that if special and remedial educators do not lead the way in these matters, others will do it for them.

Research

It is disappointing to observe how little research and researchers influence trends in the fields of special and remedial education. There has been massive political action in special and remedial education, and we do not assign fault to this fact. But it is surprising to see how rarely the findings of research and evaluation enter public policy deliberations.

It is noteworthy, in this connection, that the National Academy of Sciences, when called upon to review placement practices in special education, did so in a highly distinguished way—but their conclusions and recommendations have gone almost unnoticed in practical matters of the field (Heller, Holtzman, & Messick, 1982). One way to enhance the utility of research in these special fields may be to link it more closely with general support systems for educational research, using "earmark" procedures if necessary to assure attention by researchers to issues concerning disabled and at-risk students. In the long range, it would be disgraceful to develop a broad system for educational research, dissemination, and utilization void of disability-related work. But this is a likely prospect if we do not achieve a closer linkage of research programs across all lines. It is time now for creating an integrated research system which is attentive to needs of all students, just as practical school programs are also being integrated. Further, research systems should be fully coordinated with systemic developments now underway in relation to curriculum, school organization, teacher preparation, and assessments.

A Concluding Note

The fields of special and remedial education are entering a period of difficult change. Funding issues are large and threatening and are treated in a competitive way with general education. Cuts seem inevitable. Even more important may be the broadened range of stakeholders and activists concerning policies and practical operations in special and remedial education. There are new and powerful actors on the policy scene, and all is not well.

We believe that the newly emerging coalitions of schools, parents, and community agencies will be with us over the long term, and a chief focus of their operations will be children and youth who have special needs. It will be very difficult to accomplish the goals set for the new consortium arrangements. But they take us in a needed direction. Staff now in operating roles in special and remedial education should help lead these efforts, we believe.

It appears likely in these busy and difficult days that we will observe a mix of progressive and defensive behavior by special and remedial educators, some of them holding tightly to existing policies and programs. In our view, some will find themselves mere spectators in a changing scene that demands efficiency and that already has many new actors and energies. That story of the future may be unveiled in the next edition of this book.

References

Evans, P. (1993). Foreward. *European Journal of Special Needs Education, 8,* 191–193.

Hegarty, S. (1993). *Educating children and young people with disabilities: Principles and the review of practice.* Paris: UNESCO.

Heller, K. A., Holtzman, W. H., & Messick, S. (Eds.) (1982). *Placing children in special education: A strategy for equity.* Washington DC: National Academy Press.

United Nations: Educational, Scientific and Cultural Organization (1994). *World Conference on Special Needs Education: Access and Quality.* New York: Author.

U.S. Department of Education (1994). *Sixteenth annual report to Congress on the implementation of the Individuals with Disability Act.* Washington DC: Author.

Author index

Subject index